# Lecture Notes in Computer Science 3906

Commenced Publication in 1973
Founding and Former Series Editors:
Gerhard Goos, Juris Hartmanis, and Jan van Leeuwen

## Editorial Board

T0232281

Jens Gottlieb   Günther R. Raidl (Eds.)

# Evolutionary Computation in Combinatorial Optimization

6th European Conference, EvoCOP 2006
Budapest, Hungary, April 10-12, 2006
Proceedings

 Springer

Volume Editors

Jens Gottlieb
SAP AG
Dietmar-Hopp-Allee 16, 69190 Walldorf, Germany
E-mail: jens.gottlieb@sap.com

Günther R. Raidl
Vienna University of Technology, Institute of Computer Graphics
Algorithms and Datastructures Group
Favoritenstr. 9-11/186, 1040 Vienna, Austria
E-mail: raidl@ads.tuwien.ac.at

The cover illustration is the work of Pierre Grenier

Library of Congress Control Number: 2006922468

CR Subject Classification (1998): F.1, F.2, G.1.6, G.2.1, G.1

LNCS Sublibrary: SL 1 – Theoretical Computer Science and General Issues

ISSN        0302-9743
ISBN-10     3-540-33178-6 Springer Berlin Heidelberg New York
ISBN-13     978-3-540-33178-0 Springer Berlin Heidelberg New York

Springer is a part of Springer Science+Business Media

springer.com

© Springer-Verlag Berlin Heidelberg 2006
Printed in Germany

Typesetting: Camera-ready by author, data conversion by Scientific Publishing Services, Chennai, India
Printed on acid-free paper      SPIN: 11730095      06/3142      5 4 3 2 1 0

# Preface

Metaheuristics have often been shown to be effective for difficult combinatorial optimization problems appearing in various industrial, economical, and scientific domains. Prominent examples of metaheuristics are evolutionary algorithms, simulated annealing, tabu search, scatter search, memetic algorithms, variable neighborhood search, iterated local search, greedy randomized adaptive search procedures, estimation of distribution algorithms, and ant colony optimization. Successfully solved problems include scheduling, timetabling, network design, transportation and distribution problems, vehicle routing, the traveling salesperson problem, satisfiability, packing and cutting problems, planning problems, and general mixed integer programming.

The EvoCOP event series started in 2001 and has been held annually since then. It was the first specifically dedicated to the application of evolutionary computation and related methods to combinatorial optimization problems. Evolutionary computation involves the study of problem-solving and optimization techniques inspired by principles of natural evolution and genetics. Following the general trend of hybrid metaheuristics and diminishing boundaries between the different classes of metaheuristics, EvoCOP has broadened its scope over the last years and invited submissions on any kind of metaheuristic for combinatorial optimization problems.

This volume contains the proceedings of EvoCOP 2006, the 6th European Conference on Evolutionary Computation in Combinatorial Optimization. It was held in Budapest, Hungary, on April 10–12 2006, jointly with EuroGP 2006, the 9th European Conference on Genetic Programming, and EvoWorkshops 2006, which consisted of the following seven individual workshops: EvoBIO, the 4th European Workshop on Evolutionary Bioinformatics; EvoCOMNET, the Third European Workshop on Evolutionary Computation in Communications, Networks, and Connected Systems; EvoHOT, the Third European Workshop on Hardware Optimization; EvoIASP, the 8th European Workshop on Evolutionary Computation in Image Analysis and Signal Processing; EvoInteraction, the First European Workshop on Interactive Evolution and Humanized Computational Intelligence; EvoMUSART, the 4th European Workshop on Evolutionary Music and Art; and EvoSTOC, the Third European Workshop on Evolutionary Algorithms in Stochastic and Dynamic Environments.

EvoCOP, originally held as an annual workshop, became a conference in 2004. The events gave researchers an excellent opportunity to present their latest research and to discuss current developments and applications, besides stimulating closer future interaction between members of this scientific community. Accepted papers of previous events were published by Springer in the series *Lecture Notes in Computer Science* (LNCS – Volumes 2037, 2279, 2611, 3004, and 3448).

| EvoCOP | submitted | accepted | acceptance ratio |
|--------|-----------|----------|------------------|
| 2001 | 31 | 23 | 74.2% |
| 2002 | 32 | 18 | 56.3% |
| 2003 | 39 | 19 | 48.7% |
| 2004 | 86 | 23 | 26.7% |
| 2005 | 66 | 24 | 36.4% |
| 2006 | 77 | 24 | 31.2% |

The rigorous, double-blind reviewing process of EvoCOP 2006 resulted in a strong selection among the submitted papers; the acceptance rate was 31.2%. Each paper was reviewed by at least three members of the international Program Committee. All accepted papers were presented orally at the conference and are included in this proceedings volume. We would like to give credit to the members of our Program Committee, to whom we are very grateful for their quick and thorough work and the valuable advice on how to improve papers for the final publication.

EvoCOP 2006 covers evolutionary algorithms as well as various other meta-heuristics, like scatter search, tabu search, memetic algorithms, variable neighborhood search, greedy randomized adaptive search procedures, ant colony optimization, and particle swarm optimization algorithms. The contributions are dealing with representations, heuristics, analysis of problem structures, and comparisons of algorithms. The list of studied combinatorial optimization problems includes prominent examples like graph coloring, knapsack problems, the traveling salesperson problem, scheduling, graph matching, as well as specific real-world problems.

We would like to express our sincere gratitude to the two internationally renowned invited speakers, who gave the keynote talks at the conference: Richard J. Terrile, astronomer, Director, of the Center for Evolutionary Computation and Automated Design at NASA's Jet Propulsion Laboratory, and Stefan Voß, Chair and Director of the Institute of Information Systems at the University of Hamburg.

The success of the conference resulted from the input of many people to whom we would like to express our appreciation. We would like to thank Philipp Neuner for administrating the Web-based conference management system. The local organizers and Judit Megyery did an extraordinary job for which we are very grateful. Last but not least, the tremendous effort of Jennifer Willies and the School of Computing, Napier University, in the administration and coordination of EuroGP 2006, EvoCOP 2006, and EvoWorkshops 2006 was of paramount importance.

April 2006                                                            Jens Gottlieb
                                                                Günther R. Raidl

# Organization

EvoCOP 2006 was organized jointly with EuroGP 2006 and EvoWorkshops 2006.

## Organizing Committee

| | |
|---|---|
| Chairs | Jens Gottlieb, SAP AG, Germany |
| | Günther R. Raidl, Vienna University of Technology, Austria |
| Local chair | Anikó Ekárt, Hungarian Academy of Sciences, Hungary |
| Publicity chair | Steven Gustafson, University of Nottingham, UK |

## Program Committee

Adnan Acan, Eastern Mediterranean University, Turkey
Hernan Aguirre, Shinshu University, Japan
Enrique Alba, University of Málaga, Spain
M. Emin Aydin, London South Bank University, UK
Ruibin Bai, University of Nottingham, UK
Jean Berger, Defence Research and Development Canada, Canada
Christian Bierwirth, University Halle-Wittenberg, Germany
Christian Blum, Universitat Politècnica de Catalunya, Spain
Peter Brucker, University of Osnabrück, Germany
Edmund Burke, University of Nottingham, UK
David Corne, University of Exeter, UK
Ernesto Costa, University of Coimbra, Portugal
Carlos Cotta, University of Málaga, Spain
Peter I. Cowling, University of Bradford, UK
Bart Craenen, Napier University, Edinburgh, UK
David Davis, NuTech Solutions Inc., USA
Karl F. Dörner, University of Vienna, Austria
Anton V. Eremeev, Omsk Branch of the Sobolev Institute of Mathematics,
    Russia
David Fogel, Natural Selection, Inc., USA
Bernd Freisleben, University of Marburg, Germany
Michel Gendreau, Université de Montréal, Canada
Jens Gottlieb, SAP AG, Germany
Walter Gutjahr, University of Vienna, Austria
Jin-Kao Hao, University of Angers, France
Emma Hart, Napier University, Edinburgh, UK

Richard F. Hartl, University of Vienna, Austria
Geir Hasle, SINTEF, Norway
Jano van Hemert, University of Edinburgh, UK
Jörg Homberger, Stuttgart University of Cooperative Education, Germany
Bryant A. Julstrom, St. Cloud State University, USA
Graham Kendall, University of Nottingham, UK
Joshua D. Knowles, University of Manchester, UK
Gary A. Kochenberger, University of Colorado, USA
Gabriele Koller, Vienna University of Technology, Austria
Mario Köppen, Fraunhofer IPK, Germany
Jozef J. Kratica, Serbian Academy of Sciences and Arts, Serbia and Montenegro
Andrea Lodi, University of Bologna, Italy
Helena R. Lourenço, Universitat Pompeu Fabra, Spain
Arne Løkketangen, Molde University College, Norway
Dirk C. Mattfeld, Technical University Braunschweig, Germany
Helmut A. Mayer, University of Salzburg, Austria
Daniel Merkle, University of Leipzig, Germany
Peter Merz, University of Kaiserslautern, Germany
Zbigniew Michalewicz, University of Adelaide, Australia
Martin Middendorf, University of Leipzig, Germany
Pablo Moscato, University of Newcastle, Australia
Christine L. Mumford, Cardiff University, UK
Ibrahim H. Osman, American University of Beirut, Lebanon
Francisco J. B. Pereira, University of Coimbra, Portugal
Adam Prugel-Bennett, University of Southampton, UK
Jakob Puchinger, Vienna University of Technology, Austria
Günther R. Raidl, Vienna University of Technology, Austria
Marcus C. Randall, Bond University, Australia
Colin Reeves, Coventry University, UK
Marc Reimann, ETH Zurich, Switzerland
Mauricio G. C. Resende, AT&T Research, USA
Franz Rothlauf, University of Mannheim, Germany
Andreas Sandner, SAP AG, Germany
Marc Schoenauer, INRIA, France
Christine Solnon, University Lyon I, France
Eric Soubeiga, KPMG, UK
Giovanni Squillero, Politecnico di Torino, Italy
Thomas Stützle, Universite Libre de Bruxelles, Belgium
Eric Taillard, HEIG-VD, Switzerland
El-ghazali Talbi, INRIA Futurs – Lille, France
Edward Tsang, University of Essex, UK
Stefan Voß, University of Hamburg, Germany
Ingo Wegener, University of Dortmund, Germany
Takeshi Yamada, NTT Communication Science Laboratories, Japan

# Table of Contents

# Hybrid Genetic Algorithm Within Branch-and-Cut for the Minimum Graph Bisection Problem

Michael Armbruster[1], Marzena Fügenschuh[2], Christoph Helmberg[1], Nikolay Jetchev[2], and Alexander Martin[2]

[1] Chemnitz University of Technology,
Department of Mathematics, D-09107 Chemnitz, Germany
`michael.armbruster@mathematik.tu-chemnitz.de`
[2] Darmstadt University of Technology,
Department of Mathematics, D-64289 Darmstadt, Germany
`mfuegenschuh@mathematik.tu-darmstadt.de`

**Abstract.** We develop a primal heuristic based on a genetic algorithm for the minimum graph bisection problem and incorporate it in a branch-and-cut framework. The problem concerns partitioning the nodes of a weighted graph into two subsets such that the total weight of each set is within some lower and upper bounds. The objective is to minimize the total cost of the edges between both subsets of the partition. We formulate the problem as an integer program. In the genetic algorithm the LP-relaxation of the IP-formulation is exploited. We present several ways of using LP information and demonstrate the computational success.

## 1 The Minimum Graph Bisection Problem

We consider a weighted graph $G = (V, E)$ with edge costs $w_e \in \mathbb{R}_+$, $e \in E$, and node weights $f_i \in \mathbb{Z}_+$, $i \in V$. A pair $(V_1, V_2)$ satisfying $V_1 \cup V_2 = V$ and $V_1 \cap V_2 = \emptyset$ is called *bipartition*, if $V_1 \neq \emptyset$ and $V_2 \neq \emptyset$. $V_1$ and $V_2$ are called *clusters*. Given a real number $\tau \in [0, 1]$ we define bounds $l_\tau$ and $u_\tau$ such that

$$l_\tau = \frac{1 - \tau}{2} \sum_{i \in V} f_i \qquad \text{and} \qquad u_\tau = \frac{1 + \tau}{2} \sum_{i \in V} f_i.$$

A bipartition $(V_1, V_2)$ such that the total node weight of each cluster stays within the bounds $l_\tau$, $u_\tau$, i.e., $l_\tau \leq \sum_{i \in V_k} f_i \leq u_\tau$, $k = 1, 2$, holds, is called *bisection*. A *bisection cut* $\Delta(V_1, V_2)$ is the set of edges joining nodes in different clusters of the bisection $(V_1, V_2)$. The minimum graph bisection problem is to find a bisection $(V_1, V_2)$ with the minimum cost of $\Delta(V_1, V_2)$:

$$\sum_{e \in \Delta(V_1, V_2)} w_e.$$

This problem is known to be NP-hard [8].

J. Gottlieb and G.R. Raidl (Eds.): EvoCOP 2006, LNCS 3906, pp. 1–12, 2006.

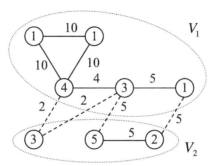

**Fig. 1.** An example of a graph bisection $(V_1, V_2)$ with $\tau = 0.1$, $l_\tau = 9$ and $u_\tau = 11$. The dashed edges build a bisection cut $\Delta(V_1, V_2)$. The numbers within the nodes and the numbers nearby the edges correspond to node and edge weights respectively.

Graph bisection and its generalizations, e.g. when $V$ is partitioned into more than two subsets [6], have considerable practical significance, especially in the areas of VLSI design and parallel computing. With state-of-the-art solution methods it is still unrealistic to obtain exact solutions for large instances. Usually such problems are tackled heuristically. Genetic algorithms are known to find good solutions to graph partitioning problems, see e.g. [5, 11, 12, 14]. Successful approaches combining methods from different classical metaheuristics, among others including evolutionary algorithms, are presented in [4, 14]. This motivated us to incorporate a primal heuristic based on a genetic algorithm in a branch-and-cut framework. In [13] a classification of methods that combine exact and heuristic procedures is given. Within this scheme our work can be categorized as a heuristic that *exactly solves relaxed problems*: We exploit the solution of the LP-relaxation for generating a start population as well as in the mutation procedures of our genetic algorithm.

Our paper is structured as follows. In Section 2 we present integer programming formulations for the minimum graph bisection problem. In Section 3 we outline the structure of the genetic algorithm based on the fractional solution of the LP-relaxation hybridized with the edge costs. In Section 4 we provide computational results.

## 2   Integer Programming Models

For an integer programming formulation of the minimum graph bisection problem we introduce binary variables $y_{ij}$ for all $ij \in E$. Each $y \in \{0, 1\}^{|E|}$ satisfying $y_{ij} = 1$ if nodes $i$ and $j$ are in different clusters, and $y_{ij} = 0$ otherwise, corresponds to an incidence vector of a bisection cut in $G$. Let

$$Y = \{y \in \{0, 1\}^{|E|} \,|\, y \text{ is an incidence vector of a bisection cut in } G\}.$$

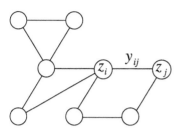
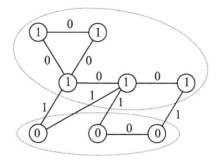

**Fig. 2.** Variables $y_{ij}$, $ij \in E$ and $z_i$, $i \in V$

**Fig. 3.** Values of variables $y_{ij}$, $ij \in E$ and $z_i$, $i \in V$ corresponding to the bisection cut presented in Fig. 1

We search for such an element of the set $Y$ that minimizes $\sum_{e \in E} w_e y_e$. It remains to describe $Y$ by linear constraints. One formulation can be derived directly from the model presented in [6] concerning a generalization of the graph bisection problem. There are introduced additional binary variables $z_i$ for each node $i \in V$. It is required that all $z$-variables corresponding to nodes assigned to one cluster have the same value, see Fig. 2 and Fig. 3.

The following constraints guarantee that $y_{ij} = 1$ if and only if $z_i \neq z_j$, i.e., nodes $i$ and $j$ are in different clusters, and $y_{ij} = 0$ otherwise.

$$\forall ij \in E \quad z_i - z_j - y_{ij} \leq 0 \tag{1}$$
$$\forall ij \in E \quad z_j - z_i - y_{ij} \leq 0 \tag{2}$$
$$\forall ij \in E \quad -z_j - z_i + y_{ij} \leq 0 \tag{3}$$
$$\forall ij \in E \quad z_i + z_j + y_{ij} \leq 2 \tag{4}$$

The constraint

$$l_\tau \leq \sum_{i \in V} f_i z_i \leq u_\tau \tag{5}$$

assures that the total weight of nodes in each cluster stays within the given lower and upper bound. The projection of the feasible set defined by constraints (1) - (5) onto the $y$-space equals the set $Y$. Note that, since our objective is to minimize a positive weighted sum of $y$-components, constraint (4) is redundant in the problem formulation.

Since the node variables $z_i$, $i \in V$ do not appear in the objective function we get rid of them in the following way. We replace the constraints (1) - (4) with a known class of valid inequalities for $P = \text{conv}(Y)$ called *odd-cycle* inequalities (see also [3]):

$$\sum_{e \in F} y_e - \sum_{e \in E_C \setminus F} y_e \leq |F| - 1, \quad \forall C \subset G, F \subset E_C, |F| \text{ is odd,}$$

where $C = (V_C, E_C)$ is a cycle in $G$. These constraints require that each cycle $C$ in $G$ must contain an even number of edges from the cut. Next, we reformulate constraint (5) by selecting a node $s \in V$ and extending $E$ so that $s$ is adjacent to all other nodes in $V$. The weights $w_{is}$ of new edges are set to zero, see Fig. 4.

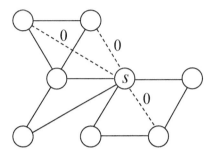

**Fig. 4.** New edges (dashed) incident to node $s$ and their costs

For all nodes $i \in V$, which are in the same cluster as $s$, holds

$$l_\tau \le f_s + \sum_{i \in V \setminus \{s\}} f_i(1 - y_{is}) \le u_\tau,$$

which is equivalent to

$$l_\tau \le \sum_{i \in V \setminus \{s\}} f_i y_{is} \le u_\tau$$

using $l_\tau + u_\tau = \sum_{i \in V} f_i$.

Thus we obtain a new integer programming model for the minimum graph bisection problem:

$$\min \sum_{e \in E} w_e y_e$$

s.t.

$$(B) \qquad \sum_{e \in F} y_e - \sum_{e \in E_C \setminus F} y_e \le |F| - 1, \quad \forall C \subset G, F \subset E_C, |F| \text{ is odd}$$

$$l_\tau \le \sum_{i \in V \setminus \{s\}} f_i y_{is} \le u_\tau, \text{ for some } s \in V$$

$$y_e \in \{0, 1\}, \forall e \in E.$$

The feasible solutions of $(B)$ are in one-to-one correspondence to the elements of $Y$. In the sequel we consider the latter model. Since the number of all possible odd-cycles is exponential in $|E|$ we do not include them all in the initial formulation. Using the polynomial time separation algorithm given in [3] we add violated odd-cycle-inequalities sequentially during separation within the branch-and-cut algorithm.

# 3 Hybrid Genetic Algorithm

Genetic algorithms solve optimization problems in an analogous manner to the evolution process of nature [9]. A solution of a given problem is coded into a string of *genes* termed *individual*. New solutions are generated by operations called *crossover* and *mutation*. In a crossover two *parents* are drawn from the current population and parts of their genes are exchanged resulting in *child* solutions. A mutation is an adequate transformation of a single individual. Individuals delivering the best objective value are selected for creating the next generation.

**Table 1.** The generic genetic algorithm

| | |
|---|---|
| *input* | $y^h \in \mathbb{R}^{|E|}$ |
| | $p$ number of individuals in population, $k$ factor of population growth, |
| | $g$ number of generations, $f$ number of fitness loops, |
| | $M$ mutation type, $m_r$ mutation rate |
| | (1)   based on $y^h$ create initial population with $p$ individuals |
| | (2)   *while*   sufficient improvement on fitness |
| |             and number of loops less than $g$ |
| |          *do* |
| |                  (2a)   perform crossover and mutation $M$ |
| |                          till population grows to $kp$ individuals |
| |                  (2b)   evaluate fitness of each individual |
| |                  (2c)   select $p$ best individuals |
| |          *done* |
| *output* | individual with the best fitness value |

In our heuristic, outlined in Table 1, we code an individual in two equivalent ways. One form is the *node representation*, i.e., as a vector $v \in \{0,1\}^n$, $n = |V|$. If $v_i = v_j$, nodes $i$ and $j$ are in one cluster. The other form is the *edge representation* as a vector $y \in \{0,1\}^E$, where

$$y_{ij} = \begin{cases} 0, & v_i = v_j \\ 1, & v_i = 1 - v_j. \end{cases}$$

To obtain a wider selection range of individuals we allow infeasible solutions in the sense that the bounds $l_\tau$ and $u_\tau$ on the total weight node of clusters do not need to be fulfilled.

The main idea of our genetic algorithm is to construct solutions to the bisection problem using the vector $y' \in [0,1]^E$, which is a fractional solution to the linear relaxation of $(B)$. As an additional criterion we use the edge cost vector $w$. Thus we hybridize the components of $y'$ and $w$ by considering $y^h = f(y', w)$. The components $y^h_{ij}$, $ij \in E$ are supposed to provide the information, if $i$ and $j$ should be assigned to the same cluster. The hybridization forms of $y^h$ vary on

subroutines of the algorithm (creating the starting population, mutation procedures) and will be explained sequentially in the next subsections.

## 3.1  Initial Population

The initial population is determined in the following way. Suppose we are given
a set $F \subseteq E$. How $F$ is constructed we explain below. We compute a spanning forest on $F$ such that for each of its components $(V^1, E^1), \ldots, (V^k, E^k) \subset$
$(V_F, E_F)$, $k \geq 2$, holds:

$$\sum_{i \in V^j} f_i \leq u_\tau, \quad 1 \leq j \leq k.$$

The two heaviest sets $V^s, V^t$ with respect to the weighted node sum form the
initial clusters. Using a bin packing heuristic we complete the clusters with the
node sets $V^j$, $1 \leq j \leq k$, $j \neq s, t$. Thus we obtain an individual $\tilde{v} \in \{0, 1\}^n$ with
$\tilde{v}_i = 0$ for all $i \in V^s$ and $\tilde{v}_i = 1$ for all $i \in V^t$. See also heuristic *Edge* in [7].

We construct $F$ due to the information delivered by the fractional LP-solution
$y'$ and the edge cost vector $w$. Concerning $y'$ the set $F$ should contain edges with
LP-values close to zero. In regard to $w$, $F$ should possibly contain edges with
high costs. We applied three methods of combining components of $y'$ and $w$. We
consider $F := F_\varepsilon \cup F_\rho$, where $F_\varepsilon = \{e \in E : y'_e < \varepsilon\}$ and $F_\rho$ takes one of the
following forms.

(F1) $F_\rho = \{e \in E : \frac{1}{1+w_e^2} \leq X_\rho\}$,

(F2) $F_\rho = \{e \in E : y'_e + \frac{1}{1+w_e^2} \leq X_\rho\}$,

(F3) $F_\rho = \{e \in E : Zy'_e + (1 - Z)\frac{1}{1+w_e^2} \leq X_\rho\}$ for a random number
    $Z \in [0, 1]$,

where $X_\rho$ is either constantly equal to $\rho$ or a random number in $[0, \rho]$. In (F1)
an edge enters $F$, if it has high cost and also a high LP-value with respect to $\varepsilon$.
In (F2) we allow the extension of $F$ by edges with high cost if their LP-value is
not too high. In (F3) we prioritize randomly the approach (F1) and (F2) using a
parameterization of the corresponding components of $y'$ and $w$. As a comparison
to the above hybridization of the set $F$ we consider also a random approach:

(R) $F_\rho = \{e \in E : y'_e < X_\rho$ and $\frac{y'_e}{X_\rho} < Z\}$ for a random $Z \in [0, 1]$.

Note that the bounds $\varepsilon$ and $\rho$ control the number of edges entering $F$. If $F_\rho \neq \emptyset$
and one of the methods (F1) - (F3) is applied, we say that the set $F$ is *hybridized*.

If $F_\rho = \emptyset$ the starting population contains $p$ duplicates of $\tilde{v}$. Otherwise, we
select a new random value $X_\rho$ in each iteration creating a new individual, and
thus obtain a differentiated initial population.

## 3.2  Mutations

To create a new generation we apply four mutation types. The parameter
$m_r$ called *mutation rate* is introduced to control the percentage of exchanged

components of vector $v$, or $y$ respectively, in one mutation round. We apply also hybridization methods to some of our mutations procedures. Concerning the implementation issues, we were able to apply the edge costs to Mutation 3 and 4 (see below) without loss on the efficiency of our algorithm. We implemented the following transformations.

**Mutation 1.** Let $v \in \{0,1\}^{|V|}$ be an individual selected for the mutation. Let $V_M \subset V$ be a randomly selected subset of $v$ components such that $|V_M| < m_r$. For all $i \in V_M$ we set $v_i := 1 - v_i$. Note that this general procedure does not make use of the specifics of the underlying bisection problem.

**Mutation 2.** This procedure works in a similar way as Mutation 1 applied to the edge representation of the individual. Let $E_M \subset E$ be a random selection of edges such that $|E_M| < m_r$. For each $ij \in E_M$ such that $y_{ij} := 1$ we set $y_{ij} := 0$. Then we accordingly update the values of $v_i$ and $v_j$. It holds $v_i \neq v_j$. Under a random decision we set either $v_i := v_j$ or $v_j := v_i$.

**Mutation 3.** Here we use the same idea as for creation of the initial population. Let $E_M \subset E$ be a random selection of edges such that $|E_M| < m_r$ and let $X_\rho \in$ be a random number. We consider the value $p_{ij} := 1 - y_{ij}^h$ as the probability that nodes $i$ and $j$ are in the same cluster. For each $ij \in E_M$, if $p_{ij} > 1 - X_\rho$ we set either $v_i := v_j$ or $v_j := v_i$ at random. If $p_{ij} < X_\rho$ then we set $v_i := 1 - v_j$ or $v_j := 1 - v_i$, again randomly.

In the non-hybridized version of this mutation we set $y_{ij}^h := y'$. To hybridize the method we apply the following mapping

$$y_{ij}^h := Z y_{ij}' + \frac{1 - Z}{1 + w_{ij}},$$

for $ij \in E_M$ and a random $Z \in [0,1]$.

**Mutation 4.** This is a kind of a neighborhood search. Let $v \in \{0,1\}^{|V|}$ be an individual selected for the mutation. Let $V_M \subset V$ be a randomly selected subset of $v$ components such that $|V_M| < m_r$. For $i \in V_M$ we count the number of nodes adjacent to $i$ in each cluster, i.e., we determine the numbers

$$c_0 = |\{j : ij \in E, \, v_j = 0\}| \text{ and } c_1 = |\{j : ij \in E, \, v_j = 1\}|.$$

If $c_0 > c_1$ we set $v_i = 0$, and 1 otherwise.

To hybridize this mutation we sum also the edge costs of joining node $i \in V_M$ with each cluster. We calculate

$$w_0 = \sum_{ij \in E : v_j = 0} w_{ij} \quad \text{and} \quad w_1 = \sum_{ij \in E : v_j = 1} w_{ij}.$$

We decide randomly, either $v_i$ takes its value depending on $\max\{c_0, c_1\}$, as in the original procedure, or depending on $\min\{w_0, w_1\}$, i.e., if $w_0 < w_1$ we set $v_i = 1$.

### 3.3  Crossover and Fitness Value

For the crossover operation we implemented the one-point crossover: we select at random two parents from the present population and a crossover point from the numbers $\{1,\ldots,n-1\}$. A new solution is produced by combining the pieces of the parents. For instance, suppose we selected parents $(u_1,u_2,u_3,u_4,u_5)$ and $(v_1,v_2,v_3,v_4,v_5)$ and crossover point 2. The child solutions are $(u_1,u_2,v_3,v_4,v_5)$ and $(v_1,v_2,u_3,u_4,u_5)$.

To each individual we assign a fitness value. If it corresponds to a feasible bisection, i.e., the total node weight in both clusters stays within the limits $l_\tau$ and $u_\tau$, we take the inverse of the objective function value in the incidence vector of the corresponding bisection cut. Otherwise we take a negated feasibility violation, i.e., the weight of the cluster which is greater than $u_\tau$. The $p$ fittest individuals are selected from the expanded population and the next generation is created. The fitness value of the best individual is stored. It defines the fitness of the generation. We consider two stopping criteria of the genetic process. One is defined by $f$, called fitness loop number. It gives the limit of loops we perform without increase in the generation fitness. The second limit is the maximal number of loops we perform in one heuristic round. The output is the fittest individual. If it corresponds to a feasible bisection cut the inverse of its fitness value gives an upper bound for the problem $(B)$.

## 4  Computational Results

All tests presented below are carried out using the settings

$$m_r = 1, \quad f = 60, \quad g = 300, \quad p = 150 \quad k = 4.$$

In addition all mutation types are selected randomly uniformly distributed.

We refer to [2] for an overview on how the heuristic performs depending on the selection of parameters $p$, $k$ and $M$. In these preliminary studies, we only considered $\varepsilon = 0.01$ and also did not apply any of the hybridization methods explained in subsections 3.1 and 3.2. We established that small instances ($|E| < 1000$) can be most efficiently solved applying just the standard mutation type (Mutation 1). In these cases the solution time was proportional to the parameters $p$ and $k$. On the contrary, on bigger instances ($3500 < |E| < 6500$) the heuristic performed better with problem related mutation types (Mutation 2,3,4) and the biggest choice of solutions, i.e., the population growth. Furthermore better solutions of the LP-relaxation lead to better primal bounds computed by our heuristic and to less time consumption.

In this paper we present our empirical investigation on three graph instances selected from the sample given in [10] varying in the number of edges between 3500 and 6500. We generally set $\tau = 0.05$. As a branch-and-cut framework we use SCIP [1] with CPLEX [15] as LP-solver. All computations are executed on a 2.6 GHz Pentium IV processor with 2 GB main storage.

In our investigations we search for the possibly best selection of the set $F$ in the sense that our heuristic constantly improves upper bounds and hence considerably contributes to the efficient termination of the branch-and-cut algorithm, i.e., the closing of the optimality gap:

$$gap = \frac{b_u - w^T y'}{w^T y'},$$

where $b_u$ is the lowest upper bound found so far. To this end, we vary the following parameters and methods. At the beginning we just increase the value of $\varepsilon$ from 0.01 to 0.03 and 0.05. The success of the methods (F1) - (F3) depends on the adequate selection of the pair $(\varepsilon, \rho)$. It turns out that too small and too large cardinalities of $F$ deteriorate the solution quality. The parameters $\varepsilon$ and $\rho$ appear to be coupled in a way that if we reduce $\varepsilon$ we need to increase $\rho$ and the other way round. For our tests we selected the pairs $(\varepsilon, \rho)$ from set

$$B_{\varepsilon, \rho} := \{(0.03,\, 0.2),\, (0.05,\, 0.1),\, (0.05,\, 0.2),\, (0.01, 0.4)\}.$$

**Table 2.** Test results on the hybridization of set $F$ and mutations 3 and 4

| Instance.$n.m$ | $\varepsilon$ | $\rho$ | $F$ | $\mathcal{M}$ | $bub$ | $aub$ | $\sigma$ |
|---|---|---|---|---|---|---|---|
| taq.334.3763 | 0.03 | ~0.2 | $\mathcal{R}$ | + | 341 | 530 | 184 |
|  | 0.01 | ~0.4 | F3 | - | 341 | 505 | 231 |
|  | 0.01 | 0.4 | F2 | - | 341 | 477 | 142 |
|  | 0.05 | ~0.2 | F2 | + | 341 | 461 | 116 |
|  | 0.01 | ~0.4 | F1 | - | 341 | 445 | 100 |
| diw.681.3752 | 0.05 | - | - | + | 1228 | 1768 | 357 |
|  | 0.03 | ~0.2 | F2 | + | 1219 | 1903 | 378 |
|  | 0.05 | ~0.1 | F1 | - | 1205 | 1552 | 253 |
|  | 0.05 | ~0.1 | F1 | + | 1199 | 1785 | 389 |
|  | 0.05 | ~0.1 | F1 | + | 1185 | 1744 | 297 |
| taq.1021.6356 | 0.05 | 0.1 | F3 | + | 3678 | 3768 | 302 |
|  | 0.01 | ~0.4 | F3 | + | 3641 | 4120 | 403 |
|  | 0.05 | - | - | + | 3425 | 3986 | 405 |
|  | 0.01 | 0.4 | $\mathcal{R}$ | + | 3403 | 3957 | 438 |
|  | 0.05 | ~0.2 | F3 | + | 3392 | 3992 | 481 |

$n$ number of nodes
$m$ number of edges
$\sim X_\rho \in [0, \rho]$ is selected randomly, otherwise $X_\rho = \rho$
$\mathcal{M} = \{+\}$ hybrid mutations 3 and 4 are applied
$bub$ best upper bound delivered by the heuristic
$aub$ average upper bound
$\sigma$ standard deviation

We investigate next the impact of the updated mutations on the solution while $F_\rho = \emptyset$ and only $\varepsilon$ vary. Then we integrate them together with all variations of $F_\rho$ described in subsection 3.1. It turns out that the simultaneous application of both, the hybrid set $F$ and the hybrid mutations, not always contributes to better solutions, in comparison to a separate one. However, improvements based on the underlying problem, i.e., either to the initial phase of the algorithm or to the mutations, still perform better than random ones.

In Table 2 we present the parameter selections for each instance, which contribute to the best upper bounds and hence to the reduction of the optimality gap. Table 3 shows the number of processed nodes of the branch-and-cut tree $(N)$, which we set as the limit for the computations for each instance, the average number of heuristic rounds $(r)$ and the average execution time of one heuristic round. None of the methods (F1) - (F3) appears to be generally

**Table 3.** Statistics on instances

| Instance.$n.m$ | $N$ | $r$ | av. CPU time |
|---|---|---|---|
| taq.334.3763 | 300 | 50 | 25 sec. |
| diw.681.3752 | 100 | 20 | 51 sec. |
| taq.1021.6356 | 30 | 7 | 82 sec. |

applicable for all instances. However, selecting an appropriate one contributes to good results. The first instance is solved most efficiently with the parameter pair $(\varepsilon, \rho) = (0.01, 0.4)$ independently from the hybridization form of the set $F$. $(\varepsilon, \rho) = (0.05, 0.1)$ and the method (F1) seems to be the best parameter selection for the second instance, while $\varepsilon = 0.05$ and method (F3) are the best for the third instance.

Finally in Tables 4, 5 and 6 we give a comparison on how the branch-and-cut performs without the genetic algorithm as well as the impact of the hybridization

**Table 4.** Branch-and-cut without the genetic algorithm

| Instance.$n.m$ | $N$ | $bub$ | $blb$ | $gap$ | $time$ |
|---|---|---|---|---|---|
| taq.334.3952 | 317 | - | 299 | $inf$ | 3600 |
| diw.681.3752 | 109 | - | 389 | $inf$ | 3600 |
| taq.1021.6356 | 42 | - | 469 | $inf$ | 3600 |

$blb$ best lower bound

$gap$ optimality gap by termination of the branch-and-cut

$time$ in CPU seconds

on the solution process. Table 4 shows the results when applying the pure branch-and-cut without the genetic heuristic. In Table 5 we show the best results, when the heuristic is applied in the non-hybrid version, and in Table 6 the best results for each instance achieved with the hybrid version. As time limit we used 3600 CPU sec.

**Table 5.** Branch-and-cut with the non-hybrid genetic algorithm

| Instance.$n.m$ | $\varepsilon$ | $N$ | $bub$ | $blb$ | $gap$ | $time$ |
|---|---|---|---|---|---|---|
| taq.334.3952 | 0.05 | 337 | 341 | 311 | 9.6 % | 3600 |
| diw.681.3752 | 0.01 | 108 | 1288 | 390 | 230.2 % | 3600 |
| taq.1021.6356 | 0.03 | 32 | 3896 | 435 | 795.6 % | 3600 |

It pays off to apply the hybridization of the genetic heuristic with the exception of the instance taq.334.3952. But even in this case the hybrid version does not deteriorate considerably the result. For the other two larger instances the hybrid genetic algorithm yields the smallest gap by improving the upper bound. This keeps the branch-and-bound tree smaller so that the best lower bound also benefits.

**Table 6.** Branch-and-cut with the hybrid genetic algorithm

| Instance.$n.m$ | $N$ | $bub$ | $blb$ | $gap$ | $time$ |
|---|---|---|---|---|---|
| taq.334.3952 | 327 | 341 | 309 | 10.1 % | 3600 |
| diw.681.3752 | 115 | 1185 | 445 | 166.2 % | 3600 |
| taq.1021.6356 | 32 | 3403 | 450 | 656.2 % | 3600 |

# 5 Conclusion

We developed a genetic algorithm as a primal heuristic routine in a branch-and-cut framework for solving a minimum graph bisection problem. Following our earlier observation that the good quality of fractional LP-solution can be a significant help for the algorithm to deliver good solutions we looked for further developments. Thus we integrated edge costs in selected subroutines of the algorithm as a kind of decision support. We were able to show that attempts of such hybridization deliver a good alternative to the standard random approach.

**Acknowledgments.** This work is supported by German Research Foundation (DFG).

# References

1. Achterberg, T.: SCIP - a framework to integrate constraint and mixed integer programming. ZIB-Report (2004)
2. Armbruster, M., Fügenschuh, M., Helmberg, C., Jetchev, N., Martin, A.: LP-based Genetic Algorithm for the Minimum Graph Bisection Problem. Operations Research Proceedings 2005 (to appear)
3. Barahona, F., Mahjoub, A. R.: On the cut polytope. Math. Programming 36(2) (1986) 157–173
4. Banos, R., Gil, C., Ortega, J., Montoya, F. G.: Multilevel heuristic algorithm for graph partitioning. Proceedings of Applications of Evolutionary Computing, EvoWorkshops 2003. Springer, LNCS 2611 (2003) 143-153
5. Bui, T. N., Moon, B. R.: Genetic algorithm and graph partitioning. IEEE Trans. Comput. 45(7) (1996) 841–855
6. Ferreira, C. E., Martin, A., de Souza, C. C., Weismantel, R., Wolsey, L. A.: Formulations and valid inequalities for the node capacitated graph partitioning problem. Math. Programming 74 (1996) 247–267
7. Ferreira, C. E., Martin, A., de Souza, C. C., Weismantel, R., Wolsey, L. A.: The node capacitated graph partitioning problem: A computational study. Math. Programmming 81(2) (1998) 229–256
8. Garey, M. R., Johnson, D. S.: Computers and Intractability. W.H. Freeman and Company (1979)
9. Hoos, H. H., Stützle, T.: Stochastic Local Search: Foundations and Applications. Morgan Kaufmann (2004) San Francisco (CA)
10. Jünger, M., Martin, A., Reinelt, G., Weismantel, R.: Quadratic 0/1 optimization and a decomposition approach for the placement of electronic circuits. Math. Programmming B 63(3) (1994) 257–279
11. Kohmoto, K., Katayaman, K., Narihisa, H.: Performance of a genetic algorithm for the graph partitioning problem. Math. Comput. Modelling 38 (11-13) (2003) 1325–1333
12. Maini, H., Mehrotra, K., Mohan, C., Ranka, S.: Genetic algorithms for graph partitioning and incremental graph partitioning. Supercomputing '94: Proceedings of the 1994 ACM/IEEE conference on Supercomputing (1994) 449–457, New York, USA, ACM Press.
13. Puchinger, J., Raidl, G. R.: Combining metaheuristics and exact algorithms in combinatorial optimization: A survey and classification. In Proceedings of the First International Work-Conference on the Interplay Between Natural and Artificial Computation, Springer LNCS 3562 (2005) 41-53
14. Soper, A.J., Walshaw, C., Cross, M.: A combined evolutionary search and multilevel optimisation·approach to graph-partitioning. J. Glob. Optim. 29(2) (2004) 225-241
15. ILOG CPLEX Division, 889 Alder Avenue, Suite 200, Incline Village, NV 89451, USA. Information available at URL http://www.cplex.com.

# The Trade Off Between Diversity and Quality for Multi-objective Workforce Scheduling[*]

Peter Cowling, Nic Colledge, Keshav Dahal, and Stephen Remde

MOSAIC Research Group, University of Bradford, Great Horton Road,
Bradford, BD7 1DP, Great Britain
{p.i.cowling, n.j.colledge, k.p.dahal,
s.m.remde}@bradford.ac.uk

**Abstract.** In this paper we investigate and compare multi-objective and weighted single objective approaches to a real world workforce scheduling problem. For this difficult problem we consider the trade off in solution quality versus population diversity, for different sets of fixed objective weights. Our real-world workforce scheduling problem consists of assigning resources with the appropriate skills to geographically dispersed task locations while satisfying time window constraints. The problem is NP-Hard and contains the Resource Constrained Project Scheduling Problem (RCPSP) as a sub problem. We investigate a genetic algorithm and serial schedule generation scheme together with various multi-objective approaches. We show that multi-objective genetic algorithms can create solutions whose fitness is within 2% of genetic algorithms using weighted sum objectives even though the multi-objective approaches know nothing of the weights. The result is highly significant for complex real-world problems where objective weights are seldom known in advance since it suggests that a multi-objective approach can generate a solution close to the user preferred one without having knowledge of user preferences.

## 1 Introduction

In collaboration with an industrial partner we have studied a workforce scheduling problem which is a resource constrained scheduling problem similar to but more complex and "messy" than many other well-studied scheduling problems like the RCPSP (Resource Constrained Project Scheduling Problem) and job shop scheduling problem, for which much work has been done [1]. The problem is based on our work with Vidus Ltd. (an @Road company) which has developed scheduling solutions for very large, complex mobile workforce scheduling problems in a variety of industries. Our workforce scheduling problem is concerned with assigning people and other resources to geographically dispersed tasks while respecting time window constraints and is like many scheduling problems NP-Hard [1] because it contains the RCPSP as a sub-problem.

---

[*] This work was funded by an EPSRC CASE Studentship in partnership with Vidus Ltd. (an @Road company).

J. Gottlieb and G.R. Raidl (Eds.): EvoCOP 2006, LNCS 3906, pp. 13–24, 2006.
© Springer-Verlag Berlin Heidelberg 2006

This paper is structured as follows: we outline a model of our workforce scheduling problem in section 2, discuss related work in section 3 and propose a multi-objective genetic algorithm for solution of the workforce scheduling problem in section 4. In section 5 we investigate multi-objective genetic algorithms and compare them to other single objective genetic algorithms in terms of solution quality and diversity. We present conclusions in section 6.

## 2   The Workforce Scheduling Problem

The workforce scheduling problem that we consider consists of four main components: Tasks, Resources, Skills and Locations. A *task* $T_i$ is a job or part of a job that needs to be completed. Each task must start and end at a specified location. Usually the start and end locations are the same but they may be different. Each task has one or more time windows. Some time windows which are an inconvenience for the customer have an associated penalty. We have a set $\{T_1, T_2, ..., T_n\}$ of tasks to be completed. Each task is undertaken by one or more resources. We have set of resources $\{R_1, R_2, ..., R_m\}$. A task requires resources with the appropriate skills. We have a set $\{S_1, S_2, ..., S_k\}$ of skills. Task $T_i$ requires skills $[TS_1^i, TS_2^i, ..., TS_{t(i)}^i]$ with work requirements $[w_1^i, w_2^i, ..., w_{t(i)}^i]$ where $w_q^i$ is the amount of skill $TS_q^i$ required. Task $T_i$ also has an associated priority $p(T_i)$. Resources are the components that undertake the work and possess skills. Resource $R_j$ possesses skills $[RS_1^j, RS_2^j, ..., RS_{r(j)}^j]$. A function $c(R,S)$ expresses the competence of resource $R$ at skill $S$, relative to an average competency. Each resource $R$ travels from location to location at speed $v(R)$. For tasks $T_1$, $T_2$, $d(T_1, T_2)$ measures the distance between the end location of $T_1$ and the start location of $T_2$. There are three main groups of constraints: task constraints, resource constraints and location constraints and they are described below.

Task constraints:

- Each task can be worked on only within specified time windows.
- Some tasks require other tasks to have been completed before they can begin (*precedence* constraints).
- Some tasks require other tasks to be started at the same time (*assist* constraints).
- Tasks may be split across breaks within a working day. No tasks may take more than one day.
- For a task to be scheduled it must have exactly one resource assigned to it for each of the skills it requires.
- All assigned resources have to be present at a task for its whole duration regardless of their skill competency and task skill work requirement.
- If a task $T_i$ with skill requirements $[TS_1^i, TS_2^i, ..., TS_{t(i)}^i]$ and amounts $[w_1^i, w_2^i, ..., w_{t(i)}^i]$ is carried out by resources $[R_1^i, R_2^i, ..., R_{t(i)}^i]$ then the time taken is

$$\max_{q \in \{1,2,...,t(i)\}} \left( \frac{w_q^i}{c(R_q^i, TS_q^i)} \right)$$

i.e. the greatest time taken for any single resource to complete a skill requirement

Resource constraints:

- A resource $R$ travels from location to location at a fixed speed $v(R)$.
- Resources may only work during specified time windows.
- Resources can only work on one task at once and only apply one skill at a time.

Location constraints:

- Resources must travel to the location of each task they work on, and are unavailable during this travel time.
- Resources must start and end each day at a specified "home" location and must have sufficient time to travel to and from their home location at the start and end of each day.

## 2.1  Objectives

When building a schedule many different and often contradictory business objectives are possible. In this paper we consider three objectives. The first objective is Schedule Priority (SP), given by

$$SP = \sum_{\{i:T_i \text{ is scheduled}\}} p(T_i)$$

Maximising Schedule Priority maximises the importance of the tasks scheduled to the user (and implicitly minimises the importance of tasks unscheduled).

The second objective measures Travel Time (TT) across all resources. Define $A=\{(i1,i2,j):$task $T_{i1}$ comes immediately before $T_{i2}$ in the schedule of resource $R_j\}$. Then,

$$TT = \sum_{(i1,i2,j) \in A} \frac{d(T_{i1}, T_{i2})}{v(R_j)}$$

Travel to and from home locations is handled by considering dummy tasks fixed at the start and end of the working day, at the home location of each resource.

The third objective measures the inconvenience associated with completing tasks or using resources at an inconvenient time, which we have labelled Schedule Cost (SC). In order to express this accurately we express the time windows for Resource $R$ using a function $\tau$ where $\tau(R,t)$ is the cost per unit time for resource $R$ working at time $t$. We introduce a variable

$$X(R,t) = \begin{cases} 1 & \text{if resource } R \text{ is busy (on a task or traveling) at time } t \\ 0 & \text{otherwise} \end{cases}$$

Similarly we introduce $\tau'$ where $\tau'(T,t)$ is the cost per unit time for task $T$ being executed at time $t$ and

$$X'(T,t) = \begin{cases} 1 & \text{if task } T \text{ is being executed at time } t \\ 0 & \text{otherwise} \end{cases}$$

Then

$$SC = \sum_{i=1}^{n} \int X'(T_i,t)\tau'(T_i,t)dt + \sum_{j=1}^{m} \int X(R_j,t)\tau(R_j,t)dt$$

Other objectives are possible but these three objectives express most of the primary concerns of the users in this case, at a high level. Considering lower level objectives at regional and resource group level could, however, give in many more objectives for this problem.

# 3 Related Work

## 3.1 Similar Problems

The RCPSP is a generalisation of several common scheduling problems including job-shop, open-shop and flow-shop scheduling problems [1]. The RCPSP consists of a set of Tasks to be performed using a set of finite capacity resources. In common with the model in the preceding section, the RCPSP has tasks which are undertaken by finite capacity resources subject to precedence constraints. However, the notion of tasks requiring skills and resources possessing multiple skills is essentially absent, as is the travel aspect. Precedence constraints are a major part of the RCPSP model and in many RCPSP problems the precedence constraints are arguably the most complex constraints with every task involved in a precedence relationship with one or more other tasks [1]. They are less significant for our model with many tasks having no predecessors or successors. Our problem has time-varying resource availability which is rarely considered in the RCPSP, however Brucker [2] discusses an RCPSP with time dependent resource profiles which model the changing availabilities of resources.

The resource constrained multiple project scheduling problem (rc-mPSP) [3] is less widely studied than the RCPSP. An rc-mPSP problem consists of several RCPSP problems that are not connected by precedence constraints, making the overall problem less time constrained than an equivalent RCPSP problem with the same tasks (and the extra precedence constraints). These precedence constraints are closer to the ones observed in our workforce scheduling problem.

A recent review of work on personnel scheduling is presented in [18]. In contrast to the RCPSP, personnel scheduling does include a notion of time-varying resource availability, in some cases resources availability is limited to certain times of day as for our problem. Some personnel scheduling problems such as crew scheduling also include travel between locations. However, personnel scheduling problems generally do not contain any notion of precedence, which is highly significant to both our problem and the RCPSP.

## 3.2 Solution Techniques for the RCPSP

Since their introduction by Bremermann and Fraser and the seminal work done by Holland [4], genetic algorithms (GAs) have been developed extensively to tackle problems including the travelling salesman problem, bin packing problems and scheduling problems (see for example [5]). [6] reviewed a range of meta-heuristics for solution of the RCPSP and showed genetic algorithms to work consistently well compared with other methods such as tabu search and simulated annealing. Many of the genetic algorithms presented in [6] worked by optimising the order in which tasks are passed to a schedule generation scheme (a greedy constructive heuristic). A serial schedule generation scheme (serial SGS) builds schedules by inserting tasks one by one into the schedule as early as possible. Most of the techniques reviewed in [6] use a serial SGS. Serial SGSs have been shown to be superior in general to parallel SGSs which works by taking a slice of time and inserting as many tasks as possible into it before moving on to another later slice of time. One such genetic algorithm is presented in [5]. Another more complex approach is presented in [7], here a genetic algorithm is used to generate a schedule in a similar fashion to [5], however once the schedule is generated, a forward backward improvement heuristic is used to improve the resulting schedule. The GA in [5] is shown to work well across a set of test RCPSP problems [6], however the genetic algorithm presented in [7] with forward backward improvement is shown to be one of the best approaches presented across the test problem set. None of the aforementioned approaches use real world problems or real problem data and so their performance under such conditions is unknown.

## 3.3 Multi-objective Scheduling Techniques

The most widely used method for combining multiple objectives in genetic algorithms is the weighted sum method. This can be problematic as practitioners estimated weights are often not a good reflection of the true requirements for a globally good solution. Simply put, a user is not used to being asked to explicitly define the relative importance of different problem goals, and the weights defined may reflect small local effects (since they are easy for the user to understand) rather than more difficult to define global ones.

Multi-objective approaches do not rank solutions directly as weighted sum approaches do; instead they use the notion of dominance and the distribution of solutions in objective space to decide the overall quality of a population of solutions. An important notion is that of a non-dominated solution, which is a solution where there are no other solutions better with respect to all the objectives. Multi-objective approaches try to maintain a set of solutions which are non-dominated and to get a good distribution of these solutions in objective space. This is useful in scheduling as the user no longer has to specify a set of specific weights representing the kind of schedule they think they are looking for, instead they can choose a schedule from a diverse set. A range of multi-objective genetic algorithms are among the most widely used approaches.

The Vector Evaluated Genetic Algorithm (VEGA) was proposed and compared with an adaptive random search technique in [8], which showed that VEGA out performed adaptive random search in terms of solution quality. VEGA finds a set of non-dominated solutions and works by splitting the population randomly in to a number of subpopulations (the number of subpopulations being equal to the number of objectives to be considered). The population evolves with a mating pool created using a proportion of individuals from each sub-population with different objective functions. The technique is simple, however, each solution is only evaluated with one objective at any time and because of this eventually all solutions converge towards one solution with respect to a single objective [9]. In real world problems this convergence towards one objectives best solution is unworkable where we seek a trade off between objectives.

The Non-dominated Sorting Genetic Algorithm (NSGA) [10], and its enhancement NSGA-II [11] sort the population into non-dominated fronts. This is done by first identifying non-dominated individuals in the population, these are on the first front, the first front is then removed from the population and then non-dominated individuals are identified again, these comprise the second front. This process is repeated until all fronts are identified. The computational complexity of the algorithm has been reduced from $O(mN^3)$ for NSGA to $O(mN^2)$ for NSGA-II per generation, where $m$ is the number of objectives and N is the population size; however the memory requirements have increased from $O(N)$ to $O(N^2)$. Elitism has been introduced in NSGA-II to force the algorithm to keep the extreme maximum and minimum solutions for all objectives, as has a new algorithm for crowding distance calculation. Crowding distance is a representation of the density of neighbouring individuals on any given front. Mating selection in NSGA-II is performed by binary tournament selection using the crowded comparison operator. If both solutions are on the same front, then the crowding distance is used as a tie breaker. The parent and child populations are combined and this new population (of size 2N) is ranked according to front. Fronts are then added to the next generation's population (starting with the non-dominated front) until the size exceeds N. Once this has been done, crowding distance assignment is applied to the last front that has been added and the crowded comparison operator is used to sort this final front and the worst individuals are removed to give the next generation's population.

The Strength Pareto Evolutionary Algorithm (SPEA) was introduced in [12], and further improved to SPEA2 in [13]. SPEA2 uses Pareto based ideas in both population selection and mating selection. Fitness assignment in SPEA2 works in two parts: domination, and distance to the $k^{th}$ nearest neighbour. The domination of an individual $i$, is calculated by counting the individuals dominated by $i$, this is the individuals *strength* value. Then, the *raw fitness R(i)* is calculated for each individual $i$ by adding all the *strength* values of individuals dominated by $i$. The *distance* value $\sigma_i^k$ is created for each individual $i$ by finding the Euclidean distance to all other individuals in objective space and sorting them in ascending order. The $k^{th}$ closest is taken where $k$ is a user defined parameter ([13] recommends $k$ parameter to be the square root of the combined size of the population and the archive population). Then a *density* value $D(i)$ is calculated by the equation $D(i) = \dfrac{1}{\left(\sigma_i^k + 2\right)}$ ensuring that (in the final calculation of fitness) the raw fitness value takes precedence over it. Finally, the total fitness for each individual ($i$) can be calculated using $F(i) = R(i) + D(i)$.

SPEA2 uses an archive to allow individuals to survive from one generation to the next. At the end of a generation all non-dominated individuals from the archive and the population are copied to a new archive. If the archive is not filled by this process the best non-dominated individuals (according to the SPEA2 fitness function) from the population can also be copied into it. If however, the archive is overfilled by non-dominated individuals it is truncated by removing individuals based on their $\sigma_i^k$ distance to other individuals in the archive.

NSGA-II and SPEA2 have been compared with SPEA and Pareto Envelope-Based Selection Algorithm (PESA) [14] on a set of test problems including the knapsack problem in [13]. NSGA-II and SPEA2 are shown to perform the best out of all tested. SPEA2 is said to be better in higher dimensional objective spaces and the solutions generated by SPEA2 are shown to dominate those generated by NSGA-II 80% of the time (on average).

# 4 Application of Multi-objective Genetic Algorithms to Our Workforce Scheduling Problem

As the review of [6] showed genetic algorithms such as [7] and [5] perform well across a set of test problem instances, we have chosen to adopt a similar approach to our workforce scheduling problem. We use a serial schedule generation scheme to encapsulate the assignment of tasks to resources and constraint handling for our problem. A permutation based genetic algorithm optimises the order in which tasks are passed to this serial schedule generation scheme. We consider two multi-objective GAs to a single, weighted sum objective GA which uses binary tournament selection and elitist replacement. The multi-objective genetic algorithms will use NSGA-II and SPEA2 multi-objective approaches for the mating and environmental selection phases.

## 4.1 Serial Schedule Generation Scheme (SGS)

The serial schedule generation scheme we have built uses a permutation of tasks to generate a schedule. As our problem is more complex than traditional RCPSP problems (due the way skills, resources and task durations are defined) the task scheduling process within the SGS is split into two sub processes, resource selection and task insertion. This is also where many of the similarities between the RCPSP serial SGS methods defined in [15] and our method end. This is unfortunate as for the RCPSP solved with an activity list serial SGS there is always an order of tasks which will generate an optimal schedule when a regular performance measure is considered [15]. However, this is not the case with our SGS that because of the complexity of allocating resources having appropriate skills to a task requiring multiple skills, and time window constraints. The schedules generated will however always be feasible.

The motivation for divorcing resource selection from task insertion in this way is mainly due to the fact that until resources are selected for a task, there is no way of knowing exactly how long the task will take. This uncertainty causes many problems in the scheduling process, and is often the cause of tasks going unscheduled. Resources are selected by finding the intersection of the periods of time they have available in common with the task time windows and the other resources already selected (here a larger amount of available time is preferable). Due to the consideration of travel and competency in resource selection ties rarely occur, however when they do they are broken randomly. Once all resources are chosen the task is then inserted as early as possible in the schedule.

## 5  The Competitiveness of Multi-objective Genetic Algorithms

Real world scheduling problems often have many and contradictory objectives making them an interesting testbed for comparing the performance of multi-objective genetic algorithms and weighted sum genetic algorithms. In our experiments we intend to investigate the competitiveness of multi-objective genetic algorithms when compared with weighted sum genetic algorithms in terms of both solution quality and diversity.

In our experiments diversity will be measured using the "Max Spread" measure from [9] and the "Morrison and De Jong" measure which is based around ideas taken from the moment of inertia in mechanical engineering [16]. Solution quality will be assessed for each population (whether from a multi-objective or weighted sum run) by several weighted sum objective functions. We do not expect that multi-objective methods will outperform all weighted sum methods when assessed by weighted sum objective functions but we are interested to see how close they come. Equally, we do not expect weighted sum methods to outperform multi-objective methods in terms of diversity but are interested to see how the diversity of the two respective populations compares. In order to reduce the effects of randomness, each method (7 weighted sum and 2 multi-objective) will be run ten times each on a set of three different problem instances with a stopping criterion of 250 generations of evolution. The problem instances were generated using the problem generator we have developed in collaboration with Vidus Ltd. (an @Road company). The test problems we used had 100 tasks, 10 resources and 6 skills, considered over a 3-day scheduling period, and correspond to a "small to medium sized" problem in practice.

These experiments were run using a master-slave approach to parallelise and thus speed up the GA runs. We used 26 2.8Ghz Pentium 4 machines (25 slaves and one master) and each of the GAs has a population size of 100 and was allowed the same number of schedule evaluations over 250 generations. Each run took around 1 (single-machine) CPU hour on this parallel architecture. Since we consider 9 solution approaches and 10 runs for each approach for 3 problem instances, a total of 270 runs were carried out.

Values for mutation and crossover rates will be taken from previous parameter tuning experiments for the weighted sum objective method with binary tournament selection and elitist replacement, these being a crossover rate of 25% and a mutation rate of 1%. The reader should note that these parameters have not been specially tuned for the multi-objective methods, particularly since such tuning is much more difficult when objective weights are not assumed.

The weighted sum methods will use the objective functions shown in Table 1 where SP is the sum of all the priority values of all scheduled tasks, SC is the total schedule cost, and TT is the total travel time on the schedule. The values in Table 1 have been chosen to provide a trade off between objectives. Here we are considering a diverse range of weights to see if we can get close to a schedule which is good for a human scheduler without knowing what the human scheduler's ideal weighting scheme is, since the human scheduler is unlikely to know these weights in practise.

**Table 1.** Objective functions for weighted sum methods

| Formula for Calculation | Name on Graphs |
|---|---|
| $f = SP$ | SP |
| $f = SP - (SC * 6)$ | SP_6SC |
| $f = SP - (TT * 6)$ | SP_6TT |
| $f = SP - (SC * 2) - (TT * 4)$ | SP_2SC_4TT |
| $f = SP - (SC * 4) - (TT * 2)$ | SP_4SC_2TT |
| $f = SP - (SC * 2) - (TT * 6)$ | SP_2SC_6TT |
| $f = SP - (SC * 6) - (TT * 2)$ | SP_6SC_2TT |

The genetic algorithm was run ten times with each weighted sum objective function shown in Table 1, as well as ten times with both NSGA-II and SPEA2 multi-objective methods. Each of these runs were assessed by each of the objective functions in Table 1 as well as the "Max Spread" [9] and "Morrison and De Jong" [16] diversity measures. Averages over the ten runs of each have been taken and plotted as a percentage of the best average found.

## 5.1 Experimental Results

Figure 1 shows the average fitness value of the best individual in the final population over thirty runs (three problem instances run ten times each). Each group of results along the x-axis represents the performance of all the different types of GA when assessed using fixed objective weights as given in table 1. For example, the first group (of 9 bars) in Figure 1 shows the performance of the multi-objective methods NSGA-II and SPEA2 as well as the single objective methods (with different objective weights) when assessed using the SP objective function.

Figure 1 shows that multi-objective algorithms find solutions that are within 2% of the best solution found by weighted sum objectives (when assessed by the weighted sum objectives) without knowing what the weights are and often find a solution within 1% of the best. This cannot be said of the weighted sum objective approaches which use the "wrong" weights whose performance is much less consistent. Figure 1 illustrates the possible effect of a poorly defined set of weights on resulting solution quality, for example if our actual global objective function (not known to the user) is SP_6SC and the user defines SP as the objective function then the solution they ended up with would be much worse in global terms than if they had used a multi-objective approach. The NSGA-II approach yielded results as good or slightly better than SPEA2 on average in all cases.

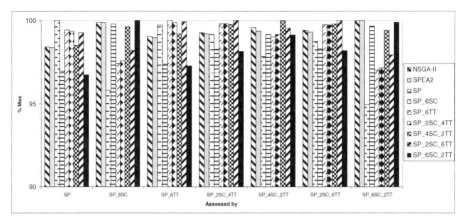

**Fig. 1.** The relative performance of multi-objective and single objective weighted sum genetic algorithms when assessed by different weighted sum objective functions

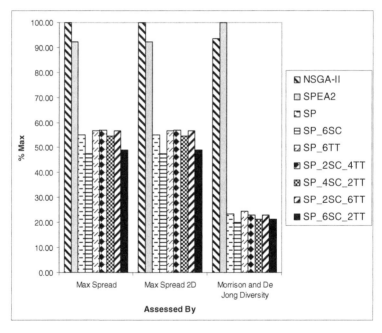

**Fig. 2.** The relative performance of multi-objective and single objective weighted sum genetic algorithms when assessed by diversity metrics

An interesting observation is that sometimes the "best" result is not found by the GA that is running the same objective function weights as is being used to assess the runs, an example of this is shown in Figure 1 when assessing the runs with the SP_6SC_2TT objective, here the GA that was run with the SP_6SC_2TT objective is outperformed by both NSGA-II and SPEA2 albeit by a small amount. In a case like this we believe that there may be a mismatch between the schedule generation scheme's task-resource allocation heuristic and the objective weights chosen, but this flaw is not exposed by the NSGA-II and SPEA2 methods which have a diverse population.

Figure 2 shows the average values of the diversity measures for the different GA populations. Here NSGA-II and SPEA2 yield a much more diverse population than weighted sum approaches.

The results of figures 1 and 2 show that where it may not be possible for a user to define weights for each objective, still at least one of the population generated using a multi-objective approach is likely to be close to the users preference, even though this preference is not known in advance. A new problem arises in this case for the user, that of selecting a good solution from a population. However, this is a problem that the user is more familiar with and which is likely to yield better results than choosing objective weights in many cases. Ways in which the users ability to choose between schedules for the RCPSP such as those considered in Shackleford and Corne [17] and there approaches could be useful here.

## 6   Conclusions

In this paper multi-objective genetic algorithms have been shown to be an effective approach allowing the user to avoid having to define weights for a set of objectives in exchange for a small decrease in solution quality. This is useful as expressing their knowledge as a set of weights for a real world problem is usually difficult for human users who are not used to being asked to explicitly define the relative importance of different problem goals.

One problem with this approach is that it replaces the problem of defining a set of weights with the problem of selecting a solution from the Pareto optimal set which is an interesting research question in itself. However, analysing schedules and choosing a good one is a much more familiar activity for a human scheduler.

In future work it will be interesting to consider problems with a larger number of objectives, and methods which integrate the users' ability to effectively select a solution into the solution process.

## References

1. Pinedo, M., Chao, X.: Operations scheduling with applications in manufacturing and services. McGraw-Hill, New York (1999)
2. Brucker, P., Knust, S.: Resource-Constrained Project Scheduling and Timetabling. In: Burke, E. Erben, W. (eds.): Practice and Theory of Automated Timetabling III (PATAT). Springer-Verlag Berlin Hidelberg New York (2000) 277-293
3. Lova, A., Maroto, C., Thomas, P.: A multicriteria heuristic method to improve resource allocation in multiproject scheduling. Eur. J. Op. Res. Vol. 127. Elsevier (2000) 408-424
4. Holland, J. H.: Adaptation in Natural and Artificial Systems. University of Michigan Press. Michigan. (1975)
5. Hartmann, S.: A Competitive Genetic Algorithm for Resource-Constrained Project Schedul-ing. Naval Research Logistics. Vol. 45. Wiley (1998) 733-750.

6. Kolisch, R., Hartmann, S.: Experimental investigation of heuristics for resource-constrained project scheduling: An update. Article to appear in the Eur. J. Op. Res. (2005)
7. Valls, V., Ballestin, F., Quintanilla, S.: A hybrid genetic algorithm for the RCPSP. Technical Report, Department of Statistics and Operations Research, University of Valencia.. (2005)
8. Schaffer J.D.: Multiple Objective Optimization with Vector Evaluated Genetic Algorithms. Proceedings of an International Conference on Genetic Algorithms and their Applications, (1985) 93-100
9. Deb, K.: Multi-objective Optimisation using Evolutionary Algorithms. John Wiley and Sons, New York.(2001)
10. Srinivas, N., Deb, K.: Multiobjective Optimisation Using Nondominated Sorting in Genetic Algorithms. J. Evolutionary Computation, Vol. 2 No. 3. (1994) 221-248
11. Deb, K., Pratap, A., Agarwal, S., Meyarivan, T.: A Fast and Elitist Multi-Objective Genetic Algorithm-NSGA-II. IEEE Trans. Evolutionary Computation, Volume 6. (2002) 849-858
12. Zitzler, E., Thiele, L.: An evolutionary algorithm for multiobjective optimization: The strength pareto approach. Technical Report 43, Computer Engineering and Networks Laboratory (TIK), Swiss Federal Institute of Technology (ETH) Zurich, Gloriastrasse 35, CH-8092 Zurich, Switzerland. (1998)
13. Zitzler, E., Laumanns, M., Thiele, L.: SPEA2: Improving the Strength Pareto Evolutionary Algorithm. Technical Report 103, Computer Engineering and Networks Laboratory (TIK), Swiss Federal Institute of Technology (ETH) Zurich, Gloriastrasse 35, CH-8092 Zurich, Switzerland. (2001)
14. Corne, D. W., Knowles, J. D., Oates, M. J.: The pareto envelope-based selection algorithm for multiobjective optimisation. In M. S et al. (Ed.) Parallel Problem Solving from Nature. PPSN VI, Berlin (2000) 839-848
15. Kolisch, R., Hartmann, S.: Heuristic algorithms for solving the resource-constrained project scheduling problem: Classification and computational analysis. In Weglarz, J. (ed): Project scheduling: Recent models, algorithms and applications. Kluwer, Amsterdam, the Netherlands (1999) 147-178
16. Landa Silva, J.D., Burke, E.K.: Using Diversity to Guide the Search in Multi-Objective Optimization. In: Coello Coello C.A., Lamont G.B. (eds.): Applications of Multi-Objective Evolutionary Algorithms, Advances in Natural Computation, Vol. 1, World Scientific. (2004) 727-751
17. Shackelford, M., Corne, D.: Collaborative Evolutionary Multi-Project Resource Scheduling. In Proceedings of the 2001 Congress on Evolutionary Computation, IEEE Press (2001) 1131-1138
18. Ernst, A.T., Jiang, H., Krishnamoorthy, M., Sier, D.: Staff scheduling and rostering: A review of applications, methods and models. European Journal of Operational Research. Vol. 153. Elsevier (2004) 3-27

# Evolving the Structure of the Particle Swarm Optimization Algorithms

Laura Dioşan and Mihai Oltean

Department of Computer Science,
Faculty of Mathematics and Computer Science,
Babeş-Bolyai University, Cluj-Napoca, Romania
{lauras, moltean}@cs.ubbcluj.ro

**Abstract.** A new model for evolving the structure of a Particle Swarm Optimization (PSO) algorithm is proposed in this paper. The model is a hybrid technique that combines a Genetic Algorithm (GA) and a PSO algorithm. Each GA chromosome is an array encoding a meaning for updating the particles of the PSO algorithm. The evolved PSO algorithm is compared to a human-designed PSO algorithm by using ten artificially constructed functions and one real-world problem. Numerical experiments show that the evolved PSO algorithm performs similarly and sometimes even better than standard approaches for the considered problems.

## 1 Introduction

Particle Swarm Optimization (PSO) is a population based stochastic optimization technique developed by Kennedy and Eberhart in 1995 [8]. Standard PSO algorithm randomly initializes a group of particles (solutions) and then searches for optima by updating all particles along a number of generations. In any iteration, each particle is updated by following some rules [16].

Standard model implies that particles are updated synchronously [16]. This means that the current position and speed for a particle is computed taking into account only information from the previous generation of particles.

A more general model allows updating any particle anytime. This basically means three things:

1. The current state of the swarm is taken into account when a particle is updated. The best global and local values are computed for each particle which is about to be updated, because the previous modifications could affect these two values. This is different from the standard PSO algorithm where the particles were updated taken into account only the information from the previous generation. Modifications performed so far (by a standard PSO) in the current generation had no influence over the modifications performed further in the current generation.
2. Some particles may be updated more often than other particles. For instance, in some cases, is more important to update the best particles several times per generation than to update the worst particles.

J. Gottlieb and G.R. Raidl (Eds.): EvoCOP 2006, LNCS 3906, pp. 25–36, 2006.

3. We will work with only one swarm. Any updated particle will replace its parent. Note that two populations/swarms are used in the standard PSO and the current swarm is filled taken information from the previous generation.

Unlike Parsopoulos' work [13] which use Differential Evolution algorithm (suggested by Storn and Price [14]) for "on fly" adaptation of the PSO parameters, our work looks for designing a new PSO algorithm by taking into account the information from the problem being solved.

Our main purpose is to evolve the structure of a PSO algorithm. This basically means that we want to find which particles should be updated and which is the order in which these particles are updated. In this respect we propose a new technique which is used for evolving the structure of a PSO algorithm. We evolve arrays of integers which provide a meaning for updating the particles within a PSO algorithm during iteration.

Our approach is a hybrid technique that works at two levels: the first (macro) level consists in a steady-state GA [7] whose chromosomes encode the structure of PSO algorithms. In order to compute the quality of a GA chromosome we have to run the PSO encoded into that chromosome. Thus, the second (micro) level consists in a modified PSO algorithm that provides the quality for a GA chromosome.

Firstly, the structure of a PSO algorithm is evolved and later, the obtained algorithm is used for solving eleven difficult function optimization problems The evolved PSO algorithm is compared to a human-designed PSO algorithm by using 10 artificially constructed functions and one real-world problem. Numerical experiments show that the evolved PSO algorithm performs similarly and sometimes even better than standard approaches for several well-known benchmarking problems.

This research was motivated by the need of answering several important questions concerning PSO algorithms. The most important question is: *Can a PSO algorithm be automatically synthesized by using only the information about the problem being solved?* And, if yes, which is the optimal structure of a PSO algorithm (for a given problem)? We better let the evolution find the answer for us.

The rules employed by the evolved PSO during a generation are not preprogrammed. These rules are automatically discovered by the evolution.

Several attempts for evolving Evolutionary Algorithms (EAs) using similar techniques were made in the past. A non-generational EA was evolved [11] using the Multi Expression Programming (MEP) technique [11]. A generational EA was evolved [12] using the Linear Genetic Programming (LGP) technique. Numerical experiments have shown [11, 12] that the evolved EAs perform similarly and sometimes even better than the standard evolutionary approaches with which they have been compared. A theoretical model for evolving EAs has been proposed in [15].

The paper is structured as follows: section 2 describes, in detail, the proposed model. Several numerical experiments are performed in section 3. Test functions are given in section 3.1. Conclusions and further work directions are given in section 4.

# 2    Proposed Model

## 2.1    Representation

Standard PSO algorithm works with a group of particles (solutions) and then searches for optima by updating them during each generation.

During iteration, each particle is updated by following two "best" values. The first one is the location of the best solution that a particle has achieved so far. This value is called *pBest*. Another "best" value is the location of the best solution that any neighbor of a particle has achieved so far. This best value is a neighborhood best and called *nBest*.

In a standard PSO algorithm all particles will be updated once during the course of iteration.

In real-world swarm (such as flock of birds) not all particles are updated in the same time. Some of them are updated more often and others are updated later or not at all. Our purpose is to simulate this, more complex, behavior. In this case we were interested to discover (evolve) a model which can tell us which particles must be updated and which is the optimal order for updating them.

We will use a GA [7] for evolving this structure. Each GA individual is a fixed-length string of genes. Each gene is an integer number, in the interval $[0...SwarmSize - 1]$. These values represent indexes of the particles that will be updated during PSO iteration.

Some particles could be updated more often and some of them are not updated at all. Therefore, a GA chromosome must be transformed so that it has to contain only the values from 0 to $Max$, where $Max$ represents the number of different genes within the current array.

**Example.** Suppose that we want to evolve the structure of a PSO algorithm with 8 particles. This means that the $SwarmSize = 8$ and all chromosomes of our macro level algorithm will have 8 genes whose values are in the $[0, 7]$ range. A GA chromosome with 8 genes can be:

$$C_1 = (2, 0, 4, 1, 7, 5, 6, 3).$$

For computing the fitness of this chromosome we will use a swarm with 8 individuals and we will perform, during one generation, the following updates:

| | |
|---|---|
| update(Swarm[2]), | update(Swarm[7]), |
| update(Swarm[0]), | update(Swarm[5]), |
| update(Swarm[4]), | update(Swarm[6]), |
| update(Swarm[1]), | update(Swarm[3]). |

In this example all 8 particles have been updated once per generation.

Let us consider another example which consists of a chromosome $C_2$ with 8 genes that contain only 5 different values.

$$C_2 = (6, 2, 1, 4, 7, 1, 6, 2)$$

In this case particles 1, 2 and 6 are updated 2 times each and particles 0, 3, 5 are not updated at all. Because of that it is necessari to remove the useless particles and to scale the genes of the GA chromosome to the interval $[0 \ldots 4]$. The obtained chromosome is:

$$C_2' = (3, 1, 0, 2, 4, 0, 3, 1).$$

The quality for this chromosome will be computed using a swarm of size 5 (5 swarm particles), performing the following 8 updates:

update(Swarm[3]),                update(Swarm[4]),
update(Swarm[1]),                update(Swarm[0]),
update(Swarm[0]),                update(Swarm[3]),
update(Swarm[2]),                update(Swarm[1]).

We evolve an array of indexes based on the information taken from a function to be optimized. Note that the proposed mechanism should not be based only on the index of the particles in the *Swarm* array. This means that we should not be interested in updating a particular position since that position can contain (in one run) a very good individual and the same position could hold a very poor individual (during another run). For instance it is easy to see that all GA chromosomes, encoding permutations, perform similarly when averaged over (let's say) 1000 runs.

In order to avoid this problem we sort (after each generation) the *Swarm* array ascending based on the fitness value. The first position will always hold the best particle at the beginning of a generation. The last particle in this array will always hold the worst particle found at the beginning of a generation. In this way we will know that *update(Swarm*[0]), will mean something: not that one of the particles is updated, but that the best particle (at the beginning of the current generation) is updated.

## 2.2   Fitness Assignment

The model proposed in this paper in divided in two levels: a macro level and a micro level. The macro-level is a GA algorithm that evolves the structure of a PSO algorithm. For this purpose we use a particular function as training problem. The micro level is a PSO algorithm used for computing the quality of a GA chromosome from the macro level.

The array of integers encoded into a GA chromosome represents the order of update for particles used by a PSO algorithm that solves a particular problem. We embed the evolved order within a modified Particle Swarm Optimization algorithm as described in sections 2 and 2.3.

Roughly speaking the fitness of a GA individual is equal to the fitness of the best solution generated by the PSO algorithm encoded into that GA chromosome. But, since the PSO algorithm uses pseudo-random numbers, it is very likely that successive runs of the same algorithm will generate completely different solutions. This problem can be handled in a standard manner: the PSO algorithm encoded by the GA individual is run multiple times (50 runs in fact) and the fitness of the GA chromosome is averaged over all runs.

## 2.3   The Algorithms

The algorithms used for evolving the PSO structure are described in this section. Because we use a hybrid technique that combines GA and PSO algorithm within a two-level model, we describe two algorithms: one for macro-level (GA) and another for micro-level (PSO algorithm).

**The Macro-level Algorithm.** The macro level algorithm is a standard GA [7] used for evolving particles order of update. We use steady-state evolutionary model as underlying mechanism for our GA implementation. The GA algorithm starts by creating a random population of individuals. Each individual is a fixed-length array of integer numbers. The following steps are repeated until a given number of generations are reached: Two parents are selected using a standard selection procedure. The parents are recombined (using one-cutting point crossover) in order to obtain two offspring. The offspring are considered for mutation which is performed by replacing some genes with randomly generated values. The best offspring $O$ replaces the worst individual $W$ in the current population if $O$ is better than $W$.

**The Micro-level Algorithm.** The micro level algorithm is a modified Particle Swarm Optimization algorithm [16] used for computing the fitness of a GA individual from the macro level.

$S_1$ Initialize the swarm of particles randomly
$S_2$ While not stop_condition
$S_3$ For each gene of the GA chromosome
    $S_{31}$ Compute fitness of the particle specified by the current gene of the GA chromosome
    $S_{32}$ Update $pBest$ if the current fitness value is better than $pBest$
    $S_{33}$ Determine $nBest$ for the current particle: choose the particle with the best fitness value of all the neighbors as the $nBest$
    $S_{34}$ Calculate particle's velocity according to eq. 1
    $S_{35}$ Update particle's position according to eq. 2
$S_4$ EndFor
$S_5$ Sort particles after fitness.

$$v_{id} = w * v_{id} + c_1 * rand() * (p_{id} - x_{id}) + c_2 * rand() * (p_{nd} - x_{id}) \qquad (1)$$

$$x_{id} = x_{id} + v_{id} \qquad (2)$$

where $rand()$ generates a random real value between 0 and 1.

The above algorithm is quite different from the standard PSO algorithm [16].

Standard PSO algorithm works on two stages: one stage that establishes the fitness, $pBest$ and $nBest$ values for each particle and another stage that determines the velocity (according to equation 1) and makes update according to equation 1 for each particle. Standard PSO usually works with two populations/swarms. Individuals are updated by computing the $pBest$ and $nBest$ value using the information from the previous population. The newly obtained individuals are added to the current population.

Our algorithm performs all operations in one stage only: determines the fitness, *pBest*, *nBest* and velocity values only when a particle is about to be updated. In this manner, the update of the current particle takes into account the previous updates in the current generation. Our PSO algorithm uses only one population/swarm. Each updated particle will automatically replace its parent.

## 3    Experiments

Numerical experiments for evolving a PSO algorithm for function optimization are performed in this section. The obtained PSO algorithm is tested against 11 difficult problems. Several numerical experiments, with a standard Particle Swarm Algorithm [16] are also performed. Finally the results are compared. We evolve the structure of a PSO algorithm and then we asses its performance by comparing it with the standard PSO algorithm.

### 3.1    Test Functions

Eleven test problems are used in order to asses the performance of the evolved EA. Functions $f_1 - f_6$ are unimodal test function. Functions $f_7 - f_{10}$ are highly multi modal (the number of the local minimum increases exponentially with problem's dimension [17]). Functions $f_1 - f10$ are given in Table 1. Function $f_{11}$ corresponds to the constrained portfolio optimization problem.

**The Portfolio Selection Problem.** Modern computational finance has its historical roots in the pioneering portfolio theory of Markowitz [10]. This theory is based on the assumption that investors have an intrinsic desire to maximize return and minimize risk on investment. Mean or expected return is employed as a measure of return, and variance or standard deviation of return is employed as a measure of risk. This framework captures the risk-return tradeoff between a single linear return measure and a single convex nonlinear risk measure. The solution typically proceeds as a two-objective optimization problem where the return is maximized while the risk is constrained to be below a certain threshold. The well-known risk-return efficient frontier is obtained by varying the risk target and maximizing on the return measure.

The Markowitz mean-variance model [10] gives a multi-objective optimization problem, with two output dimensions. A portfolio $p$ consisting of $N$ assets with specific volumes for each asset given by weights $w_i$ is to be found, which minimizes the variance of the portfolio:

$$\sigma_p = \sum_{i=1}^{N} \sum_{j=1}^{N} w_i w_j \sigma_{ij} \tag{3}$$

maximizes the return of the portfolio:

$$\mu_p = \sum_{i=1}^{N} w_i \mu_i \text{ subject to: } \sum_{i=1}^{N} w_i = 1, 0 \leq w_i \leq 1, \tag{4}$$

**Table 1.** Test functions used in our experimental study. The parameter $n$ is the space dimension ($n = 5$ in our numerical experiments) and $f_{min}$ is the minimum value of the function. All functions should be minimized.

| Test function | Domain | $f_{min}$ |
|---|---|---|
| $f_1(x) = \sum\limits_{i=1}^{n} (i \cdot x_i^2).$ | $[-10, 10]^n$ | 0 |
| $f_2(x) = \sum\limits_{i=1}^{n} x_i^2.$ | $[-100, 100]^n$ | 0 |
| $f_3(x) = \sum\limits_{i=1}^{n} |x_i| + \prod\limits_{i=1}^{n} |x_i|.$ | $[-10, 10]^n$ | 0 |
| $f_4(x) = \sum\limits_{i=1}^{n} \left( \sum\limits_{j=1}^{i} x_j \right)^2.$ | $[-100, 100]^n$ | 0 |
| $f_5(x) = \max_i\{x_i, 1 \leq i \leq n\}.$ | $[-100, 100]^n$ | 0 |
| $f_6(x) = \sum\limits_{i=1}^{n-1} 100 \cdot (x_{i+1} - x_i^2)^2 + (1 - x_i)^2.$ | $[-30, 30]^n$ | 0 |
| $f_7(x) = 10 \cdot n + \sum\limits_{i=1}^{n} (x_i^2 - 10 \cdot \cos(2 \cdot \pi \cdot x_i))$ | $[-5, 5]^n$ | 0 |
| $f_8(x) = -a \cdot e^{-b\sqrt{\frac{\sum\limits_{i=1}^{n} x_i^2}{n}}} - e^{\frac{\sum \cos(c \cdot x_i)}{n}} + a + e.$ | $[-32, 32]^n$ $a = 20,\, b = 0.2,\, c = 2\pi.$ | 0 |
| $f_9(x) = \frac{1}{4000} \cdot \sum\limits_{i=1}^{n} x_i^2 - \prod\limits_{i=1}^{n} \cos(\frac{x_i}{\sqrt{i}}) + 1.$ | $[-500, 500]^n$ | 0 |
| $f_{10}(x) = \sum\limits_{i=1}^{n} (-x_i \cdot \sin(\sqrt{|x_i|}))$ | $[-500, 500]^n$ | $-n* 418.98$ |
| $f_{11} = $ The Portfolio Selection Problem | $[0, 1]^n$ | 0 |

where $i = 1...N$ is the index of the asset, $N$ represents the number of assets available, $\mu_i$ the estimated return of asset $i$ and $\sigma_{ij}$ the estimated covariance between two assets. Usually, $\mu_i$ and $\sigma_{ij}$ are to be estimated from historic data. While the optimization problem given in (3) and (4) is a quadratic optimization problem for which computationally effective algorithms exist, this is not the case if real world constraints are added. In this paper we treat only the cardinality constraints problem.

**Cardinality constraints** restrict the maximal number of assets used in the portfolio

$$\sum_{i=1}^{N} z_i = K, \text{ where } z_i = sign(w_i). \tag{5}$$

Let $K$ be the desired number of assets in the portfolio, $\epsilon_i$ be the minimum proportion that must be held of asset $i$, ($i = 1, ..., N$) if any of asset $i$ is held, $\delta_i$ be the maximum proportion that can be held of asset $i$, ($i = 1, ..., N$) if any of asset $i$ is held, where we must have $0 \leq \epsilon_i \leq \delta_i \leq 1 (i = 1, ..., N)$. In practice, $\epsilon_i$ represents a "min-buy" of "minimum transaction level" for asset $i$ and $\delta_i$ limits the exposure of the portfolio to asset $i$.

$$\epsilon_i z_i \leq w_i \leq \delta_i z_i, i = 1, ..., N \qquad (6)$$
$$w_i \in [0, 1], i = 1, ..., N. \qquad (7)$$

Equation (5) ensures that exactly $K$ assets are held. Equation (6) ensures that if any of asset $i$ is held ($z_i = 1$) its proportion $w_i$ must lie between $\epsilon_i$ and $\delta_i$, whilst if none of asset is held ($z_i = 0$) its proportion $w_i$ is zero. Equation (7) is the integrality constraint.

The objective function (equation (3)), involving as it does the covariance matrix, is positive semi-definite and hence we are minimizing a convex function.

The chromosome representation (within a GA algorithm) supposes (conform to [2]) a set $Q$ of $K$ distinct assets and $K$ real numbers $s_i$, ($0 \leq s_i \leq 1$), $i \in Q$.

Now, given a set $Q$ of $K$ assets, a fraction $\sum_{j \in Q} \epsilon_j$ of the total portfolio is already accounted for and so we interpret $s_i$ as relating to the share of the *free* portfolio proportion $(1 - \sum_{j \in Q} \epsilon_j)$ associated with asset $i \in Q$.

So, our GA chromosome will encode real numbers $s_i$ and the proportion of asset $i$ from $Q$ in portfolio will be:

$$w_i = \epsilon_i + \frac{s_i}{\sum_{j \in Q} s_j} (1 - \sum_{j \in Q} \epsilon_j) \qquad (8)$$

For this experiment we have used the daily rate of exchange for a set of assets quoted to Euronext Stock [6].

**Experiment 1.** The structure of a PSO algorithm is evolved in this experiment. We use function $f_1$ as training problem.

For GA we run during 50 generations a population with 50 individuals, each individual having 10 genes. We perform a binary tournament selection, one cutting point recombination (applied with probability 0.8) and weak mutation (applied with probability 0.1). The parameters of the PSO algorithm (micro level) are given in Table 2. The *SwarmSize* is not included in this table because different PSOs may have different number of particles. However, the number of function evaluations/generation is equal to 10 for all evolved PSO.

Our algorithm uses a randomized inertia weight, selected in the spirit of Clerc's constriction factor [3], [5] (many reports use a linearly decreasing inertia weight which starts at 0.9 and ends at 0.4, but we want to not restrict our

**Table 2.** The parameters of the PSO algorithm (the micro level algorithm) used for computing the fitness of a GA chromosome

| Parameter | Value |
|---|---|
| Number of generations | 50 |
| Number of function evaluations/generation | 10 |
| Number of dimensions of the function to be optimized | 5 |
| Learning factor $c_1$ | 2 |
| Learning factor $c_2$ | 1.8 |
| Inertia weight | 0.5 + rand() / 2 |

inertia to a fix model: decreasing or increasing function). Learning factors are not identical. Initial we have used same values for this parameters, but recent work [1] reports that it might be even better to choose a larger cognitive parameter, $c_1$, than a social parameter, $c_2$, but with $c_1 + c_2 < 4$.

We performed 50 independent runs for evolving order for particles. The results obtained in one of the runs (randomly selected from the set of 50 runs) are presented in Figure 1.

Different orders of particles have been evolved. Two of these orders are represented by the chromosomes: $C_1 = (1112020113)$ and $C_2 = (1502033043)$. The second chromosome ($C_2$) will be used in the numerical experiments performed in the next section.

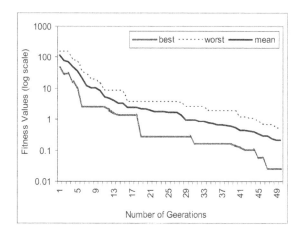

**Fig. 1.** The evolution of the fitness of the best/worst GA individual, and the average fitness (of all GA individuals in the population) in a particular run

**Experiment 2.** For assessing the performance of the evolved PSO we will compare it with a standard PSO algorithm. For this comparison we use the test functions given in Table 1.

In order to make a fair comparison we have to perform the same number of function evaluations in both evolved PSO and standard PSO. The evolved PSO has 6 particles, but because some of them are updated more times, it will perform 10 function evaluations / generation (it means that in each generation will be processed 10 updates - some particles will be updated more times). The standard PSO [16] has 10 particles and thus it will perform 10 function evaluations/generation (for standard PSO each particle will be updated one time - 10 updates in total). The other parameters of the standard PSO are similar to those used by the evolved PSO and are given in Table 2.

Taking into account the averaged values we can see in Table 3 that the evolved PSO algorithm performs better than the standard PSO algorithm in 7 cases (out of 11). When taking into account the solution obtained in the best run,

**Table 3.** The results obtained by the evolved PSO algorithm and the standard PSO algorithm for the considered test functions. *Best/Worst* stands for the fitness of the best individual in the best/worst run. The results are averaged over 500 runs.

| Func-tions | Evolved PSO | | | | Standard PSO | | | |
|---|---|---|---|---|---|---|---|---|
| | Worst | Best | Mean | StdDev | Worst | Best | Mean | StdDev |
| $f_1$ | 2.717 | 0.001 | 0.526 | 0.723 | 7.123 | 0.282 | 2.471 | 1.517 |
| $f_2$ | 3.413 | 0.070 | 0.757 | 0.718 | 3.908 | 0.077 | 1.495 | 0.995 |
| $f_3$ | 2.332 | 0.340 | 1.389 | 0.638 | 1.639 | 0.395 | 0.846 | 0.344 |
| $f_4$ | 632.667 | 5.549 | 151.433 | 134.168 | 547.968 | 2.360 | 81.114 | 127.010 |
| $f_5$ | 3.106 | 0.457 | 0.766 | 0.410 | 9.298 | 0.457 | 1.278 | 1.268 |
| $f_6$ | 1883.010 | 11.585 | 281.541 | 407.174 | 1625.180 | 4.254 | 88.371 | 230.074 |
| $f_7$ | 23.295 | 6.790 | 12.766 | 4.884 | 29.973 | 5.743 | 15.619 | 5.084 |
| $f_8$ | 3.639 | 0.165 | 1.601 | 1.248 | 3.491 | 0.469 | 1.960 | 0.681 |
| $f_9$ | 13.987 | 8.749 | 11.911 | 1.252 | 41.282 | 3.233 | 18.044 | 9.370 |
| $f_{10}$ | -773.118 | -910.367 | -853.202 | 30.990 | -512.661 | -1563.899 | -880.475 | 207.074 |
| $f_{11}$ | 3.412 | 0.001 | 0.785 | 1.168 | 3.689 | 0.180 | 1.285 | 0.801 |

the evolved PSO algorithm performs better than the standard PSO algorithm in 5 cases (out of 11) and tied in 1 case. When taking into account the solution obtained in the worst run, the evolved PSO algorithm performs better than the standard PSO algorithm in 7 cases.

We have also compared the evolved PSO to another PSO algorithm that updates all particles one by one (i.e. the order of update is 0, 1, ... $SwarmSize - 1$). In 9 cases the evolved PSO performed better (on average) than the other algorithm.

In order to determine whether the differences between the evolved PSO algorithm and the standard PSO algorithm are statistically significant, we use a t-test with a 0.05 level of significance. Before applying the T-test, an F-test has been used for determining whether the compared data have the same variance. The P-values of a two-tailed T-test with 499 degrees of freedom are given in Table 4. Table 4 shows that the differences between the results obtained by standard PSO and by the evolved PSO are statistically significant ($P < 0.05$) in 9 cases (out of 11).

**Table 4.** The results of F-test and T-test

| Functions | F-test | T-test | Functions | F-test | T-test |
|---|---|---|---|---|---|
| $f_1$ | 7.40E-07 | 5.13E-13 | $f_7$ | 7.80E-01 | 2.58E-03 |
| $f_2$ | 2.43E-02 | 2.41E-05 | $f_8$ | 4.06E-05 | 3.87E-02 |
| $f_3$ | 3.14E-05 | 3.63E-07 | $f_9$ | 4.13E-30 | 6.64E-06 |
| $f_4$ | 7.03E-01 | 4.18E-03 | $f_{10}$ | 8.55E-28 | 1.80E-01 |
| $f_5$ | 5.76E-13 | 3.90E-03 | $f_{11}$ | 9.57E-03 | 7.12E-03 |
| $f_6$ | 1.06E-04 | 2.17E-03 | | | |

# 4    Conclusion and Further Work

A new hybrid technique for evolving the structure of a PSO algorithm has been proposed in this paper. The model has been used for evolving PSO algorithms for function optimization. Numerical experiments have shown that the evolved PSO algorithm performs similarly and sometimes even better than the standard PSO algorithm for the considered test functions.

Note that according to the No Free Lunch theorems [18] we cannot expect to design a perfect PSO which performs the best for all the optimization problems. This is why any claim about the generalization ability of the evolved PSO should be made only based on some numerical experiments.

Further work will be focused on: finding patterns in the evolved structures. This will help us design PSO algorithms that use larger swarms, evolving better PSO algorithms for optimization, evolving PSO algorithms for other difficult problems.

# References

1. A. Carlisle, G. Dozier, "An Off-the-shelf PSO", *Proceedings of the Particle Swarm Optimization Workshop*, pp. 1-6, 2001.
2. T. -J. Chang, (et al.), "Heuristics for cardinality constrained portfolio optimisation" Comp. & Opns. Res. 27, pp. 1271-1302, 2000.
3. M. Clerc, "The swarm and the queen: towards a deterministic and adaptive particle swarm optimization", *Proceedings of the IEEE Congress on Evolutionary Computation (CEC 1999)*, pp. 1951-1957, 1999.
4. R. C. Eberhart, Y. Shi, "Comparison Between Genetic Algorithms and Particle Swarm Optimization", *Evolutionary Programming VII: Proceedings of the Seventh International Conference*, pp. 611-616, 1998.
5. R. C. Eberhart, Y. Shi, "Particle swarm optimization: developments, applications and resources", *Proceedings of the CEC*, 2001.
6. http://www.euronext.com
7. D. Goldberg, *Genetic algorithms in search, optimization and machine learning*, Addison-Wesley, Boston, USA, 1989.
8. J. Kennedy, R. C. Eberhart, "Particle Swarm Optimization", *Proceedings of the 1995 IEEE International Conference on Neural Networks*, pages 1942-1948, 1995.
9. J. R. Koza, *Genetic programming, On the programming of computers by means of natural selection*, MIT Press, Cambridge, MA, 1992.
10. H. Markowitz, "Portfolio Selection", *Journal of Finance*, 7, pp. 77-91,1952.
11. M. Oltean, C. Groşan, "Evolving EAs using Multi Expression Programming", *Proceedings of the European Conference on Artificial Life*, pp. 651-658, 2003.
12. M. Oltean, "Evolving evolutionary algorithms using Linear Genetic Programming", *Evolutionary Computation*, MIT Press, Cambridge, Vol. 13, Issue 3, 2005.
13. K. E. Parsopoulos, M. N. Vrahatis, "Recent approaches to global optimization problems through Particle Swarm Optimization", *Natural Computing*, Vol. 1, pp. 235-306, 2002.
14. R. Storn, K. Price, "Differential evolution-a simple and efficient heuristic for global optimization over continuous spaces", *Global Optimization*, Vol. 11, pp. 341-359, 1997.

15. J. Tavares, (et al.), "On the evolution of evolutionary algorithms", in Keijzer, M. (et al.) editors, *European Conference on Genetic Programming*, pp. 389-398, Springer-Verlag, Berlin, 2004.
16. H. Xiaohui, S. Yuhui, R. Eberhart, "Recent Advances in Particle Swarm", *Proceedings of the IEEE Congress on Evolutionary Computation*, pp. 90 - 97, 2004.
17. X. Yao, Y. Liu, and G. Lin, "Evolutionary Programming Made Faster", *IEEE Transaction on Evolutionary Computation*, pp. 82-102, 1999.
18. D. H. Wolpert and W. G. McReady, "No Free Lunch Theorems for Optimization", *IEEE Transaction on Evolutionary Computation*, Nr. 1, pp. 67-82, IEEE Press, NY, USA, 1997.

# A Tabu Search Algorithm for Optimization of Gas Distribution Networks

Herbert de Mélo Duarte, Elizabeth F. Gouvêa Goldbarg, and Marco César Goldbarg

Dep. de Informática e Matemática Aplicada, Universidade Federal do Rio Grande do Norte,
Campus Universitário, 59072-970 Natal, Brazil
herbert@ppgsc.ufrn.br, {beth, gold}@dimap.ufrn.br

**Abstract.** In this paper a tabu search algorithm is proposed for the optimization of constrained gas distribution networks. The problem consists in finding the least cost combination of diameters, from a discrete set of commercially available ones, for the pipes of a given gas network, satisfying the constraints related to minimum pressure requirements and upstream pipe conditions. Since this is a nonlinear mixed integer problem, metaheuristic approaches seem to be more suitable and to provide better results than classical optimization methods. In this work, a tabu search heuristics is applied to the problem and the results of the proposed algorithm are compared with the results of a genetic algorithm and two other versions of tabu search algorithms. The results are very promising, regarding both quality of solutions and computational time.

## 1 Introduction

Gas distribution networks are a very important part of cities infrastructure, serving both houses and industries, and usually imply an enormous cost for their implementation. For such reasons, many companies which provide this kind of service need to use computational tools in order to determine the best design for such networks. The design involves the definition of a layout for the network and the dimensioning of pipes. To define the layout, one has to take into account a number of factors such as the streets, topology, environmental risks, economics and network reliability. Once a layout is defined, then the sizes of the diameters of the pipes to be laid have to be determined.

Pipes are produced and commercialized only in a certain number of materials and in certain fixed diameters. Their costs per unit of length usually vary with the kind of material they are made of and with diameters. For larger diameters, more expensive pipes are expected. Pressure also drops along the pipes in a decreasing rate with larger diameters for a constant flow, meaning that larger diameters imply in more reliable networks, with better guarantee of demand supply.

In this paper, it is assumed that a layout is previously defined. Thus the goal is to select the diameters of the pipes such that they are large enough to guarantee that demand requirements are met and the overall cost of pipes is minimized. An additional constraint is that each pipe, not incident to a source node, should have at least one upstream pipe of the same or greater diameter. Given a layout for the network, the objective is to minimize the sum of the costs of the pipes. In this paper fixed tree

J. Gottlieb and G.R. Raidl (Eds.): EvoCOP 2006, LNCS 3906, pp. 37–48, 2006.
© Springer-Verlag Berlin Heidelberg 2006

structures are considered for the network layout and the term "pipe" can be utilized to refer to an arc between two nodes of a given tree. Although the assumption of tree-like structures is made, the proposed method applies to looped distribution networks as well.

The henceforth denominated pipe dimensioning problem is formulated as a constrained optimization problem. Since restrictions are implicit and observe non-linear equations, the feasibility of a solution may only be determined by solving the flow equations of the network. For such reasons, analytical investigation and exact computational methods seem not to be promising on solving this kind of problem and the use of metaheuristics is the natural alternative for dealing with it.

Next section presents a short review of previous works on problems related to distribution networks optimization. A mathematical formulation of the pipe dimensioning problem as a mixed integer non-linear optimization problem is then given in the following section. Section 4 introduces some basic concepts of tabu search algorithms. Details of the proposed algorithm are presented in section 5. Section 6 shows the computational results obtained by the new approach and the last section summarizes the main contributions of this work.

## 2   Previous Works on Related Problems

In one of the earliest investigations of a related problem, Rothfarb et al. [10] explored, among other design issues of gas distribution networks from offshore fields to by-products separation plants, the problem of selection of pipe diameters. Their goal was to minimize the sum of investment and operational costs. On that paper, the considered networks were those which contain the fewest number of pipelines that can deliver gas from the fields to the separation plants or, in other words, networks with fixed tree structures.

Boyd et al. [1] developed a genetic algorithm for the pipe dimensioning problem and used a penalty function to take both the minimum pressure and upstream pipe restrictions in account. The solutions were represented by a sequence of $n$ integers, where $n$ is the number of pipe segments (arcs) in the network, each integer indicating the index of the diameter to be chosen for a given pipeline. This representation allows all the search space to be coded and eases the use of genetic operators, like the creep mutation and the uniform crossover, both utilized in the genetic algorithm. An unstructured population of 100 networks was used through 100 generations and allowed the authors to report a 4% better result for a 25 pipes real instance when compared to a heuristic utilized by the British Gas company. The latter heuristic is simple and serves as a basic idea for other methods proposed to solve the problem. It consists in guessing some initial pipe sizes which lead to a valid network and locally optimize the current solution repeatedly trying to reduce the diameter of one of the pipes until no further reduction can generate a still feasible solution.

The COMOGA method (Constrained Optimization by Multi-Objective Genetic Algorithms) [11] treats the constraints of the problem, either explicit or implicit, as a separate criterion in a multi-objective formulation of the problem. The method has the advantage of diminish the number of free parameters to be tuned when compared with the traditional penalty function formulation and to remain low sensitive to these. It consists in reducing the problem to a bi-criterion one, condensing all the constraints in a single criterion. By calculating a Pareto ranking for each solution of the population (number of solutions which dominate it) regarding the level of violation of all the original constraints, the method minimizes both the cost and the degree of unfeasibility of solutions through a reverse annealing scheme that adjust the probability of selecting and replacing individuals of the population according to one (cost) or another criterion (feasibility). The authors argue that the COMOGA method needs considerably less experimentation and is much less sensitive to the parameters involved in its design.

In the same year, Osiadacz and Góreki [8] present a reasonably comprehensive survey of the optimization of both water and gas distribution networks by means of heuristic methods, exact procedures which consider the availability of continuous diameters, and discrete optimization methods. They propose to solve the problem with the use of an iterative method which minimizes at each step a quadratic approximation to a lagrangian function subject to sequentially linearized approximations to the constraints. After all, diameters are corrected to the closest available discrete diameter size.

On another line of investigation, Boyd et al. [2] study the fuel cost minimization problem and generate useful lower bounds to evaluate the quality of the solutions provided by the pipeline optimization algorithms. A study of the solution space and objective function is carried out, however the specific problem they deal with is more concerned about the compressor stations fuel minimization and the developed lower bounds don't directly apply to the pipe dimensioning problem.

Distribution networks composed basically of pumps and pipes (once accessories can be modeled only by its diameters) are considered in the paper of Castillo and González [3]. The decision variables are then defined to be pumps Q-H characteristics (pressure and flow rates) and pipes and accessories diameters, taken from a discrete commercial set of available diameters. A formulation of the problem is presented in which the velocity of the flow comprehends the most important constraint and a genetic algorithm with specialized crossover operators is proposed. Their genetic algorithm also makes use of a penalty function to disfavor unfeasible solutions, adding up the highest possible cost of each element of the network on which the constraints are violated. They apply the procedure to two theoretical test cases and to a real one arising from activities of a dairy products company. The problem-specific genetic operators demonstrate to be better in practice than the standard ones.

Finally, Cunha and Ribeiro [6] propose a tabu search algorithm to find the least-cost design of looped water distribution networks. Even dealing with a different kind of fluid to be distributed, the mathematical structure of this problem is identical to the gas distribution networks optimization problem, including the concern with similar minimum design pressure and upstream pipe constraints. Besides providing a vast survey on previous approaches to the problem, the suggested algorithm seems to be both simple and effective to tackle this optimization problem. It starts from a trivial feasible solution where all pipes are set to the largest available diameter and for each successive step it decreases the diameter of one of the pipes, while maintaining the feasibility of the current solution with respect to all the constraints. When the algorithm finds a solution where no additional diameter decrease leads to a still feasible solution, it makes a diversification move by increasing the diameter of one of the pipes, selected according to one of the following criteria:

- the pipe whose diameter is increased is the one that presents the lowest value of a parameter given by multiplying the number of changes in the diameter of the pipe during the search procedure by the number of times that the pipe was ascribed the increased diameter;
- the chosen pipe is the one which size has maintained the same diameter for the longest period in previous iterations.

These criteria define two different versions of the algorithm, called TS1 and TS2, respectively. In either case, the modification of the size of the pipe whose diameter has been changed becomes a tabu move and enters the tabu list. The authors experiment different fixed and variable tabu tenure parameters and, comparing their algorithms results to best known solutions of five benchmark instances of the problem, they demonstrate the effectiveness of the proposed method.

## 3 Mathematical Model Formulation

The pipe dimensioning problem can be stated as follows: to select, from a discrete set of commercially available pipe diameters, the combination of diameters that gives the least cost network able to supply a set of demand nodes with at least a minimum design pressure and satisfying pipe upstream conditions. The corresponding model can be written as:

$$\min \sum_{k \in NP} c(D_k) \cdot L_k \tag{1}$$

subject to:

$$P_i \geq PD_i, \forall i \in N - \{n_0\} \tag{2}$$

$$D_k \leq DU_k, \forall k \in NP \tag{3}$$

$$D_k \in D, \forall k \in NP \tag{4}$$

where $NP$ is the pipe set, $N$ is the node set, with $n_0$ representing the gas source, $D$ is the set of commercial diameters, $D_k$ is the diameter of the pipe $k$, taken as the decision variables, $L_k$ is the length of pipe $k$, $c : D \to \Re^+$ is a function from the set of diameters to positive real numbers which represents the cost of the pipe per unit of length, $P_i$ is the pressure obtained on node $i$ for the assigned pipes diameters, $PD_i$ is the required pressure on node $i$, and $DU_k$ is the diameter of the upstream pipe of pipe $k$.

The objective function (1) represents the minimization of total gas distribution network cost, expressed as the sum of the costs of each pipe section which compose it. The pipe cost, by its turn, is a function of the pipe diameters (taken as the decision variables) and the length of the section.

The set of constraints (2) represents the minimum design pressure requirements on each node. They can only be verified by the successive solution of the non-linear equation in (5) for obtaining the pressure drop since the source for each pipe section.

$$Q = C \frac{T_b}{P_b} D^{2.5} e \left( \frac{P_1^2 - P_2^2}{LGT_a Z_a f} \right)^{0.5}$$

(5)

where $Q$ is the flow rate, $C$ is a constant for systems of units conversion, $T_b$ is the base temperature, $P_b$ is the base pressure, $D$ is the diameter of the pipe, $e$ is the pipe efficiency, $P_1$ and $P_2$ are the input and output pressure, respectively, $L$ is the length of the pipe, $G$ is the gas specific gravity, $T_a$ is the gas temperature, $Z_a$ is the gas compressibility factor and $f$ is the friction factor of the pipe.

The set of constraints (3) indicates the upstream pipe conditions, meaning that each pipe must have a diameter less than or equal to the diameter of its upstream pipe. Once the networks we are considering in this paper have fixed tree structures, the upstream pipe can be easily determined as the only one which directly connects to the given pipe and is closer to the source node. If looped distribution networks were considered, flow equations would have to be solved to determine flow directions and the upstream pipes. Finally, the constraints in (4) restrict the possible diameters of the pipes to the commercially available ones.

## 4  Tabu Search Algorithms

Among the numerous metaheuristic methods proposed in the past few decades for the solution of complex optimization problems, tabu search distinguishes as a successful heuristic in many fields of application. Its main features lay down on analogies to the human memory process and are fully explained by Glover and Laguna [7].

The tabu search method proposes to iteratively search the neighborhood of the current solution in such an ordering that allows the best moves to be explored and at the same time prevents the repetition of previously visited solutions. To achieve this goal a certain neighborhood structure and a function to generate neighbor solutions (by means of structured moves) have to be defined. The exploration of the neighborhood can be done entirely or restricted to a certain number of solutions.

A further part of tabu search functionality is preventing from repeatedly visiting the same solution. This is done by setting some possible moves to tabu status, meaning that they cannot be executed for a number of iterations. It can avoid cycling on visited solutions. Moreover the tabu tenure parameter defines the number of iterations the elements added to a tabu list maintain their tabu status, making the search more or less restrictive. A tabu search implementation may or may not make use of an aspiration criterion. This criterion allows a tabu move to be done in a given iteration if it will lead to a solution that satisfies some condition, e.g. if the new solution is better than any of the previously encountered solutions.

Eventually the search will lead to a local optimal solution and no move will be able to bring improvements to the process. In such cases, the algorithm shall use a diversification procedure which will permit the exploration of regions of the search space that haven't been considered yet. The ability of determining those unexplored regions is supported by the use of long term memory structures that keep register of the already visited vicinities. Finally the search is concluded when some termination criteria are achieved. It's evident that this metaheuristic technique incorporates strategies for efficiently avoiding local optima while keeping the implementation fairly simple and this can partially explains the popularity of the method.

## 5 Algorithm Implementation

In this work, we propose a tabu search algorithm for the solution of the pipe dimensioning problem. The implementation considers only feasible networks, which means that moves never generate solutions that violate the problem constraints. It simplifies the objective function in the sense that there is no need of a penalty function. The goal is to minimize the total cost of the pipes of a gas distribution network.

This section describes the basic features involved in the implementation of the proposed algorithm, such as: the initial solution, the neighborhood exploration (including the neighbor generation move, the diversification procedure, the aspiration criterion, the tabu moves and the tabu tenure parameter), and termination criteria. A pseudo-code of the described algorithm is presented in the following.

**Pseudo-code of the proposed algorithm**

```
curr_sol = generate_initial_sol ();
tabu_list = {};
repeat
  if (no diameter decrease is possible) then
    curr_sol = diversification (curr_sol);
  else
    new_sol = decrease_diameter (curr_sol);
    if (reduced pipe is not tabu or
        aspiration criterion holds) then
        curr_sol = new_sol;
  end_if
  for each pipe i in tabu list
    alpha = random value between 0.0 and 1.0;
    if (alpha < p) remove pipe i from tabu list;
  end_for
  insert last reduced pipe in tabu list;
until termination criteria are accomplished
```

## 5.1  Initial Solution

Once we are concerned in maintaining the feasibility of solutions through all the iterations, it is necessary to consider a method to generate initial solutions that guarantees the satisfaction of pressure and upstream requirements. Therefore, the strategy adopted in this work was to utilize the safest possible network (though the most expensive) as the initial solution. This network is the one where the largest possible diameter is assigned to each pipe. This strategy brings the extra benefit of being able to determine on the very beginning whether the instance has or not a single solution. If the proposed initial solution is infeasible, no other solution will be. A randomized initial solution would not be adequate because the density of feasible solutions in the search space is extremely low and it could take a large computational time to randomly find one of them.

## 5.2  Neighbor Generation Moves

To detail how the algorithm explores the neighborhood of the current solution, the underlying neighbor generation move is explained. First, consider that the solutions are represented as a list of pipes and corresponding diameters, ordered by depth of incident nodes. The basic move consists in decreasing pipe diameters to the minimum possible value keeping the feasibility of the solution. It is done from the leaves up to the root of the tree to favor the observance of upstream restrictions. The bottom-up style of pipe reductions may allow that the decrease of a pipe diameter enables its upstream pipe to have its diameter reduced too. However, due to the complexity of the constraints involved, the pressure at each node has to be recalculated for every diameter reduction to ensure that the network is still meeting the requirements.

## 5.3  Diversification Procedure

When a solution is found for which no pipe diameter reduction is possible, a local optimum was reached. To proceed with the search, the diversification procedure is applied. The neighbor solution is then generated by increasing pipe diameters. A random pipe (whose diameter is not the largest available one) is selected and its diameter is increased to the next available size. This move always keeps the network valid regarding the pressure constraints, but may cause the selected pipe to violate the upstream requirement. This is easily fixed by recursively increasing the diameter of the upstream pipe of the altered pipes while necessary.

## 5.4  Tabu Moves and Aspiration Criterion

Either if the pipe diameter is decreased or increased, the branch which diameter was changed receives a "tabu" status. Then, the modification of its diameter is prohibited for a certain number of iterations. The tabu tenure parameter is implemented to be randomized. For all iteration each move has a probability $p$ of leaving the tabu list. Some experimentation was carried out and a value of $p=0.05$ showed to be effective for the studied problem. This value was used in all tests we report in this paper. Also, an aspiration criterion was utilized to allow a tabu status to be removed if the correspondent move leads to a new best solution.

## 5.5 Termination Criteria

Two termination criteria were used: a maximum runtime and a maximum number of iterations without improvement of the best solution. Let $n$ be the number of pipes in a given instance. The maximum time allowed for the tabu search algorithm execution in computational experiments was $n$ seconds. The maximum number of iterations without improvement of the best solution was $2n$. It can be verified on the following section through the tests carried out that in general the second criterion was the one which interrupted the search.

# 6   Computational Experiments and Results

For purposes of validation of the proposed algorithm, it was applied to 58 theoretical instances with sizes varying from 51 to 1432 nodes and, consequently, from 50 to 1431 pipe sections. These instances are based on instances of the TSPLIB – a traveling salesman problem bank of instances [9] – and have been generated as follows. First, the minimum spanning tree of the complete graph obtained from the TSP instances was calculated, the first node fixed as the source node. Then, values for flow on each pipe and minimum pressure required by each node were randomly generated in a previously defined range. Finally, a set of commercial diameters was chosen from real-world available ones [4][5]. Each instance of the pipe dimensioning problem is referred to with the same name of the TSP instance that originated it.

The two versions of the tabu search algorithm from Cunha and Ribeiro [6] and the multi-objective genetic algorithm from Surry et al. [11] were applied to the problem – maintaining all their basic features unchanged. The proposed algorithm is compared with them. The termination criteria for these algorithms was set to be same as the proposed algorithm, a maximum time of $n$ seconds and a maximum of $2n$ iterations without improvement of the best already found solution, where $n$ is the number of nodes of the instance. For every instance, 30 independent runs of the algorithms were done on a Pentium IV 2.8 GHz computer with 512MB of RAM, except for the genetic algorithm on instances with more than 1000 nodes, for the reasons explained below.

Table 1 shows the results the algorithms obtained for the smaller instances (less than 1000 nodes). The best encountered solution by any of the tested algorithms for each instance along with the minimum and average gap in percentage and the standard deviation of the gap for each algorithm is presented. 14 small easy instances (up to 264 nodes) are omitted since all the methods achieved the same results. Only the results of TS1 [6] are shown because the performance of the second version was inferior to it. TS2 was able to find a cheaper solution than TS1 for only 1 instance and its running time was equal to or worse than TS1 running time in 55 of the cases. The genetic algorithm [11] didn't find any feasible solution for pr299 or u724 instances. Proposed TS stands for the proposed algorithm while TS-C and GA-S refer to the compared tabu search [6] and genetic algorithm [11], respectively.

**Table 1.** Solutions quality comparison for 34 small and medium networks

| Instance | Best Sol. | Proposed TS | | | TS-C | | | GA-S | | |
|---|---|---|---|---|---|---|---|---|---|---|
| | | Min. Gap | Avg. Gap | SD Gap | Min. Gap | Avg. Gap | SD Gap | Min. Gap | Avg. Gap | SD Gap |
| pr76 | 913473.10 | 0.00 | 0.00 | 0.00 | 0.00 | 0.00 | 0.00 | 3.50 | 4.56 | 0.58 |
| kroA100 | 193351.60 | 0.00 | 0.00 | 0.00 | 0.00 | 0.00 | 0.00 | 2.65 | 4.43 | 0.95 |
| kroE100 | 197996.90 | 0.00 | 0.00 | 0.00 | 0.00 | 0.00 | 0.00 | 1.23 | 2.52 | 1.45 |
| lin105 | 134466.50 | 0.00 | 0.00 | 0.00 | 0.00 | 0.00 | 0.00 | 5.05 | 5.98 | 0.74 |
| pr107 | 392422.10 | 0.00 | **0.00** | 0.00 | 0.00 | 5.09 | 27.88 | 0.34 | *0.94* | 0.23 |
| pr124 | 532125.00 | 0.00 | 0.00 | 0.00 | 0.00 | 0.00 | 0.00 | 8.06 | 11.06 | 2.36 |
| bier127 | 975471.80 | 0.00 | 0.00 | 0.00 | 0.00 | 0.00 | 0.00 | 0.79 | 1.54 | 1.07 |
| pr136 | 932218.70 | 0.00 | 0.00 | 0.00 | 0.00 | 0.00 | 0.00 | 5.05 | 7.54 | 1.58 |
| pr144 | 539456.30 | 0.00 | **0.00** | 0.00 | 0.00 | 5.40 | 29.57 | 5.62 | 9.10 | 1.50 |
| kroA150 | 253311.10 | 0.00 | **0.00** | 0.00 | 0.00 | 6.64 | 5.14 | 12.73 | 18.45 | 3.73 |
| kroB150 | 234850.30 | 0.00 | 0.00 | 0.00 | 0.00 | 0.00 | 0.00 | 2.02 | 2.92 | 0.64 |
| pr152 | 658830.80 | 0.00 | **0.00** | 0.00 | 0.00 | 14.38 | 11.13 | 9.14 | *12.51* | 3.09 |
| u159 | 391681.80 | 0.00 | 0.00 | 0.00 | 0.00 | 0.00 | 0.00 | 8.05 | 12.21 | 3.20 |
| d198 | 128528.90 | 0.00 | 0.00 | 0.00 | 0.00 | 0.00 | 0.00 | 5.49 | 8.54 | 2.37 |
| kroA200 | 274397.80 | 0.00 | 0.00 | 0.00 | 0.00 | 0.00 | 0.00 | 5.29 | 8.42 | 3.13 |
| kroB200 | 273240.10 | 0.00 | 0.00 | 0.00 | 0.00 | 0.00 | 0.00 | 4.47 | 8.25 | 2.40 |
| ts225 | 1198600.00 | 0.00 | **0.00** | 0.00 | 0.00 | 5.67 | 31.08 | 7.83 | 23.82 | 11.55 |
| tsp225 | 36647.40 | 0.00 | 0.00 | 0.00 | 0.00 | 0.00 | 0.00 | 3.98 | 4.83 | 0.49 |
| pr226 | 763475.40 | 0.00 | 0.00 | 0.00 | 0.00 | 0.00 | 0.00 | 14.71 | 25.60 | 6.73 |
| gil262 | 21516.70 | 0.00 | 0.00 | 0.00 | 0.00 | 0.00 | 0.00 | 0.31 | 0.48 | 0.23 |
| a280 | 25080.50 | 0.00 | 0.00 | 0.00 | 0.00 | 0.00 | 0.00 | 0.32 | 0.50 | 0.11 |
| pr299 | 479309.90 | 0.00 | **0.00** | 0.00 | 5.25 | 5.25 | 0.00 | - | - | - |
| lin318 | 405353.80 | **0.00** | **0.00** | 0.01 | 0.55 | 8.64 | 3.68 | 7.09 | 19.90 | 13.04 |
| rd400 | 140256.80 | 0.15 | 0.15 | 0.00 | 0.15 | 0.15 | 0.00 | **0.00** | 1.47 | 5.03 |
| fl417 | 107799.80 | 0.00 | 0.00 | 0.00 | 0.00 | 0.00 | 0.00 | 5.46 | 7.83 | 1.66 |
| pr439 | 1053042.50 | **0.00** | **0.09** | 0.15 | 0.02 | 21.02 | 7.12 | 59.34 | 59.34 | 0.00 |
| pcb442 | 504545.90 | **0.00** | **0.00** | 0.00 | 0.00 | 0.49 | 1.85 | 35.98 | 35.98 | 0.00 |
| d493 | 331245.30 | **0.00** | **0.00** | 0.00 | 27.20 | 27.20 | 0.00 | 60.83 | 60.83 | 0.00 |
| u574 | 348626.20 | **0.00** | **0.00** | 0.02 | 8.24 | 8.24 | 0.00 | 42.62 | 42.62 | 0.00 |
| rat575 | 65209.40 | 0.00 | 0.00 | 0.00 | 0.00 | 0.00 | 0.00 | 6.45 | 8.21 | 0.46 |
| p654 | 361242.50 | **0.00** | **0.43** | 0.11 | 26.76 | 27.13 | 2.03 | 48.40 | 48.40 | 0.00 |
| d657 | 460791.80 | **0.00** | **0.02** | 0.11 | 5.73 | 10.28 | 24.61 | 67.83 | 67.83 | 0.00 |
| u724 | 414143.80 | **0.00** | **0.00** | 0.01 | 10.55 | 10.55 | 0.00 | - | - | - |
| rat783 | 86522.50 | **0.00** | **0.00** | 0.00 | 3.56 | 3.56 | 0.00 | 38.98 | 38.98 | 0.00 |

The effectiveness of the proposed algorithm is clearly demonstrated in this table. Our tabu search finds the best solution for all but one instance (rd400), for which the solution found was only 0.15% above the genetic algorithm solution. Besides, for the four instances in which the best solution was not found in all executions, the average gap didn't reached 0.5%. On the other hand, Cunha and Ribeiro tabu search algorithm [6] also showed a consistent performance, but wasn't able to find the best known solution in 8 instances and its average gap was worse than the average gap of the proposed algorithm for 15 of these instances. The results also indicate that TS-C performance is superior to GA-S, once that for only 2 instances the former presented an average gap worse than the latter and GA-S found only one best known solution.

As the performance of GA-S seriously deteriorate for the last instances in Table 1, the proposed algorithm was only compared to the TS-C for the largest instances.

Table 2 shows the results these two algorithms presented for 10 instances with more than 1000 nodes in the same format of Table 1. It is evident that the proposed tabu search overcomes the reference method once it obtains the best minimum and average solutions for all the networks. Additionally, the robustness of the algorithm is visible because the average gap from the best found configuration is at most 1.83% and the standard deviation exceeds 1.0% for only one case. It can be seen that TS-C fails to obtain competitive solutions for large instances, perhaps because of its running time dependency and the employed termination criteria.

**Table 2.** Comparison of the solutions of the 10 largest networks

| Instance | Best Sol. | Proposed TS | | | TS-C | | |
|----------|-----------|-------------|-------------|------------|-------------|-------------|------------|
| | | Min. Gap | Avg. Gap | SD Gap | Min. Gap | Avg. Gap | SD Gap |
| pr1002 | 2662871.80 | **0.00** | **1.15** | 0.80 | 10.17 | 12.68 | 11.96 |
| u1060 | 2302421.10 | **0.00** | **1.55** | 0.94 | 11.65 | 19.75 | 22.03 |
| vm1084 | 2324953.00 | **0.00** | **1.09** | 0.84 | 11.31 | 12.53 | 6.71 |
| pcb1173 | 591012.00 | **0.00** | **1.28** | 0.73 | 30.06 | 41.68 | 20.27 |
| d1291 | 528294.10 | **0.00** | **0.69** | 0.73 | 78.77 | 84.34 | 11.24 |
| rl1304 | 2701706.20 | **0.00** | **1.83** | 0.75 | 47.15 | 58.57 | 16.07 |
| rl1323 | 3134737.40 | **0.00** | **1.15** | 0.70 | 6.53 | 6.53 | 0.00 |
| nrw1379 | 561381.10 | **0.00** | **0.25** | 0.28 | 70.30 | 71.09 | 0.46 |
| fl1400 | 202637.50 | **0.00** | **1.70** | 1.07 | 87.48 | 87.54 | 0.04 |
| u1432 | 1584001.50 | **0.00** | **0.32** | 0.20 | 113.51 | 114.24 | 0.42 |

The good solutions quality presented by the proposed algorithm in previous tables can consistently point to its usefulness. This observation is ratified in Table 3, where average running time is given for each instance. Once more, the proposed algorithm showed good results, spending in smaller instances less than one tenth of the time spent by TS-C and significantly less time in the other cases. It's relevant to note that the termination criterion of maximum execution time was hardly necessary for the proposed algorithm. However, TS-C shows poor quality results for the 12 largest instances, probably because of the maximum allowed running time, which was reached in all executions for these instances. GA-S average running times are omitted because this method utilized the whole available time for all executions of all instances.

Since a global optimal solution is not known for these problems, it is very important to evaluate different heuristic methods to find the most suitable one. Through the analysis carried out, the proposed tabu search algorithm demonstrates to be very effective in tackle the problem difficulties. Moreover, the compared algorithms are of recognized value on this class of problems, so the analysis may state that this is in fact one adequate method for this kind of problem. It is reasonably expected since tabu search approach has presented good results in other related problems as well, like the water distribution network optimization problem, studied by Cunha and Ribeiro [6].

**Table 3.** Average running time (in seconds)

| Instance | Proposed TS | TS-C |
|----------|-------------|------|
| pr76 | 0.119 | 1.076 |
| kroA100 | 0.039 | 0.852 |
| kroE100 | 0.039 | 0.865 |
| lin105 | 0.043 | 0.895 |
| pr107 | 0.313 | 3.620 |
| pr124 | 0.406 | 4.403 |
| bier127 | 0.063 | 1.801 |
| pr136 | 0.494 | 5.736 |
| pr144 | 0.634 | 8.450 |
| kroA150 | 0.923 | 10.774 |
| kroB150 | 0.090 | 2.730 |
| pr152 | 0.966 | 9.238 |
| u159 | 0.744 | 7.889 |
| d198 | 1.149 | 17.115 |
| kroA200 | 1.027 | 23.772 |
| kroB200 | 1.200 | 38.465 |
| ts225 | 1.694 | 31.311 |
| tsp225 | 0.202 | 9.578 |
| pr226 | 2.158 | 17.576 |
| gil262 | 0.271 | 14.643 |
| a280 | 0.312 | 17.470 |
| pr299 | 8.923 | 86.404 |
| lin318 | 10.880 | 65.950 |
| rd400 | 1.560 | 49.649 |
| fl417 | 5.026 | 174.955 |
| pr439 | 62.854 | 218.321 |
| pcb442 | 16.048 | 213.174 |
| d493 | 18.130 | 354.374 |
| u574 | 36.154 | 504.909 |
| rat575 | 6.559 | 464.208 |
| p654 | 72.481 | 654.269 |
| d657 | 26.071 | 642.325 |
| u724 | 91.008 | 724.510 |
| rat783 | 26.242 | 783.501 |
| pr1002 | 557.313 | 1002.387 |
| u1060 | 918.610 | 1060.277 |
| vm1084 | 944.113 | 1084.526 |
| pcb1173 | 1032.390 | 1173.583 |
| d1291 | 456.914 | 1292.244 |
| rl1304 | 1227.100 | 1304.494 |
| rl1323 | 1291.161 | 1323.406 |
| nrw1379 | 1117.667 | 1379.547 |
| fl1400 | 462.311 | 1401.127 |
| u1432 | 1251.755 | 1433.146 |

# 7  Conclusions

A tabu search algorithm for the pipe dimensioning problem was proposed. Its specialized features allowed high quality solutions to be found in much reduced computational time. It demonstrates the capacity of the metaheuristics to deal with the complex constraints of this combinatorial problem. However, more experimentation is needed to permit a final conclusion to be drawn.

In future works an adaptation of the parallel and serial merge method [10] will be investigated as part of heuristic algorithms. Other line of research will consider looped networks and adding new devices to the distribution model, such as pumps and compressors.

## Acknowledgements

This research was partially funded by the program PRH 22 of the National Agency of Petroleum.

## References

1. Boyd, I. D., Surry, P. D., Radcliffe, N. J. (1994), Constrained gas network pipe sizing with genetic algorithms, EPCC-TR94-11, University of Edinburgh.
2. Boyd, E. A., Scott, L. R., Wu, S. (1997), Evaluating the quality of pipeline optimization algorithms, http://www.psig.org/paper/1997/9709.pdf.
3. Castillo, L., González, A. (1998), Distribution network optimization: finding the most economic solution by using genetic algorithms, European Journal of Operational Research, 108 (3), pp. 527-537.
4. Castro, M. P. (2004), Algoritmos evolucionários para o problema de dimensionamento dos dutos de uma rede urbana de distribuição de gás natural. Dissertation (Master's degree in Systems and Computers). Universidade Federal do Rio Grande do Norte.
5. Costa, W. E. (2004), Um estudo algorítmico para o problema do dimensionamento de dutos em uma rede urbana de distribuição de gás natural. Dissertation (Master's degree in Systems and Computers). Universidade Federal do Rio Grande do Norte.
6. Cunha, M. C., Ribeiro, L. (2004), Tabu search algorithms for water network optimization, European Journal of Operational Research, 157 (3) pp. 746-758.
7. Glover, F., Laguna, M. (1997). Tabu search. Kluwer Academic Publishers, Dordrecht.
8. Osiadacz, A. J., Górecki, M. (1995), Optimization of pipe sizes for distribution gas network design, http://www.psig.org/papers/1987/9511.pdf.
9. Reinelt, G. (1991), A traveling salesman problem library, ORSA Journal on Computing 3, pp. 376-384.
10. Rothfarb, B., Frank, H., Rosenbaum, D., Steiglitz, K., Kleitman, D. (1970), Optimal design of offshore natural-gas pipeline systems, Operations Research, Vol. 6, 18, pp. 922-1020.
11. Surry, P. D., Radcliffe, N. J., Boyd, I. D. (1995), A multi-objective approach to constrained optimization of gas supply networks: the COMOGA method, in: Evolutionary Computing. AISB Workshop, T. Fogarty (Ed.), Springer-Verlag.

# Design of a Retail Chain Stocking Up Policy with a Hybrid Evolutionary Algorithm

Anna I. Esparcia-Alcázar[1], Lidia Lluch-Revert[2], Manuel Cardós[2],
Ken Sharman[1], and Carlos Andrés-Romano[2]

[1] Instituto Tecnológico de Informática,
Edif. 8G Acc. B, Ciudad Politécnica de la Innovación,
46022 Valencia, Spain
{anna, ken}@iti.upv.es
http://www.iti.upv.es

[2] Departamento de Organización de Empresas, Econ. Financiera y Contabilidad,
Universidad Politécnica de Valencia,
Camino de Vera s/n,
46022 Valencia, Spain
{mcardos, candres}@omp.upv.es, lillure@iti.upv.es
http://www.upv.es

**Abstract.** In this paper we address the joint problem of minimising both the transport and inventory costs of a retail chain that is supplied from a central warehouse. We propose a hybrid evolutionary algorithm where the delivery patterns are evolved for each shop, while the delivery routes are obtained employing the multistart sweep algorithm. The experiments performed show that this method can obtain acceptable results consistently and within a reasonable timescale. The results are also of a lower cost than those obtained by other strategies employed in previous research. Furthermore, they confirm the interest of addressing the optimisation problem jointly, rather than minimising separately inventory and transport.

## 1 Introduction

The design of a retail chain usually involves finding a compromise between the minimisation of the transport costs (from the central warehouse's point of view) and the minimisation of the inventory costs (from the shops' point of view). Retail chain shops are usually stocked periodically, with the inventory costs being given by the delivery days established for every shop. Once these are established, the transportation costs can be calculated by solving the vehicle routing problem (VRP).

Although there is substantial literature on both transportation and inventory management, this is not so for the joint problem[1] [3]. In [4] the authors study this problem by considering a single product controlled by an order-point system and

---

[1] This problem arises when both the warehouse and the shops are owned by the same company.

J. Gottlieb and G.R. Raidl (Eds.): EvoCOP 2006, LNCS 3906, pp. 49–60, 2006.

a cost per unit transportation cost. In [2] an approximate analytical method for minimizing inventory and transportation costs under known demand is developed, based on an estimation of the travel distance. [1] analyzes production, inventory and transportation costs in a capacitated network, yet once again transportation costs are assumed to be proportional to the units moved. [9] focuses on a multiproduct production system on a single link. [5] presents a (s, Q)-type inventory policy for a network with multiple suppliers by replenishing a central depot, which in turn distributes to a large number of retailers; his paper considers transportation costs, but only as a function of the shipment size. [7] deals with an inbound material-collection problem so that decisions for both inventory and transportation are made simultaneously; however vehicle capacity is assumed to be unlimited so that it is solved as a traveling salesman problem (TSP). Finally, [10] presents a modified economic ordering quantity for a single supplier-retailer system in which production, inventory and transportation costs are all considered.

In this paper we propose a hybrid evolutionary algorithm that aims at minimising the sum of the inventory and transportation costs. Initially a population of chromosomes is generated, each representing a set of delivery patterns (and hence delivery frequencies) to shops. The inventory cost for each shop is a function of the delivery frequency to that shop. Then for each chromosome we employ the multistart sweep algorithm to calculate the five sets of routes (one per working day of the week) that have minimal transportation cost. This depends on the type of vehicle used, as well as the distances traveled. The fitness that guides the evolutionary algorithm is calculated as the sum of the inventory and transportation costs.

We test our algorithm on real problem data with three objectives in mind. Firstly, we want to show that our method can be a useful tool to design stocking up policies. Secondly, we intend to confirm whether the joint approach proposed in[3], is more suitable than addressing the problem in two stages (first calculating a solution that provides minimum inventory costs and then the minimum transport cost for that solution). Finally, we want to see if we can share their conclusions as per the optimum type of vehicle to use.

The rest of the paper is laid out as follows. The problem of designing a stocking up policy is explained in section 2; in section 2.1 previous approaches are described, while section 2.2 lays out our evolutionary approach. The experiments are described in section 3, where a statistical comparison is made, and finally conclusions and further work are given in section 4.

## 2   Design of a Stocking Up Policy

Given a warehouse that supplies a chain of shops, we have to determine not only the *optimal frequency*, $f$ (number of days a week), with which to serve each shop, but also what is the *optimal pattern*, $p$, for that frequency that minimises the cost. A pattern $p$ represents a set of days in which the shop is served. For each frequency and shop there is an associated inventory cost (the higher the frequency the lower the cost) so to obtain the total inventory cost we add up the individual inventory costs per shop.

Once the patterns are known for all shops, the next step is to obtain the transportation costs. For this we obtain solutions for the Euclidean VRP for each day of the week and add up the associated costs.

We consider the following restrictions, as imposed by the customer:

- The fleet is homogeneous, so although we will consider three types of vehicles (van, small truck and medium truck) only one type of vehicle is used per experiment.
- The transport cost only includes the cost per km, not the time of use of the vehicle. Nor there is a fixed cost per vehicle.
- The capacity of the vehicles is limited but the number of vehicles is unlimited.
- The load is containerised. The amount to deliver to each shop varies with the shop and the frequency of delivery.
- No single shop can consume more than the capacity of any one vehicle, i.e. a shop is served by one and only one vehicle.
- The shops have no time windows.
- The working day of a driver is 8 hours, including driving time and stops.
- Not all frequencies are admissible for all shops.
- Not all patterns are admissible for a given frequency.

The problem data can be found in Tables 4, 5, 6 and 7.

## 2.1   Previous Approaches

The number of possible delivery frequencies combinations is given by $\prod_{i=1}^{N} F_i$, where $N$ is the number of shops and $F_i$ is the number of admissible delivery frequencies for shop $i$. For our problem, this value is of the order of $5.44 \cdot 10^9$. The total number of possible combinations of patterns is $\prod_{j=1}^{N} \sum_{i=1}^{F_i} P_{ij}$, where $P_{ij}$ is the number of admissible delivery patterns for frequency $i$ in shop $j$. In our case, this value is $1.89 \cdot 10^{17}$, which rules out exhaustive search.

In [3] the authors develop an approximate search procedure, involving partial enumeration of solutions and solving five VRPs (one per working day of the week) to determine the delivery patterns. For this, the algorithm employed in [3] was the sweep, or daisy, algorithm as described in [6], but modified to sweep both clockwise and counter-clockwise and employing multiple starting points. This bi-directional multistart sweep, or **daisy algorithm** will also be used in our experiments as a first approach to solve the VRP. It provides $2 * n_d$ sets of routes per weekday (with $n_d$ being the number of shops to be visited on that particular day) of which we will select the one that provides the minimal transportation cost. The final transportation cost will be obtained by adding the minimal costs of all five days of the week.

**Drawbacks.** One disadvantage of this method is the computational expense, but as the authors point out themselves, this is not an important one, since the calculations are done at a strategic phase and hence they're not time-critical.

A more relevant shortcoming is that it is a limited exploration carried out in order to understand the shape and behaviour of the cost functions and hence it only provides a rough approximation to the possible optima. Furthermore, from a business point of view, it is not a practical tool to use. Clearly, some other method must be found. The scene is then set for evolutionary algorithms.

### 2.2 The Evolutionary Approach

**Evolutionary Set-up.** We will start by generating a random population of individuals or chromosomes representing solutions to the problem at hand. The chromosomes are vectors of length equal to the number of shops in which the components, or genes, are integers representing a particular pattern, out of the set of admissible ones (see Table 5). For instance, the chromosome $(4\,3\,6\,8\,9\,7\,5\ldots 10\,1)$ represents a solution in which shop 1 is served with pattern 4, shop 2 with pattern 3 and so on. The fitness of such a chromosome, which we intend to *minimise*, is calculated as the sum of the associated inventory and transportation costs,

$$f = InventoryCost(\text{\euro}) + TransportCost(\text{\euro}) \tag{1}$$

To calculate the former, given the patterns for each shop, we know the associated delivery frequency and with this we can look up the inventory cost per shop (see Table 6). To obtain the transport cost, we run the daisy algorithm as explained in the previous section.

Evolution proceeds employing the following operators:

- `1-point crossover` : Two parents are selected and the genes of each are swapped from the (random) crossover point to the end of the chromosome
- `2-point crossover` : As above but swapping the genes between two random crossover points
- `n-point mutation` : Pick a random value of $n$, $1 \leq n \leq N$, and mutate $n$ positions in the chromosome selected randomly. The upper and lower bounds of $n$ are varied when the fitness has remaind unchanged for a given number of iterations

## 3    Experiments

The main aim of the experiments is to show how our hybrid evolutionary algorithm can be employed to design stocking up policies, and in this way demonstrate the relevance of taking into account the inventory and transport costs jointly. We want to show that the algorithm is robust, in that it obtains consistently good results, and also that it can be done within a reasonable timescale. Additionally, we want to see whether there is one type of vehicle that provides significantly better results than the others.

In order to do this we run the algorithm 30 times per type of vehicle. The termination criterion in all cases was a number of iterations equal to 5000, which corresponds to approximately 9 minutes per run in the computers employed. The algorithm is laid out in Figure 1.

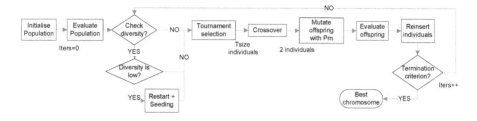

**Fig. 1.** The evolutionary algorithm employed in the experiments

**Table 1.** Parameters of the evolutionary algorithm. The probability of mutation and the value of $n$ are variable. Every $10^{th}$ iteration the diversity is measured and if 80% or more individuals are the same, they are replaced by newly created ones.

| Parameters of the evolutionary algorithm | |
|---|---|
| Population size | 200 |
| Selection method | Tournament |
| Tournament size | 5 |
| Crossover operators | 1 and 2-point Xover |
| Mutation operators | n-point mutation |
| Mutation probability | Variable from 0.2 to 1 |
| Measure diversity every | 10 iterations |
| Diversity preservation mechanism | Restart + seeding |
| apply when | 80% individuals are equal |
| Termination criterion | 5000 iterations |

### 3.1 Van

The best chromosome obtained in the 30 runs was:

$$chrom = \left( 11\ 10\ 10\ 9\ 11\ 9\ 9\ 10\ 11\ 6\ 7\ 11\ 8\ 11\ 11\ 10\ 9\ 2\ 7\ 1 \right)$$

with a total cost of 13782 €, of which 11707 € are inventory and 2062 € are transport costs.
The routes for each day of the week were:

| | | | | |
|---|---|---|---|---|
| Mon | 0 13 9 4 1 | 0 2 3 5 8 | 0 7 6 15 12 14 16 | 0 |
| Tues | 0 19 20 11 17 | 0 9 4 5 8 7 | 0 6 10 15 12 14 | 0 |
| Wed | 0 14 19 16 13 | 0 18 9 4 1 | 0 2 3 5 8 | 0 7 6 10 15 12 0 |
| Thurs | 0 16 20 11 17 | 0 9 1 2 3 | 0 5 8 7 15 12 14 | 0 |
| Fri | 0 14 19 16 13 11 | 0 18 9 4 1 | 0 2 3 5 8 | 0 7 6 10 15 12 0 |

### 3.2 Small Truck

The optimal chromosome obtained in the 30 runs was:

$$chrom = \left( 11\ 10\ 10\ 11\ 11\ 10\ 11\ 11\ 7\ 4\ 7\ 11\ 6\ 11\ 11\ 10\ 7\ 4\ 1\ 3 \right)$$

with a total cost of 13599 € , of which 11740 € are inventory and 1859 € are transport costs. The routes for each day were:

| Mon | 0 4 1 2 3 5 8 | 0 7 6 10 15 12 14 16 13  0 | |
| Tues | 0 7 6 15 12 14 19 | 0 16 20 11 17 18 9 | 0 4 1 2 3 5 8  0 |
| Wed | 0 4 1 2 3 5 8 | 0 7 6 10 15 12 14 16 13  0 | |
| Thurs | 0 15 12 14 19 20 11 17 | 0 9 4 1 5 8 7 | 0 |
| Fri | 0 7 6 10 15 12 14 | 0 16 13 11 17 18 9 | 0 4 1 2 3 5 8  0 |

These routes are also depicted in Figure 2

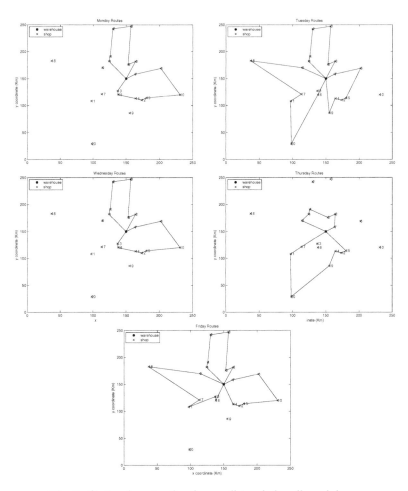

**Fig. 2.** Optimal routes for the small truck for all weekdays

### 3.3   Medium Truck

The optimal chromosome was:

$$chrom = \left( 10\ 10\ 10\ 10\ 10\ 10\ 11\ 6\ 11\ 8\ 1\ 11\ 8\ 8\ 10\ 8\ 1\ 3\ 1\ 1 \right)$$

with a total cost of 13715 €, of which 11760 €  are inventory and 1955 €  are transport costs. The routes for each day were:

| Mon | 0 18 9 4 1 2 3 5 | 0 7 6 10 15 12 14 16 13 | 0 |
|---|---|---|---|
| Tues | 0 15 12 14 19 20 11 17 | 0 4 1 2 3 5 8 7 | 0 |
| Wed | 0 9 4 1 2 3 5 | 0 7 6 10 15 12 14 16 13 | 0 |
| Thurs | 0 18 9 4 1 5 8 | 0 7 6 15 12 19 20 11 17 | 0 |
| Fri | 0 9 4 1 2 3 5 8 | 0 7 6 10 15 12 14 19 16 13 0 | |

## 3.4    Comparison and Analysis

A point to make about this problem is that the inventory cost is much higher than the transport cost. This is due to the nature of the products stocked (perfumes, cosmetics) which have a high density value (in terms of € $/m^3$) and also a high inventory discount rate (around 30%). This discount rate is justified by the stock deterioration, thefts and changes in fashion, as well as the cost of opportunity of the capital, the renting of the premises and the cost of personnel.

Using the kruskalwallis function in MATLAB we run a Kruskal-Wallis test on the best results obtained by all the runs to establish whether the costs with any type of vehicle were significantly different from the others. The results of the test are shown in Table 2.

**Table 2.**  Results of the Kruskal-Wallis test

| Kruskal-Wallis ANOVA table | | | | | |
|---|---|---|---|---|---|
| Source | SS | degrees of freedom | MS | Chi-sq | Prob > Chi-sq |
| Columns | 52598.4 | 2 | 26299.2 | 77.07 | 0 |
| Error | 8143.6 | 87 | 93.6 | | |
| Total | 60742 | 89 | | | |

**Table 3.**  Summary of the best results obtained in 30 runs for the three vehicle types. The inventory cost is always much higher than the transport cost due to the nature of the products stocked (perfumes, cosmetics) which have a high density value (in terms of € $/m^3$) and also a high inventory discount rate (around 30%).

| Type | Transport cost (€) | Inventory cost (€) | Total cost |
|---|---|---|---|
| Van | 2062 | 11707 | 13782 |
| Small truck | 1859 | 11740 | 13599 |
| Medium truck | 1955 | 11760 | 13715 |

**Table 4.** Vehicle data

| Vehicle type | Van | Small Truck | Medium Truck |
|---|---|---|---|
| Capacity (roll containers) | 8 | 12 | 15 |
| Transportation cost(€/Km) | 0.55 | 0.6 | 0.67 |
| Average speed | | 90 km/h | |
| Unloading time | | 15 min | |
| Maximum working time | | 8h | |

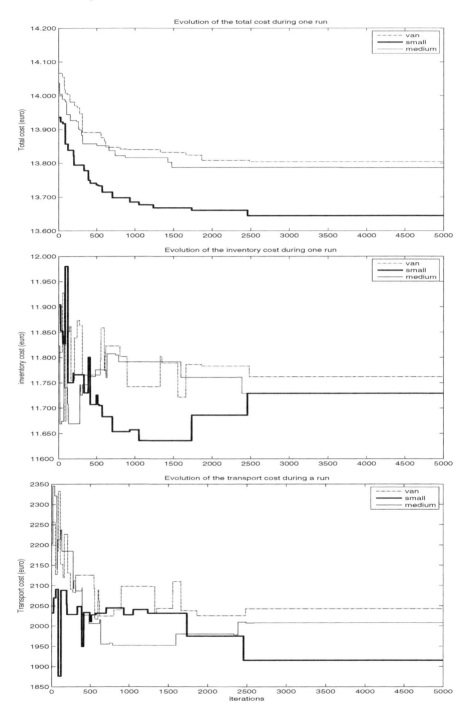

**Fig. 3.** Comparison of the evolution of the costs during a run for each type of vehicle. It can be noticed that the global minimum doesn't correspond to a minimum inventory cost, especially for the case of the small truck.

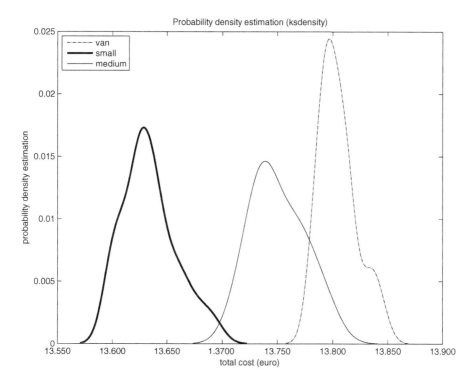

**Fig. 4.** Estimation of the probability density functions of the results for each type of vehicle, obtained with the MATLAB function `ksdensity`. The peaks obtained are relatively narrow, especially for the case of the van; the differences between the maximum and minimum values obtained in the 30 runs of the three experiments are never higher that 1.3% of the median; this means the algorithm is robust, in the sense that any run will not fall very far from the median, which is important in a business context.

Because the probability value is zero, this casts doubt on the null hypothesis and suggests that at least one sample median is significantly different from the others, i.e. the costs of at least one vehicle differ significantly. We then employed a multiple comparisons procedure (the `multcompare` function in MATLAB) to determine which one provided the lower cost. The results of the procedure showed that the cost for the small truck was lower than the rest, followed by the medium truck and the van. This can also be seen when estimating the probability density functions (pdf), see Figure 4. This figure also provides an interesting conclusion. When calculating the maximum width of the estimated pdf peaks, the result is never bigger than 1.3% of the peak median. From this we conclude that the algorithm is robust, in the sense that any run will not fall far from the median, which is important in a business context.

When our results are compared to those obtained by [3] it can be noticed that the costs obtained are much lower in all three cases. Furthermore, our conclusion differs from theirs in that our preferred vehicle is always the small truck, while in their case the best results where obtained by the van.

**Table 5.** Admissible patterns per frequency. The value 1 represents that the shop is served on that day, 0 that it isn't. We will only consider 11 patterns out of the 31 that are possible, a restriction imposed by the customer (the retail chain). For instance, there are no patterns of frequency 1, and only 2 patterns are admissible out of the 5 existing of frequency 4: (Monday, Wednesday, Thursday, Friday) and (Monday, Tuesday, Wednesday, Friday).

| Pattern | Freq | Mon | Tues | Wed | Thu | Fri |
|---------|------|-----|------|-----|-----|-----|
| 1 | 2 | 1 | 0 | 0 | 1 | 0 |
| 2 | 2 | 1 | 0 | 0 | 0 | 1 |
| 3 | 2 | 0 | 1 | 0 | 1 | 0 |
| 4 | 2 | 0 | 1 | 0 | 0 | 1 |
| 5 | 2 | 0 | 0 | 1 | 0 | 1 |
| 6 | 3 | 1 | 0 | 1 | 0 | 1 |
| 7 | 3 | 0 | 1 | 0 | 1 | 1 |
| 8 | 3 | 0 | 1 | 1 | 0 | 1 |
| 9 | 4 | 1 | 0 | 1 | 1 | 1 |
| 10 | 4 | 1 | 1 | 1 | 0 | 1 |
| 11 | 5 | 1 | 1 | 1 | 1 | 1 |

**Table 6.** Shop coordinates, inventory cost (in €) and size of the deliveries per shop depending on the delivery frequency. Shop number 0 corresponds to the warehouse.

| Shop | Location (km) | | Inventory cost (€) | | | | | Delivery size (roll containers) | | | | |
|------|-----|-----|-------|-------|-------|-------|-------|---|---|---|---|---|
|      | x   | y   | 1     | 2     | 3     | 4     | 5     | 1 | 2 | 3 | 4 | 5 |
| 0 | 150 | 150 | - | - | - | - | - | - | - | - | - | - |
| 1 | 126 | 191 | 871.3 | 725.9 | 701.6 | 671.9 | 650.8 | 9 | 5 | 3 | 2 | 2 |
| 2 | 130 | 242 | 862.7 | 718.8 | 694.7 | 665.4 | 644.4 | 9 | 4 | 3 | 2 | 2 |
| 3 | 157 | 247 | 854.2 | 711.7 | 687.8 | 658.8 | 638.0 | 9 | 4 | 3 | 2 | 2 |
| 4 | 124 | 182 | 845.6 | 704.6 | 680.9 | 652.2 | 631.6 | 9 | 4 | 3 | 2 | 2 |
| 5 | 153 | 176 | 837.1 | 697.5 | 674.1 | 645.6 | 625.2 | 9 | 4 | 3 | 2 | 2 |
| 6 | 202 | 169 | 828.6 | 690.3 | 667.2 | 639.0 | 618.9 | 9 | 4 | 3 | 2 | 2 |
| 7 | 163 | 158 | 806.1 | 668.9 | 630.1 | 614.8 | 611.3 | 7 | 4 | 2 | 2 | 1 |
| 8 | 164 | 182 | 791.1 | 656.5 | 618.4 | 603.4 | 600.0 | 7 | 4 | 2 | 2 | 1 |
| 9 | 114 | 170 | 776.2 | 644.1 | 606.7 | 592.0 | 588.7 | 7 | 3 | 2 | 2 | 1 |
| 10 | 231 | 120 | 761.3 | 631.7 | 595.1 | 580.7 | 577.3 | 7 | 3 | 2 | 2 | 1 |
| 11 | 97 | 108 | 746.4 | 619.4 | 583.4 | 569.3 | 566.0 | 7 | 3 | 2 | 2 | 1 |
| 12 | 173 | 110 | 731.4 | 607.0 | 571.7 | 557.9 | 554.7 | 7 | 3 | 2 | 2 | 1 |
| 13 | 137 | 127 | 716.5 | 594.6 | 560.1 | 546.5 | 543.4 | 6 | 3 | 2 | 2 | 1 |
| 14 | 164 | 113 | 701.6 | 582.2 | 548.4 | 535.1 | 532.1 | 6 | 3 | 2 | 2 | 1 |
| 15 | 180 | 114 | 686.6 | 569.8 | 536.7 | 523.7 | 520.7 | 6 | 3 | 2 | 2 | 1 |
| 16 | 138 | 120 | 671.7 | 557.4 | 525.1 | 512.3 | 509.4 | 6 | 3 | 2 | 1 | 1 |
| 17 | 113 | 121 | 670.0 | 555.2 | 523.7 | 512.0 | 513.3 | 5 | 2 | 2 | 1 | 1 |
| 18 | 37 | 183 | 663.4 | 549.7 | 518.6 | 507.0 | 508.3 | 5 | 2 | 2 | 1 | 1 |
| 19 | 155 | 86 | 656.9 | 544.3 | 513.4 | 502.0 | 503.2 | 4 | 2 | 1 | 1 | 1 |
| 20 | 98 | 29 | 650.3 | 538.8 | 508.3 | 496.9 | 498.2 | 4 | 2 | 1 | 1 | 1 |

**Table 7.** Admissible frequencies per shop, as imposed by the customer (the retail chain). Note that frequency 1 (one delivery day a week) is not admissible for any shop.

| Freq. | Shop | | | | | | | | | | | | | | | | | | | |
|---|---|---|---|---|---|---|---|---|---|---|---|---|---|---|---|---|---|---|---|---|
|  | 1 | 2 | 3 | 4 | 5 | 6 | 7 | 8 | 9 | 10 | 11 | 12 | 13 | 14 | 15 | 16 | 17 | 18 | 19 | 20 |
| 1 | 0 | 0 | 0 | 0 | 0 | 0 | 0 | 0 | 0 | 0 | 0 | 0 | 0 | 0 | 0 | 0 | 0 | 0 | 0 | 0 |
| 2 | 0 | 0 | 0 | 0 | 0 | 0 | 1 | 1 | 1 | 1 | 1 | 1 | 1 | 1 | 1 | 1 | 1 | 1 | 1 | 1 |
| 3 | 0 | 0 | 0 | 0 | 0 | 0 | 1 | 1 | 1 | 1 | 1 | 1 | 1 | 1 | 1 | 1 | 1 | 1 | 1 | 1 |
| 4 | 1 | 1 | 1 | 1 | 1 | 1 | 1 | 1 | 1 | 1 | 1 | 1 | 1 | 1 | 1 | 1 | 1 | 1 | 1 | 1 |
| 5 | 1 | 1 | 1 | 1 | 1 | 1 | 1 | 1 | 1 | 1 | 1 | 1 | 1 | 1 | 1 | 1 | 0 | 0 | 0 | 0 |

On a more interesting note, the optimum found does not correspond to a minimum inventory cost. This can be seen in Figure 3, which shows the evolution of the global, inventory and transport costs throughout one run (chosen at random) and for each type of vehicle. For all vehicles, but more noticeably for the small truck, the global cost is lower by the end of the run, as should be expected. On the other hand, the inventory cost goes through a minimum value at earlier stages of the run, to increase by the end of it. This justifies the convenience of handling the inventory and transport problems in a joint manner.

## 4   Conclusions and Future Work

We have described a hybrid evolutionary algorithm that can be successfully employed in the design of a stocking up policy for a chain of shops in a reasonable timescale. The algorithm is robust in the sense that any run will provide results that are not too different from the median (less than 0.65% away). We have also established the importance of addressing the problem as the joint minimisation of the inventory and transport costs, rather than tackling each problem independently.

Future work will involve:

- More complex topologies, including more shops.
- Using real (map) not Euclidean distances, in which for instance distances are different one way and back.
- Improving the daisy algorithm with local learning (or even employing a completely different heuristic). This is as suggested by [8] with respect to simulated annealing. For instance, in figure 2 it can be noticed that the route for Friday can be improved by reordering shops (16, 13, 11, 17) as (13, 16, 17, 11). This reordered route is approximately 27 km shorter than the one given by the daisy algorithm. In monetary terms and for the small truck this represents a saving of 16.13 € . From a business point of view, the improvement is too minute for the effort that would entail incorporating a learning algorithm. However, it could be worthy if applied in a real non-Euclidean space. Further research is necessary on this matter.

## Acknowledgement

This work was part of the BioLog project and is supported by the Acciones de Articulación del Sistema Valenciano de Ciencia-Tecnología-Empresa. $N^0$ Expte. IIARCO/2004/17.

The data employed in the experiments was supplied by DRUNI SA, a major regional cosmetics retail chain.

## References

1. Benjamin, J.: An Analysis of Inventory and Transportation Costs in a Constrained Network, *Transportation Science*, 23(3): 177-183 (1989)
2. Burns, L.D., Hall, R.W., Blumenfeld, D.E. and Daganzo, C.F.: Distribution Strategies that Minimize Transportation and Inventory Costs, *Operations Research*, 33(3):469-490 (1985)
3. Cardós, M. and García-Sabater, J.P.: Designing a consumer products retail chain inventory replenishment policy with the consideration of transportation costs. International Journal of Production Economics. In Press. doi:10.1016/j.ijpe.2004.12.022
4. Constable, G.K. and Whybark, D.C.: The interaction of transportation and inventory decisions, *Decision Sciences*, 9(4):688-699 (1978)
5. Ganeshan, R.: Managing supply chain inventories: A multiple retailer, one warehouse, multiple supplier model. *International Journal of Production Economics*, 59(1-3):341-354 (1999)
6. Gillet, B.E. and Miller, L.R.. A Heuristic Algorthm for the Vehicle Dispatch Problem. Operations Research 22: 340-349 (1974)
7. Qu, W.W., Bookbinder, J.H. and Iyogun, P.: An integrated inventory-transportation system with modified periodic policy for multiple products. *European Journal of Operational Research*, 115(2): 254-269 (1999)
8. Robusté, F., Daganzo, C.F. and Souleyrette, R.: Implementing vehicle routing models. Transportation Research B, 24(4): 263-286 (1990).
9. Speranza, M.G. and Ukovich, W.: Minimizing Transportation and Inventory Costs for Several Products on a Single Link, *Operations Research*, 42(5):879-894 (1994)
10. Zhao, Q.H., Wang, S.Y., Lai, K.K. and Xia, G.P.: Model and algorithm of an inventory problem with the consideration of transportation cost. *Computers and Industrial Engineering*, 46(2): 398-397 (2004)

# Parametrized GRASP Heuristics for Three-Index Assignment

Armin Fügenschuh and Benjamin Höfler

Darmstadt University of Technology,
Schlossgartenstr. 7, 64289 Darmstadt, Germany

**Abstract.** Constructive greedy heuristics are algorithms that try to iteratively construct feasible solutions for combinatorial optimization problems from the scratch. For this they make use of a greedy scoring function, which evaluates the myopic impact of each possible element with respect to the solution under construction. Although fast, effective, and even exact for some problem classes, greedy heuristics might construct poor solution when applied to difficult (NP-hard) problems. To avoid such pitfalls we suggest the approach of parametrizing the scoring function by including several different myopic aspects at once, which are weighted against each other. This so-called pgreedy approach can be embedded into the metaheuristic concept of GRASP. The hybrid metaheuristic of GRASP with a parametrized scoring function is called parametrized GRASP heuristic (PGRASP). We present a PGRASP algorithm for the axial three index assignment problem (AP3) and computational results comparing PGRASP with the classical GRASP strategy.

## 1   Introduction

Greedy-type construction heuristics are used in many special-purpose optimization software packages, where a good feasible solution to a given instance of some combinatorial problem is required after a very short amount of time (typically, a few seconds). They construct feasible solutions from scratch by step-by-step inserting always the best immediate, or local, solution while finding an answer to the given problem instance. To obtain in fact good solutions, the crucial point within every greedy algorithm is having a proper criterion that selects these local solutions and thus is responsible for the search direction. For some optimization problems greedy algorithms are able to find the globally optimal solution. For example, Prim's or Kruskal's algorithms actually are greedy algorithms which find a minimum spanning tree in a given graph. On the other hand, there is no known greedy algorithm that finds a minimum Hamiltonian path, i.e., a solution to the traveling salesman problem (TSP). The aim of this article is to extend classical greedy construction heuristics by using parametrized scoring function, and to embed such parametrized greedy heuristic into the GRASP framework, in order to find better solutions in shorter time.

J. Gottlieb and G.R. Raidl (Eds.): EvoCOP 2006, LNCS 3906, pp. 61–72, 2006.

## 1.1   The Three Index Assignment Problem

As a demonstrator example we selected the axial three index assignment problem (AP3), which was first stated in [13] (see also [14]). It can be formulated as follows: Given $n \in \mathbb{N}$, three sets $I := J := K :- \{1, \ldots, n\}$, and associated costs $c(i, j, k) \in \mathbb{R}$ for all ordered triples $(i, j, k) \in I \times J \times K$. A feasible solution for the AP3 is a set of $n$ triples, such that each pair of triples $(i_1, j_1, k_1), (i_2, j_2, k_2)$ has different entries in every component, i.e., $i_1 \neq i_2, j_1 \neq j_2, k_1 \neq k_2$. The aim of the AP3 is to find a feasible solution $S$ with minimal costs $\sum_{(i,j,k) \in S} c(i, j, k)$. It was shown by Garey and Johnson that AP3 is NP-hard [7].

AP3 has a lot of real-world applications. Several exact and heuristic algorithms have been proposed for its solution, see Aiex et al. [1] and the references therein. AP3 is also an example for the large class of combinatorial optimization problems. Most of the ideas developed in this article apply not only for AP3 but also for many other problems. Thus we now take a more general point of view, and come back to AP3 later.

## 1.2   Combinatorial Optimization Problems

An instance of a combinatorial optimization problem $(E, \mathcal{F}, f)$ is given by a finite basic set $S$, a set $\mathcal{F} \subseteq 2^E$ of feasible solutions, and an objective function $f : \mathcal{F} \to \mathbb{Q}$. A combinatorial optimization problem $(E, \mathcal{F}, f)$ is called linear, if there exists a function $c : E \to \mathbb{Q}$ such that for all feasible solutions $S \in \mathcal{F}$ we have $f(S) = \sum_{e \in S} c(e)$. The aim of combinatorial optimization problems is to compute a feasible solution $S \in \mathcal{F}$ that minimizes the objective function.

# 2   Greedy Algorithms and Extensions

We start with a discussion of the probably most simple algorithmic framework to solve such combinatorial optimization problems: the constructive greedy heuristic.

## 2.1   Greedy

A *constructive greedy heuristic* is defined as a procedure that tries to construct a feasible solution for a given combinatorial optimization problem by stepwise selecting the currently most promising element in $E$ (with respect to some kind of scoring function). For this, the following steps are iteratively repeated: In the $i$-th step the element $e_i = \operatorname{argmin}\{s_i(e) < \infty : e \in E\}$ is chosen. Here $s_i : E \to \mathbb{Q} \cup \{\infty\}$ is the so-called $i$-th scoring function, which yields a score $s_i(e)$ for the selection of $e \in E$ in step $i$. In the sequel we assume that it is computationally easy to identify some $e \in E$ that minimizes the scoring function $s_i$. Moreover we assume a deterministic rule to break ties (in the case of more than one element with minimal score). After the selection of element $e_i$, $i$ is increased by one. The greedy algorithm terminates after at most $|E|$ steps when there is no element with finite score left, i.e., $s_i(e) = \infty$ for all elements $e \in E$. In the end, the set $S := \{e_1, \ldots, e_{i-1}\} \in 2^E$ called *solution*, is returned.

A solution $S$ is either *feasible* if $S \in \mathcal{F}$, or infeasible otherwise. As one can imagine the actual selection of a proper scoring function $s$ is essential for constructing a greedy heuristic that leads to good feasible solutions, or, for some hard problems, to any feasible solution at all. A feasible solution is called *optimal* if for every other feasible solution $T \in \mathcal{F}$ we have $f(S) \leq f(T)$. If a greedy heuristic always terminates with a feasible solution it is called *reliable*. If a reliable greedy heuristic always produces optimal solutions it is called *exact*. Another special case is the *static* greedy algorithm, where every two consecutive scoring functions $s_i, s_{i+1}$ fulfill the property $s_{i+1}(e) - s_i(e) \in \{0, \infty\}$ for all $e \in E$. That means, the scoring function is basically the same, except that some elements from the set $E$ are prevented from being selected in step $i + 1$ (and hence in all later steps). A non-static greedy algorithm is also called *adaptive*. The difference between static and adaptive greedy algorithms is now demonstrated by two well-known algorithms, Kruskal's and Prim's, for the construction of a graph's minimum weight spanning tree.

Given is an undirected, simple graph $G = (V, E)$ with edge weights $c(e) \geq 0$ for all $e \in E$. The weight of subgraph $S$, denoted by $c(S)$, is the total sum of the edge weights of the edges in $E_S$, i.e., $c(S) := \sum_{e \in E_S} c(e)$. A spanning tree $T$ is an acyclic subgraph of $G$ that connects all of the vertices of $G$. The minimum weight spanning tree problem asks for a spanning tree $T$ with minimum weight, that is, for any other spanning tree $T'$ of $G$ we have $c(T) \leq c(T')$.

Kruskal's algorithm [11] is an example for an exact, static greedy algorithm. The first scoring function is defined as $s_1(e) := c(e)$. Hence in the first step an edge $e_1 \in E$ with minimum weight is selected. In general, in the $i$-th step, the scoring function $s_i$ is defined as

$$s_i(e) := \begin{cases} \infty & ; e \in \{e_1, \ldots, e_{i-1}\}, \\ \infty & ; \text{subgraph } \{e_1, \ldots, e_{i-1}, e\} \text{ is cyclic}, \\ c(e) & ; \text{otherwise.} \end{cases} \tag{1}$$

Note that if subgraph $\{e_1, \ldots, e_i, e\}$ is cyclic, then subgraph $\{e_1, \ldots, e_i, e_{i+1}, e\}$ will also be cyclic. Thus if $s_i(e) = \infty$ for some $e \in E$, then also $s_{i+1}(e) = \infty$. Hence we have $s_{i+1}(e) - s_i(e) \in \{0, \infty\}$ for all $e \in E$. A proof for the exactness of Kruskal's algorithm can be found in [11].

Prim's greedy algorithm [15] on the other hand is an example for an exact, adaptive greedy algorithm. As above, the first scoring function is defined as $s_1(e) := c(e)$. Now in the $i$-th step, Prim's algorithm uses the following scoring function:

$$s_i(e) := \begin{cases} \infty & ; e \in \{e_1, \ldots, e_{i-1}\}, \\ c(e) & ; v \in e_k \text{ for some } k \in \{1, \ldots, i-1\}, \\ & \quad w \notin e_k \text{ for all } k \in \{1, \ldots, i-1\}, \text{ with } e = \{v, w\}, \\ \infty & ; \text{otherwise.} \end{cases}$$

To see that a greedy heuristic based on this family of scoring functions is adaptive, consider the following example. Let $\{e_1, \ldots, e_{i-1}\}$ be the set of edges selected so far. Furthermore let $u, v, w \in V$ be three pairwise distinct nodes with $\{u, w\}, \{v, w\} \in E, v \in e_k$ for some $k \in \{1, \ldots, i-1\}$ and $u, w \notin e_k$ for all $k \in \{1, \ldots, i-1\}$. Then $s_i(\{u, w\}) = \infty$. Suppose now that edge $\{v, w\}$ is selected in the $i$-th step. Then $s_{i+1}(\{u, w\}) < \infty$. Hence $s_{i+1}(\{u, w\}) - s_i(\{u, w\}) = -\infty$, thus the greedy algorithm is non-static. For a proof of the exactness of Prim's algorithm we again refer to the literature [15].

Kruskal's scoring functions (1) have a structure which can be considered as typical or canonical for linear combinatorial problems:

1. The setting $s_i(e) := \infty$ for all $e \in \{e_1, \ldots, e_{i-1}\}$ prevents from selecting again an element that was already chosen in a previous step of the construction heuristic.

2. The setting $s_i(e) := \infty$ for those $e \in E$ where $\{e_1, \ldots, e_{i-1}, e\}$ is cyclic prevents from selecting an element that would trivially lead to an infeasible solution.

3. The setting $s_i(e) := c(e)$ for all other $e \in E$ is an adaptation of the objective function to the scoring function.

Using such scoring function, the solution is iteratively built up by those elements that have the most promising objective function value and don't lead trivially to an infeasible solution. It is probably the easiest and perhaps most natural way to obtain a scoring function from the problem's objective function. However, especially when solving an $NP$-hard problem, greedy algorithms with scoring functions that are constructed along this methodology can easily be trapped by constructing an instance where they perform very badly. For instance, we consider the AP3. As abbreviation we denote $e = (u, v, w) \in E := I \times J \times K$ and $e_k = (u_k, v_k, w_k)$ for $k \in \mathbb{N}$. A canonical scoring function family, similar to Kruskal's, is the following: $s_1(e) := c(e)$ for all $e \in E$, and

$$
s_i(e) := \begin{cases} \infty & ; e \in \{e_1, \ldots, e_{i-1}\}, \\ \infty & ; u = u_k \text{ or } v = v_k \text{ or } w = w_k \text{ for some } k \in \{1, \ldots, i-1\}, \\ c(e) & ; \text{otherwise}, \end{cases} \quad (2)
$$

for $i > 1$. Now we apply this greedy algorithm to an instance having the following input data: $I := J := K := \{1, 2\}$,

$$
\begin{pmatrix} c_{111} & c_{112} \\ c_{121} & c_{122} \end{pmatrix} = \begin{pmatrix} 1 & 10 \\ 10 & 10 \end{pmatrix}, \quad \begin{pmatrix} c_{211} & c_{212} \\ c_{221} & c_{222} \end{pmatrix} = \begin{pmatrix} 10 & 10 \\ 10 & 100 \end{pmatrix}. \quad (3)
$$

In the first step of the heuristic, element $e_1 := (1, 1, 1)$ is selected, because $e_1$ is the (unique) minimum of the scoring function $s_1$. In the second (and last) step, only element $e_2 := (2, 2, 2)$ has a finite score and is therefore selected. The entire solution is $S := \{(1, 1, 1), (2, 2, 2)\}$, and its objective function value is $f(S) = c(e_1) + c(e_2) = 1 + 100 = 101$. Thus the greedy heuristic returned the (unique) maximum. A minimum is, for instance, the solution $S^{\mathrm{opt}} := \{(1, 1, 2), (2, 2, 1)\}$. Here the objective function value is $f(S^{\mathrm{opt}}) = 10 + 10 = 20$.

## 2.2   Parametrized Greedy

A *parametrized scoring function* with $p$ parameters $\lambda \in \mathbb{Q}^p$ is a mapping $s_i :$ $E \times \mathbb{Q}^p \to \mathbb{Q} \cup \{\infty\}$. A greedy heuristic that makes use of scoring functions $s_i(\cdot, \lambda)$ for a given $\lambda \in \mathbb{Q}^p$ is hence called a *parametrized greedy heuristic* or *pgreedy*, for short. PGreedy was initially developed to solve the vehicle routing problem with coupled time windows (see Fügenschuh [6] for the details).

The local criterion which element to select, and hence the entire solution found by the pgreedy heuristic, depends on the actual choice of $\lambda \in \mathbb{Q}^p$. We write $S(\lambda)$ for the solution found when the pgreedy algorithm is called with parameter $\lambda$, and $z(\lambda)$ for the corresponding objective function value, i.e., $z(\lambda) = f(S(\lambda))$. We are now faced with the problem to find a vector $\lambda$ with $z(\lambda) \leq z(\mu)$ for all $\mu \in \mathbb{Q}^p$ and hence to search for

$$z^{\text{pgreedy}} := \inf\{z(\lambda) : \lambda \in \mathbb{Q}^p\}.$$

First we remark that the infimum is in fact a minimum, and that the search can be restricted to a compact subset of the unbounded $\mathbb{Q}^p$. Let $\|\cdot\|$ be an arbitrary norm on the space $\mathbb{Q}^p$. The most prominent examples are, for instance, the 1-norm $\|x\|_1 := \sum_{i=1}^p |x_i|$, the euclidean norm $\|x\|_2 := \sum_{i=1}^p x_i^2$, or the max-norm $\|x\|_\infty := \max\{|x_i| : 1 \leq i \leq p\}$.

**Theorem 1.** *There exists a sufficiently large number $R > 0$ such that*

$$z^{\text{pgreedy}} = \min\{z(\lambda) : \lambda \in \mathbb{Q}^p, \|\lambda\| \leq R\}.$$

In the sequel we restrict our discussion to the case of scoring functions that are linear in the parameters, i.e., for all $e \in E, \lambda, \mu \in \mathbb{Q}^p$, and $t, u \in \mathbb{Q}$ we have $s(e, t \cdot \lambda + u \cdot \mu) = t \cdot s(e, \lambda) + u \cdot s(e, \mu)$. In the case of linear scoring functions, the search space for the parameters can be further restricted.

**Theorem 2.** *Let $\lambda, \lambda' \in \mathbb{Q}^p$. If there is a positive scalar $t \in \mathbb{Q}_+$ such that $\lambda' = t \cdot \lambda$ then $S(\lambda) = S(\lambda')$ and hence $z(\lambda) = z(\lambda')$.*

**Corollary 1.** *For every solution $S(\lambda)$ with $\lambda \in \mathbb{Q}^p \backslash \{0\}$ there is a $\lambda' \in \mathbb{Q}^p$ with $\|\lambda'\| = 1$ such that $S(\lambda) = S(\lambda')$.*

In particular, we obtain the following result.

**Corollary 2.**
$$z^{\text{pgreedy}} = \min\{z(\lambda) : \|\lambda\| = 1\}.$$

Informally speaking, in the case of linear scoring functions, one can reduce the parameter search space by one dimension.

For example, consider a search space with $p$ parameters and the euclidean norm $\|\cdot\|_2$. The set of parameters $\lambda$ with $\|\lambda\| = 1$ is the $p$-dimensional unit hypersphere $\mathbb{S}^{p-1}$. A parametrization of this sphere is given by polar coordinates:

$$\Phi_p(\varphi) := \lambda_\varphi := \begin{pmatrix} \cos\varphi_{p-1} \cdot \cos\varphi_{p-2} \cdot \ldots \cdot \cos\varphi_2 \cdot \cos\varphi_1 \\ \cos\varphi_{p-1} \cdot \cos\varphi_{p-2} \cdot \ldots \cdot \cos\varphi_2 \cdot \sin\varphi_1 \\ \cos\varphi_{p-1} \cdot \cos\varphi_{p-2} \cdot \ldots \cdot \cos\varphi_3 \cdot \sin\varphi_2 \\ \vdots \\ \cos\varphi_{p-1} \cdot \sin\varphi_{p-2} \\ \sin\varphi_{p-1} \end{pmatrix}$$

for $\varphi \in B^{p-1} := [-\pi, \pi[ \times [-\frac{\pi}{2}, \frac{\pi}{2}[^{p-2}$ (see Kaballo [10] for the details). Hence the search can be restricted to the bounded, rectangular set $B^{p-1}$.

But even in this lower-dimensional setting, the problem of finding these parameters remains. To this end, we now discuss several parameter selection strategies. In the sequel, we denote the $p-1$ dimensional, bounded search space always as $B^{p-1} \subset \mathbb{Q}^{p-1}$, and the corresponding parameter for the scoring function as $\lambda_\varphi$, where $\varphi \in B^{p-1}$.

**Enumeration.** Since the parameter search space is bounded, the optimal parameter vector $\varphi$ can in principle be found by sampling over a bounded subset of the regular, sufficiently dense grid $\Lambda_d^{p-1} := \{\frac{k}{d} \cdot e_i : k \in \mathbb{Z}, 1 \leq i \leq p-1\}$, where $e_i$ is the $i$-th unit vector, i.e., $e_i := (0, \ldots, 0, 1, 0, \ldots, 0)$ with the 1 at position $i$, and $d \in \mathbb{N}$ is a parameter taking control of the grid's density (the greater $d$, the finer the grid). For each $\varphi \in B^{p-1} \cap \Lambda_d^{p-1}$ the pgreedy heuristic has to be called and the best $\varphi$ (i.e., the $\varphi$ with the lowest objective function value $z(\lambda_\varphi)$) is kept. However, in practice this approach turns out to be inefficient, even for a relative few number of parameters.

**Pure Random Search.** The simplest idea besides enumeration is to select a candidate $\varphi$ parameters randomly (PRS, for short), which in practice also leads to a high number of runs.

**Improving Hit-and-Run.** This strategy was introduced by Zabinsky et al. [21] (see also [20]) to solve general global optimization problem. Improving hit-and-run (or IHR, for short) is a randomized (Monte-Carlo) algorithm that automatically selects parameters which lead to good, possibly optimal solutions when used in a pgreedy heuristic. In a hybrid algorithm of pgreedy and IHR, a combination of a parametrized greedy algorithm and improving hit-and-run, IHR is used to compute the weights $\lambda$ that take control of the parametrized scoring function and calls the pgreedy algorithm as a black-box to obtain a new objective function value. The basic idea behind improving hit-and-run is to use hit-and-run to generate candidate points randomly and accept only those that are an improvement with respect to the objective function. (See Smith [19] for more details on hit-and-run.) The IHR algorithm works as follows: We start with an initial $\varphi^0 \in B^{p-1}$, and set $k := 0$. The following steps are now repeated until a stopping criterion is met, for example, if the number of iterations $k$ reaches a certain limit. Generate a random direction vector $d_k$ uniformly distributed on the boundary of the unit hypersphere $\mathbb{S}^{p-1} := \{\varphi \in \mathbb{Q}^{p-1} : \|\varphi\|_2 = 1\}$. Generate a candidate point $w^{k+1} := \varphi^k + t \cdot d^k$, where $t$ is generated by sampling uniformly over the line set $L_k := \{\varphi \in B^{p-1} : \varphi = \varphi^k + t \cdot d^k, t \in \mathbb{Q}_+\}$. If the candidate point is improving, i.e., $z(\lambda_{w^{k+1}}) < z(\lambda_{\varphi^k})$, we set $\varphi^{k+1} := w^k$, otherwise $\varphi^{k+1} := \varphi^k$. Finally, increase $k$ by 1.

**Hide-and-Seek.** On some practical applications of IHR it has been observed that IHR sometimes gets trapped in local optima [20]. To avoid this, one can use a generalization of IHR, called Hide-and-Seek (HAS, for short), proposed by Romeijn and Smith [18]. HAS is a hybridization of IHR and Simulated Annealing. As in IHR, in case of an improving candidate point $w^{k+1}$, this point is always accepted. If the candidate point is not improving, then it still might be accepted with a certain probability. This probability is adjusted after each iteration, so that the acceptance of non-improving points will become less likely in later iterations. In detail, we set

$$\varphi^{k+1} := \begin{cases} w^{k+1} & \text{; with probability } P_{T_k}(\varphi^k, w^{k+1}), \\ \varphi^k & \text{; otherwise,} \end{cases}$$

where $P_{T_k}(\varphi^k, w^{k+1}) := \min\{1, \exp((z(\lambda_{\varphi^k}) - z(\lambda_{w^{k+1}}))/T)\}$, and $T_k$ is the temperature in the $k$-th iteration. The temperature is updated according to a cooling schedule. For example, one can select a starting temperature $T_0 \in [0, \infty]$ and a cooling factor $q \in [0, 1]$, and use the update rule $T_{k+1} := q \cdot T_k$. Note that for $T_0 := 0$ the whole HAS algorithm passes into the IHR algorithm.

We now extend the greedy heuristic using the scoring function (2) for the AP3 to a parametrized greedy heuristic. The crucial point for pgreedy is that more than a single criterion is needed. Finding those additional criteria is in general more of an art than science. For example, if in the $i$-th step of the greedy heuristic an element $e = (u, v, w) \in E$ is selected, one can consider $\sum_{e' \in N(e)} c(e')$ with $N(u, v, w) := \{(\alpha, \beta, \gamma) \in E : u = \alpha \text{ or } v = \beta \text{ or } w = \gamma\}$, i.e., the total weight of all neighboring elements which are not selectable anymore if $e$ is selected, as an additional term in the scoring function. Moreover, one can consider the smallest coefficient $\min\{c(e') : e' \in N(e)\}$ or the largest coefficient $\max\{c(e') : e' \in N(e)\}$ in the neighborhood of $e$ as additional terms. We set

$$\sigma(e, \lambda) := \lambda_1 \cdot c(e) + \lambda_2 \cdot \sum_{e' \in N(e)} c(e') +$$
$$\lambda_3 \cdot \min\{c(e') : e' \in N(e)\} + \lambda_4 \cdot \max\{c(e') : e' \in N(e)\}$$

and define

$$s_i(e, \lambda) := \begin{cases} \infty & \text{; } e \in \{e_1, \ldots, e_{i-1}\}, \\ \infty & \text{; } u = u_k \text{ or } v = v_k \text{ or } w = w_k \\ & \text{for some } k \in \{1, \ldots, i-1\}, \\ \sigma(e, \lambda) & \text{; otherwise} \end{cases} \qquad (4)$$

as $i$-th parametrized scoring function, where $\lambda = (\lambda_1, \ldots, \lambda_4) \in \mathbb{Q}^4$ are some scalar parameters. Note that for $\lambda = (1, 0, 0, 0)$ the classical scoring function (2) is regained. Hence the best solution found by an algorithm using (4) is expected to be at least as good as when using (2) as scoring function.

We apply pgreedy to the instance (3) from above. Consider the parameter vector $\lambda = (1, -1, 0, 0)$. Then we obtain $s_1((1, 1, 2), \lambda) = -131$ as scoring function value for $(1, 1, 2)$, which is minimal and hence $e_1 := (1, 1, 2)$ is selected. (Note that this selection is not unique, since other elements have the same minimal score.) Now for the selection of the second element, only $e_2 := (2, 2, 1)$ (also with score $s_2((2, 2, 1), \lambda) = -131$) remains. Hence $\{e_1, e_2\} = S^{\text{opt}}$.

From this small example we learn that parametrized greedy heuristics are able to construct better (even optimal) solutions than their classical counterparts. For larger instances, the solution quality can further be improved by embedding pgreedy into a GRASP search scheme.

## 2.3   GRASP

GRASP is an abbreviation of *Greedy Randomized Adaptive Search Procedure*. This metaheuristic for solving combinatorial optimization problems was first introduced by Feo and Resende for the set covering problem [5]. Since its first description in 1989, GRASP has been successfully applied to many difficult combinatorial optimization problems, see [17] for a recent survey. In its basic form it is a multi-start, iterative process where each iteration consists of two phases: A construction phase, in which a feasible solution is produced, and a local search phase, where the neighborhood of this solution is investigated to find a better solution.

In the construction phase a feasible solution $S \in \mathcal{F}$ is built step-by-step using a greedy scoring function, similar as it is done in the pure greedy heuristics described above. The difference is that not the element with the best (lowest) scoring value is chosen automatically. Instead a restricted candidate list (RCL), i.e., set of candidates with a sufficiently good greedy function value, is built and some element from this list is chosen randomly.

There are basically two approaches to build a RCL. First, there is the cardinality-based approach, where a fixed number $maxRCL$ is given, and the best $maxRCL$ candidates (with respect to the scoring function) are put into the RCL. The other approach is the value-based one, where a threshold value $g_{thresh}$ is lying between $g_{min}$, the best and $g_{max}$, the worst (but finite) scoring function value of the candidates. For example, the threshold is given by the formula $g_{thresh} := g_{min} + \alpha \cdot (g_{max} - g_{min})$, where $\alpha \in [0, 1]$ is a parameter taking control of the size of the list. Now all candidates having a scoring function value lower than $g_{thresh}$ are placed in the RCL. In both cases, the cardinality-based and the value-based, an element $e$ from the RCL is randomly chosen and put into the solution $S$ under construction.

The parameter $\alpha$ is usually chosen randomly in the interval $[0, 1]$, and its value remains constant during the whole construction phase. This parameter plays a decisive role for the quality of the constructed solution. It is a measure for the greediness of the choice of the elements. For example, $\alpha = 0$ means that only elements with the best greedy function value of all candidates can be chosen. In this case the construction phase is equivalent to the pure greedy algorithm with a random choice in the case of ties. On the other hand, $\alpha = 1$ means pure random choice among all candidates.

As a local search post-optimization routine one can use path relinking (PR). PR was proposed by Glover [8] as an approach to enhance intensification and diversification in the tabu search heuristic, but it can as well be integrated into other heuristics. The first hybrid metaheuristic of GRASP and path relinking was implemented by Laguna and Marti [12]. Aiex et al. [1] give a detailed description of an adaptation of PR for the AP3.

## 2.4   Parametrized Greedy and GRASP

The two strategies of parametrizing a greedy scoring function on the one hand and randomizing the choice among the best candidates on the other hand can be naturally combined into a single approach for solving difficult combinatorial optimization problems. We propose the term *parametrized GRASP*, or *PGRASP* for short, for this hybrid metaheuristic.

One iteration of PGRASP consists of three phases: 1) the choice of $\lambda$ parameters by means of an enumerative or stochastic procedure (such as PRS, IHR, or HAS), 2) a construction phase, which is equivalent to the construction phase of a GRASP with the additional property that a parametrized greedy function is used, and 3) a local search phase.

In contrast to the pgreedy heuristic (with a deterministic rule to break ties), the outcome of the PGRASP construction phase with a fixed parameter $\lambda$ is stochastic. In GRASP the objective function value of the solution depends crucially on the parameter $\alpha$ (see [17]), whereas in pgreedy it depends on $\lambda$. Hence it is not reasonable to change $\lambda$ after just one constructed solution. Instead, there is a given number of iterations $l$, where the same $\lambda$ but different values for $\alpha$ are used. After $l$ iterations, a new $\lambda$ is selected based on one of the rules presented above.

# 3   Computational Results

All computational experiments were carried out on a dual Intel-Pentium III computer running at 500MHz with 256MB RAM, 512kB cache memory and 1GB swap. The operating system was Debian Linux 2.4.26. The code was written in C and compiled using gnuC2.95.4. Random numbers are generated using the **rand** function of the C standard library. In the beginning of a program run, the function **srand** is once applied with a seed chosen by the user.

## 3.1   Test Instances

We considered four different classes of test instances. Two classes are generated following the outlines found in the literature. Altogether we used a pool consisting of 18 test instances.

1. Balas and Saltzman [2] describe instances with sizes ranging from $n = 12$ to $n = 26$. The cost coefficients $c_{ijk}$ are integer and uniformly generated in the interval $[0, 99]$. We generated four instances of this class, bs26, bs40, bs54, and bs66, with $n = 26, 40, 54, 66$, respectively.

2. Crama and Spieksma [4] reformulate the AP3 as an optimization problem in the tripartite graph $K_{n,n,n} := (I \cup J \cup K, (I \times J) \cup (I \times K) \cup (J \times K))$. A triple $(i, j, k)$ of the original formulation of the AP3 corresponds then to the triangle $(i, j, k)$ in $K_{n,n,n}$. The costs $c_{ijk}$ are associated with triangles, and the problem consists in finding a minimum cost collection of $n$ disjoint triangles. A length $d_{uv} > 0$ is assigned to each edge of $K_{n,n,n}$, and the cost $c_{ijk}$ of triangle $(i, j, k) \in I \times J \times K$ is its circumference $c_{ijk} := d_{ij} + d_{ik} + d_{jk}$. Crama and Spieksma suggest three types of randomly assigning a length to an edge (see [4] for the details). For each type we took two instances with $n = 33$ and $n = 66$ from [16], which gives six instances altogether. These instances are referred to as cs33t1, cs33t2, cs33t3, cs66t1, cs66t2, and cs66t3.

3. The instances of Burkard, Rudolf, and Woeginger [3] have decomposable integer cost coefficients $c_{ijk} = \alpha_i \cdot \beta_j \cdot \gamma_k$ where $\alpha_i, \beta_j, \gamma_k$ are uniformly distributed in the interval $[1, 10]$. We generated four instances according this guideline called brw26, brw40, brw 54, and brw66 with $n = 26, 40, 54, 66$, respectively.

4. The instances fh26, fh40, fh54, and fh66 (for $n = 26, 40, 54, 66$, respectively) have cost coefficients $c_{ijk} := \lfloor 10000 \cdot z^2 \rfloor$, where $z \in [0, 1]$ is randomly generated.

## 3.2   GRASP, PGRASP, and the Global Optimum

We implemented GRASP with path relinking according to the paper of Aiex et al. [1]. Our implementation computes approximately twice as much feasible solutions per hour compared to the one of Aiex et al. For each of the 18 instances we first computed the global optimum using ILOG CPLEX 9.0.1 as a MIP solver [9]. On some of the larger instances this took several hours up to a few days. For the two heuristics we set $|\mathcal{P}| = 20$ as limit for the path relinking solution pool $\mathcal{P}$. In the PGRASP algorithm $\lambda$ was chosen by IHR (i.e., $T_0 = 0$). We set $l = 20$, that is, after 20 GRASP iterations a new $\lambda$ is selected using IHR. Both heuristics, GRASP and PGRASP, were limited to a three minute search per instance. Each heuristic was then called 20 times for each instance, and we measured the average objective function value of all 20 solutions (which took one hour per entry in the table below).

In 13 out of 18 runs, PGRASP was (slightly) ahead of GRASP, that means, the average solution quality from 20 runs was better. GRASP was (also slightly) better than PGRASP in two case (for brw54 and fh40), and both were equal in three cases (for cs33t1, brw26 and brw66). Both heuristics were only twice (for cs33t1 and brw26) able to identify a globally optimal solution in all 20 runs.

## 3.3    Conclusions

The results presented in Table 1 indicate that PGRASP is able to compensate the fewer number of iterations compared to the pure GRASP heuristic in almost 4 out of 5 runs. This shows that for finding good solutions for the AP3 it might be advantageous to take multiple greedy functions into account. From additional computations (not presented in detail here) we can furthermore conclude that the more sophisticated variants of parameter selection (like IHR and HAS) clearly outperform the simple PRS.

**Table 1.** Comparison of CPLEX, GRASP and PGRASP

| Instance | CPLEX | GRASP | PGRASP |
|---|---|---|---|
| bs26 | 0.00 | 9.05 | 8.45 |
| bs40 | 0.00 | 13.25 | 11.35 |
| bs54 | 0.00 | 16.35 | 15.55 |
| bs66 | 0.00 | 19.50 | 17.15 |
| cs33t1 | 1401.00 | 1401.00 | 1401.00 |
| cs33t2 | 5067.00 | 5068.20 | 5068.15 |
| cs33t3 | 131.00 | 131.30 | 131.20 |
| cs66t1 | 2449.00 | 2458.80 | 2458.35 |
| cs66t2 | 8944.00 | 8957.85 | 8957.50 |
| cs66t3 | 286.00 | 286.45 | 286.40 |
| brw26 | 2544.00 | 2544.00 | 2544.00 |
| brw40 | 4903.00 | 4903.50 | 4903.45 |
| brw54 | 5181.00 | 5181.85 | 5182.05 |
| brw66 | 5323.00 | 5327.30 | 5327.30 |
| fh26 | 1.00 | 87.05 | 70.05 |
| fh40 | 0.00 | 123.05 | 123.45 |
| fh54 | 0.00 | 191.30 | 174.95 |
| fh66 | 0.00 | 289.25 | 280.20 |

# References

1. Aiex R.M., Resende M.G.C., Pardalos P.M., Toraldo G. (2005), GRASP with Path Relinking for Three-Index Assignment. INFORMS Journal on Computing 17(2): 224 – 247.
   Software available at URL http://www.research.att.com/~mgcr/
2. Balas E., Saltzman M.J. (1991), An Algorithm for the Three-Index Assignment Problem. Oper. Res. 39: 150 – 161.
3. Burkard R.E., Rudolf R., Woeginger G.J. (1996), Three-dimensional axial assignment problems with decomposable cost coefficients. Discrete Applied Mathematics 65: 123 – 139.
4. Crama Y., Spieksma F.C.R. (1992), Approximation algorithms for three-dimensional assignment problems with triangle inequalities. European Journal of Oper. Res. 60: 273 – 279.

5. Feo T.A., Resende M.G.C. (1989), A probabilistic heuristic for a computationally difficult set covering problem. Operations Research Letters 8: 67 – 71.
6. Fügenschuh A. (2005), The Integrated Optimization of School Starting Times and Public Transport. PhD Thesis, Logos Verlag Berlin, 165 pages.
7. Garey M.R., Johnson D.S. (1979), Computers and Intractability: A Guide to the Theory of NP-Completeness. W.H. Freeman and Company, New York.
8. Glover F. (1996), Tabu Search and Adaptive Memory Programming - Advances, Applications and Challenges. In: R.S. Barr, R.V. Helgason, J.L. Kennington (Eds.), Interfaces in Computer Science and Operations Research, Kluwer Academic Publishers: 1 – 75.
9. ILOG CPLEX, ILOG, Inc., 1080 Linda Vista Ave., Mountain View, CA 94043. Information available at URL http://www.ilog.com/products/cplex/
10. Kaballo W. (1998), Einführung in die Analysis III. Spektrum Akademischer Verlag, Heidelberg. (In German).
11. Kruskal J.B. (1956), On the shortest spanning subtree of a graph and the traveling salesman problem. Proc. Amer. Math. Soc. 7: 48 – 50.
12. Laguna M., Marti R. (1999), GRASP and Path Relinking for 2-Layer Straight Line Crossing Minimization. INFORMS Journal on Computing 11: 44 – 52.
13. Pierskalla W.P. (1967), The Tri-Substitution Method for the Three-Dimensional Assignment Problem. Canadian Operational Research Society Journal 5(2): 71 – 81.
14. Pierskalla W.P. (1968), The Multi-Dimensional Assignment Problem. Operations Research 16(2): 422 – 431.
15. Prim R.C. (1957), Shortest connection networks and some generalizations. Bell System Tech. J. 36: 1389 – 1401.
16. Resende M.G.C., Test Data for the Three Index Assignment Problem, online available at URL http://www.research.att.com/~mgcr/data/
17. Resende M.G.C, Ribeiro C.C. (2003), Greedy randomized adaptive search procedures. In: F. Glover, G. Kochenberger (Eds.), Handbook of Metaheuristics, Kluwer Academic Publishers: 219 – 249.
18. Romeijn H.E., Smith R.L. (1994), Simulated Annealing for Constrained Global Optimization. Journal of Global Optimization 5: 101 – 126.
19. Smith R.L. (1984), Efficient Monte Carlo Procedures for Generating Points Uniformly Distributed over Bounded Regions. Operations Research 32: 1296 – 1308.
20. Zabinsky Z.B. (2003), Stochastic Adaptive Search for Global Optimization. Nonconvex Optimization and its Applications. Kluwer Academic Publishers, Boston.
21. Zabinsky Z.B., Smith R.L., McDonald J.F., Romeijn H.E., Kaufman D.E. (1993), Improving Hit-and-Run for Global Optimization. Journal of Global Optimization 3: 171 – 192.

# A Memetic Algorithm with Bucket Elimination for the Still Life Problem

José E. Gallardo, Carlos Cotta, and Antonio J. Fernández

Dept. Lenguajes y Ciencias de la Computación, ETSI Informática,
University of Málaga, Campus de Teatinos, 29071 - Málaga, Spain
{pepeg, ccottap, afdez}@lcc.uma.es

**Abstract.** Bucket elimination (BE) is an exact technique based on variable elimination, commonly used for solving constraint satisfaction problems. We consider the hybridization of BE with evolutionary algorithms endowed with tabu search. The resulting memetic algorithm (MA) uses BE as a mechanism for recombining solutions, providing the best possible child from the parental set. This MA is applied to the maximum density still life problem. Experimental tests indicate that the MA provides optimal or near-optimal results at an acceptable computational cost.

## 1 Introduction

The game of life [1] consists of an infinite checkerboard in which the only player places checkers on some of its squares. Each square is a cell that has eight neighbors: the eight cells that share one or two corners with it. A cell is alive if there is a checker on it, and dead otherwise. The state of the board evolves iteratively according to three rules: (i) if a cell has exactly two living neighbors then its state remains the same in the next iteration, (ii) if a cell has exactly three living neighbors then it is alive in the next iteration and (iii) if a cell has fewer than two or more than three living neighbors, then it is dead in the next iteration. An interesting extension of this game is the *maximum density still life problem* (MDSLP) that consists of finding board configurations with a maximal number of living cells not changing along time. These stable configurations are called *maximum density stable patterns* or simply *still lifes*. In this paper we are concerned with the MDSLP and finite patterns, i.e., finding $n \times n$ still lifes. No polynomial method is known for this problem.

Our interest in this problem is manifold. Firstly, it must be noted that the patterns resulting in the game of life are very interesting. For example, by clever placement of the checkers and adequate interpretation of the patterns, it is possible to create a Turing-equivalent computing machine [2]. From a more applied point of view, it is interesting to consider that many aspects of discrete dynamical systems have been developed or illustrated by examples in life game [3, 4]. In this sense, finding stable patterns can be regarded as a mathematical abstraction of a standard issue in discrete systems control. Finally, the MDSLP is a prime example of weighted constrained optimization problem, and as such constitutes an excellent test bed for different optimization techniques.

J. Gottlieb and G.R. Raidl (Eds.): EvoCOP 2006, LNCS 3906, pp. 73–85, 2006.

The still life problem has been recently included in the CSPLib repository and a dedicated web page[1] maintains up-to-date results. This problem has been tackled using different approaches. Bosch and Trick [5] used a hybrid approach mixing integer programming and constraint programming to solve the cases for $n = 14$ and $n = 15$ in about 6 and 8 days of CPU time respectively. Smith [6] considered a pure constraint programming approach to tackle the problem and proposed a formulation of the problem as a constraint satisfaction problem with 0-1 variables and non-binary constraints. A dual formulation of the problem was also considered, and it was proved that this dual representation outperformed the initial one (although it could only solve instances up to $n = 10$). In any case, this dual encoding was particularly useful to find (90° rotational) symmetric solutions (e.g., it found the optimal solution for $n = 18$). Later on, Larrosa *et al.* [7, 8] showed the usefulness of variable elimination techniques, namely bucket elimination (BE), on this problem. Their basic approach could solve the problem for $n = 14$ in about $10^5$ seconds. Further improvements pushed the solvability boundary forward to $n = 20$ in about the same time. At any rate, it is clear that these exact approaches are inherently limited for increasing problem sizes, and their capabilities as anytime algorithms are unclear. Furthermore, to the best of our knowledge no heuristic approaches to this problem have been attempted.

In this work, we consider the hybridization of evolutionary algorithms with the BE approach. We will show that memetic algorithms (MAs) endowed with BE can provide optimal or near-optimal solutions at an acceptable computational cost. To do so, we will firstly introduce the essentials of BE in next section.

## 2   WCSPs and Bucket Elimination

A *Weighted constraint satisfaction problem* (WCSP) [9] is a constraint satisfaction problem (CSP) in which the user can express preferences among solutions. A WCSP is defined by a tuple $(X, D, F)$, where $X = \{x_1, \cdots, x_n\}$ is a set of variables taking values from their finite domains ($D_i \in D$ is the domain of $x_i$) and $F$ is a set of cost functions (also called soft constraints). Each $f \in F$ is defined over a subset of variables $var(f) \subseteq X$, called its *scope*. For each assignment $t$ of all variables in the scope of a soft constraint $f$, $t \in f$ (i.e., $t$ is *permitted*) if, and only if, $t$ is allowed by the soft constraint. A complete assignment that satisfies every soft constraint represents a solution to the WCSP. The valuation of an assignment $t$ is defined as the sum of costs of all functions whose scope is assigned by $t$. Permitted assignments receive finite costs that express their degree of preference and forbidden assignments receive cost $\infty$. The optimization goal consists of finding the solution with the lowest valuation.

### 2.1   The Bucket Elimination Approach

Bucket elimination [10] is a generic algorithm suitable for many automated reasoning and optimization problems, in particular for WCSP solving.

---

[1] http://www.ai.sri.com/~nysmith/life/

```
function BE(X, D, F)
1:      for i := n downto 1 do
2:          B_i := {f ∈ F | x_i ∈ var(f)}
3:          g_i := (∑_{f∈B_i} f) ⇓ i
4:          F := (F ⋃{g_i}) − B_i
5:      end for
6:      t := ∅
7:      for i := 1 to n do
8:          v := argmin_{a∈D_i}{(∑_{f∈B_i} f)(t · (x_i, a))}
9:          t := t · (x_i, v)
10:     end for
11:     return(F, t)
end function
```

**Fig. 1.** The general template of Bucket Elimination for a WCSP $(X, D, F)$

BE is based upon the following two operators over functions:

- The sum of two functions $f$ and $g$ denoted $(f + g)$ is a new function with scope $var(f) \cup var(g)$ which returns for each tuple the sum of costs of $f$ and $g$ defined as $(f + g)(t) = f(t) + g(t)$;
- The elimination of variable $x_i$ from $f$, denoted $f \Downarrow i$, is a new function with scope $var(f) − \{x_i\}$ which returns for each tuple $t$ the minimum cost extension of $t$ to $x_i$, defined as $(f \cdot i)(t) = min_{a \in D_i}\{f(t \cdot (x_i, a))\}$ where $t \cdot (x_i, a)$ means the extension of $t$ to the assignment of $a$ to $x_i$. Observe that when $f$ is a unary function (i.e., arity one), eliminating the only variable in its scope produces a constant.

Fig. 1 shows an operational schema of the BE algorithm for solving a certain WCSP. The displayed algorithm returns the optimal cost in $F$ and one optimal assignment in $t$. Note that BE has exponential space complexity because in general, the result of summing functions or eliminating variables cannot be expressed intensionally by algebraic expressions and, as a consequence, intermediate results have to be collected extensionally in tables.

As it can be seen in Fig. 1, BE works in two phases. In the first phase (lines 1-5), the algorithm eliminates variables one at a time in reverse order according to an arbitrary variable ordering $o$ (without loss of generality, here we assume lexicographical ordering for the variables in $X$, i.e, $o = (x_1, x_2, \cdots, x_n)$). In the second phase (lines 6-10), the optimal assignment is computed processing variables in increasing order. The elimination of variable $x_i$ is done as follows: initially (line 2), all cost functions in $F$ having $x_i$ in their scope are stored in $B_i$ (the so called *bucket of $x_i$*). Next (line 3), BE creates a new function $g_i$ defined as the sum of all functions in $B_i$ in which variable $x_i$ has been eliminated. Then (line 4), this function is added to $F$ that is also updated by removing the functions in $B_i$. The consequence is that the new $F$ does not contain $x_i$ (all functions mentioning $x_i$ were removed) but preserves the value of the optimal cost. The elimination of the last variable produces an empty scope function (i.e., a constant) which is the optimal cost of the problem. The second phase (lines

6-10) generates an optimal assignment of variables. It uses the set of buckets that were computed in the first phase: starting from an empty assignment $t$ (line 6), variables are assigned from first to last according to $o$. The optimal value for $x_i$ is the best value regarding the extension of $t$ with respect to the sum of functions in $B_i$ (lines 8,9). We use $argmin_a\{f(a)\}$ to denote the value of $a$ producing minimum $f(a)$.

The complexity of BE depends on the problem structure (as captured by its constraint graph $G$) and the ordering $o$. According to [8], the complexity of BE along ordering $o$ is time $\Theta(Q \times n \times d^{w^*(o)+1})$ and space $\Theta(n \times d^{w^*(o)})$, where $d$ is the largest domain size, $Q$ is the cost of evaluating cost functions (usually assumed $\Theta(1)$), and $w^*(o)$ is the maximum width of nodes in the induced graph of $G$ relative to $o$ (check [8] for details).

## 2.2   Bucket Elimination for the Still Life Problem

The general template presented above can be readily applied to the MDSLP. To this end, let us first introduce some notation. A board configuration for a $n \times n$ instance will be represented by a $n$-dimensional vector $(r_1, r_2, \ldots, r_n)$. Each vector component encodes (as a binary string) a row, so that the $j$-th bit of row $r_i$ (noted $r_{ij}$) indicates the state of the $j$-th cell of the $i$-th row (a value of 1 represents a live cell and a value of 0 a dead cell). Let $Zeroes(r)$ be the number of zeroes in binary string $r$ and let $Adjacents(r)$ be the maximum number of adjacent living cells in row $r$. If $r_i$ is a row and $r_{i-1}$ and $r_{i+1}$ are the rows above and below $r$, then $Stable(r_{i-1}, r, r_{i+1})$ is a predicate satisfied if, and only if, all cells in $r$ are stable.

The formulation has $n$ cost functions $f_i$ ($i \in \{1..n\}$). For $i \in \{2..n-1\}$, $f_i$ is ternary with scope $var(f_i) = \{r_{i-1}, r_i, r_{i+1}\}$ and is defined as[2]:

$$f_i(a,b,c) = \begin{cases} \infty & : \ \neg Stable(a,b,c) \\ \infty & : \ a_1 = b_1 = c_1 = 1 \\ \infty & : \ a_n = b_n = c_n = 1 \\ Zeroes(b) & : \ \text{otherwise} \end{cases} \quad (1)$$

As to $f_1$, it is binary with scope $var(f_1) = \{r_1, r_2\}$ and is specified as:

$$f_1(b,c) = \begin{cases} \infty & : \ \neg Stable(0,b,c) \\ \infty & : \ Adjacents(b) > 2 \\ Zeroes(b) & : \ \text{otherwise} \end{cases} \quad (2)$$

Likewise, the scope of $f_n$ is $var(f_n) = \{r_{n-1}, r_n\}$ and its definition is:

$$f_n(a,b) = \begin{cases} \infty & : \ \neg Stable(a,b,0) \\ \infty & : \ Adjacents(b) > 2 \\ Zeroes(b) & : \ \text{otherwise} \end{cases} \quad (3)$$

---

[2] Notice in these definitions that stability is not only required within the pattern, but also in the surrounding cells (assumed dead).

```
        function BE(n, D)
1:          for a, b ∈ D do
2:              gₙ(a, b) := min_{c∈D}{fₙ₋₁(a, b, c) + fₙ(b, c)}
3:          end for
4:          for i := n − 1 downto 3 do
5:              for a, b ∈ D do
6:                  gᵢ(a, b) := min_{c∈D}{fᵢ₋₁(a, b, c) + gᵢ₊₁(b, c)}
7:              end for
8:          end for
9:          (r₁, r₂) := argmin_{a,b∈D}{g₃(a, b) + f₁(a, b)}
10:         opt := g₃(r₁, r₂) + f₁(r₁, r₂)
11:         for i := 3 to n − 1 do
12:             rᵢ := argmin_{c∈D}{fᵢ₋₁(rᵢ₋₂, rᵢ₋₁, c) + gᵢ₊₁(rᵢ₋₁, c)}
13:         end for
14:         rₙ := argmin_{c∈D}{fₙ₋₁(rₙ₋₂, rₙ₋₁, c) + fₙ(rₙ₋₁, c)}
15:         return (opt, (r₁, r₂, . . . , rₙ))
        end function
```

**Fig. 2.** Bucket Elimination for the MDSLP

Due to the sequential structure of the corresponding constraint graph, the model can be easily solved with BE. Fig. 2 shows the corresponding algorithm. Function BE takes two parameters: $n$ is the size of the instance to be solved, and $D$ is the domain for each variable (row) in the solution. If domain $D$ is set to $\{0..2^n - 1\}$ (i.e., a set containing all possible rows) the function implements an exact method that returns the optimal solution for the problem instance (as the number of dead cells) and a vector corresponding to rows representing that solution.

Note that the complexity of this method is time $\Theta(n^2 \times 2^{3n})$ and space $\Theta(n \times 2^{2n})$. On the other hand, a basic search-based solution to the problem could be implemented with worst case time complexity $\Theta(2^{(n^2)})$ and polynomial space. Observe that the time complexity of BE is therefore an exponential improvement over basic search algorithms, although its high space complexity makes the approach unpractical for large instances.

# 3   A Memetic Algorithm for the MDSLP

WSCPs are very amenable for being tackled with evolutionary metaheuristics. The quality of the results will obviously depend on how well the structure of the soft constraints is captured by the search mechanisms used in the optimization algorithm. To this end, problem-aware algorithmic components are essential. In the particular case of the MDSLP, we will use tabu search (TS) and BE for this purpose, integrating them into a memetic approach. Before detailing these two components, let us describe the basic underlying evolutionary algorithm (EA).

## 3.1   Representation and Fitness Calculation

The natural representation of MDSLP solutions is the binary encoding. Configurations will be represented as a binary $n \times n$ matrix $r$. Clearly, not all such binary matrices will correspond to stable patterns, i.e., infeasible solutions can be represented. We have opted for using a penalty-based fitness function in order to deal with such infeasible solutions. To be precise, the fitness (to be minimized) of a configuration $r$ is computed as:

$$f(r) = n^2 - \sum_{i=1}^{n}\sum_{j=1}^{n} r_{ij} + K \sum_{i=0}^{n+1}\sum_{j=0}^{n+1} \left[ r'_{ij}\phi_1(\eta_{ij}) + (1 - r'_{ij})\phi_0(\eta_{ij}) \right] \qquad (4)$$

where $r'$ is an $(n+2) \times (n+2)$ binary matrix obtained by embedding $r$ in a frame of dead cells (i.e., $r'_{ij} = r_{ij}$ for $i, j \in \{1..n\}$, and $r'_{ij} = 0$ otherwise – recall that stability is not only required within the $n \times n$ board, but also in its immediate neighborhood), $K$ is a constant, $\eta_{ij}$ is the number of live neighbors of cell $(i, j)$, and $\phi_0, \phi_1 : \mathbb{N} \longrightarrow \mathbb{N}$ are two functions defined as:

$$\phi_0(\eta) = \begin{cases} 0 & \text{if } \eta \neq 3 \\ K' + 1 & \text{otherwise} \end{cases} \qquad \phi_1(\eta) = \begin{cases} 0 & \text{if } 2 \leqslant \eta \leqslant 3 \\ K' + 2 - \eta & \text{if } \eta < 2 \\ K' + \eta - 3 & \text{if } \eta > 3 \end{cases} \qquad (5)$$

where $K'$ is another constant. The first double sum in Eq. (4) corresponds to the basic quality measure for feasible solutions, i.e., the number of active cells. As to the last term, it represents the penalty for infeasible solutions. The strength of penalization is controlled by constants $K$ and $K'$. We have chosen $K = n^2$ and $K' = 5n^2$. With this setting, given any two solutions $r$ and $s$, the one that violates less constraints is preferred; if two solutions violate the same number of constraints, the one whose overall degree of violation (i.e., distance to feasibility) is lower is preferred. Finally, if the two solutions are feasible, the penalty term is null and the solution with the higher number of live cells is better.

## 3.2   A Local Improvement Strategy Based on Tabu Search

The fitness function defined above provides a stratified notion of gradient that can be exploited by a local search strategy. Moreover, notice that the function is quite decomposable, since interactions among variables are limited to adjacent cells in the board. Thus, whenever a configuration is modified, the new fitness can be computed just considering the cells located in adjacent positions to changed cells. To be precise, assume that cell $(i, j)$ is modified in solution $r$, resulting in solution $s$; the new fitness $f(s)$ can be computed as:

$$f(s) = f(r) + K \left[ \Delta f_1(r_{ij}, \eta_{ij}) + \sum_{i',j'} \Delta f_2(r_{i'j'}, \eta_{i'j'}, r_{ij}) \right] \qquad (6)$$

where the sum in the last term ranges across all cells $(i', j')$ adjacent to $(i, j)$, and functions $\Delta f_1$ and $\Delta f_2$ are defined as:

$$\Delta f_1(c, \eta) = \begin{cases} 0 & \eta = 2 \\ (-1)^{(1-c)}\phi_0(\eta) & \eta = 3 \\ (-1)^c\phi_1(\eta) & \text{otherwise} \end{cases} \tag{7}$$

$$\Delta f_2(c', \eta, c) = (1 - c')\Delta f_{2,0}(\eta, c) + c'\Delta f_{2,1}(\eta, c) \tag{8}$$

$$\Delta f_{2,0}(\eta, c) = \begin{cases} K'+1 & (\eta = 2 \wedge c = 0) \vee (\eta = 4 \wedge c = 1) \\ -(K'+1) & \eta = 3 \\ 0 & \text{otherwise} \end{cases} \tag{9}$$

$$\Delta f_{2,1}(\eta, c) = \begin{cases} K'+1 & (\eta = 2 \wedge c = 1) \vee (\eta = 3 \wedge c = 0) \\ -(K'+1) & (\eta = 1 \wedge c = 0) \vee (\eta = 4 \wedge c = 1) \\ 1 & (\eta = 1 \wedge c = 1) \vee (\eta \geqslant 4 \wedge c = 0) \\ -1 & (\eta = 0) \vee (\eta \geqslant 5 \wedge c = 1) \\ 0 & \text{otherwise} \end{cases} \tag{10}$$

Using this efficient fitness re-computation mechanism, our local search strategy explores the neighborhood $\mathcal{N}(r) = \{s \mid \text{Hamming}(r, s) = 1\}$, i.e., the set of solutions obtained by flipping exactly one cell in the configuration. This neighborhood comprises $n^2$ configurations, and it is fully explored in order to select the best neighbor. In order to escape from local optima, a tabu-search scheme is used: up-hill moves are allowed, and after flipping a cell, it is put in the tabu list for a number of iterations (randomly drawn from $[n/2, 3n/2]$ to hinder cycling in the search). Thus, it cannot be modified in the subsequent iterations unless the aspiration criterion is fulfilled. In this case, the aspiration criterion is improving the best solution found in that run of the local search strategy. The whole process is repeated until a maximum number of iterations is reached, and the best solution found is returned.

## 3.3 Optimal Recombination with BE

In the context of the fitness function that we have considered, the binary representation used turns out to be freely manipulable: any configuration can be evaluated, and therefore any standard recombination operator for binary strings could be utilized in principle. For example, we could consider the two-dimensional version of single-point crossover, depicted in Fig. 3. While feasible from a computational point of view, such a blind operator would perform poorly though: it would be more similar to macromutation than to a sensible recombination of information. To fulfill this latter goal, we can resort to BE.

In section 2.2 it was shown how BE could be used to implement an exact method to solve the MDSLP. Although the resulting algorithm was better than basic search-based approaches, the corresponding time and space complexity were very high. In the following we describe how BE can be used to implement a recombination operator that explores the dynastic potential [11] (possible

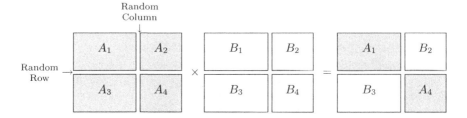

**Fig. 3.** Blind recombination operator for the MDSLP

children) of the solutions being recombined, providing the best solution that can be constructed without introducing implicit mutation (i.e., exogenous information).

Let $x = (x_1, x_2, \cdots, x_n)$ and $y = (y_1, y_2, \cdots, y_n)$ be two board configurations for a $n \times n$ instance of the MDSLP. Then, $BE(n, \{x_1, x_2, \cdots, x_n, y_1, y_2, \cdots, y_n\})$ calculates the best feasible configuration that can be obtained by combining rows in $x$ and $y$ without introducing information not present in any of the parents. Observe that we are just restricting the domain of variables to take values corresponding to the configurations being recombined, so that the result of function BE is the best possible recombination.

In order to analyze time complexity for this recombination operator, the critical part of the algorithm is the execution of lines 4-8 in Figure 2. In this case, line 6 has complexity $O(n^2)$ (finding the minimum of at most $2n$ alternatives, the computation of each being $\Theta(n)$). Line 6 has to be executed $n \times 2n \times 2n$ times at most, making a global complexity of $O(n^5) = O(|x|^{2.5})$, where $|x| \in \Theta(n^2)$ is the size of solutions. Notice also that the recombination procedure can be readily made to further exploit the symmetry of the problem, extending variable domains to column values in addition to row values. The complexity bounds remain the same in this case.

It must be noted that the described operator can be generalized to recombine any number of board configurations like $BE(n, \bigcup_{x \in S} \{x_i \mid i \in \{1..n\}\})$ where $S$ is a set comprising the solutions to be recombined. In this situation, the time complexity is $O(k^3 n^5)$ (line 6 is $O(kn^2)$, and it is executed $O(k^2 n^3)$ times), where $k = |S|$ is the number of configurations being recombined. Therefore, finding the optimal recombination from a set of MDSLP configurations is fixed-parameter tractable [12] when the number of parents is taken as a parameter.

## 4   Experimental Results

In order to assess the usefulness of the described hybrid recombination operator, a set of experiments for different problem sizes ($n = 12$ up to $n = 20$) has been realized. The experiments were done in all cases using a steady-state evolutionary algorithm (*popsize* $= 100$, $p_m = 1/n^2$, $p_X = 0.9$, binary tournament selection). With the aim of maintaining some diversity, duplicated individuals were not allowed in the population. All algorithms were run until an optimal solution was

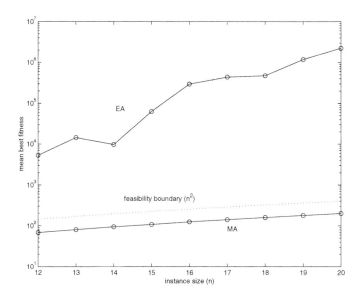

**Fig. 4.** Comparison of a plain EA, and a MA incorporating tabu search for different problem sizes. Results are averaged for 20 runs.

found or a time limit was exceeded. This time limit was set to 3 minutes for problem instances of size 12 and were gradually incremented by 60 seconds for each size increment. For each algorithm and each instance size, 20 independent executions were run. The experiments have been performed in a Pentium IV PC (2400MHz and 512MB of main memory) under SuSE Linux.

First of all, experiments were done with a plain EA. This EA did not use local search, utilized the blind recombination operator described in Sect. 3.3, and performed mutation by flipping single cells. This algorithm was compared with a MA that utilized tabu search for local improvement ($maxiter = n^2$), and the same recombination operator. Since simple bit-flipping moves were commonly reverted by the local search strategy, a stronger perturbation was considered during mutation, namely performing a cyclic rotation (by shifting bits one position to the right) in a random row (or column). Fig. 4 shows the results of this comparison. As it can be seen, the EA performs poorly, and is easily beaten by the MA. While the former cannot even find a single feasible solution in most runs, the MA finds not just feasible solutions in a consistent way, but solutions between 0.73% and 5.29% from the optimum (the optimal solution is found in at least one run for $n < 15$).

Subsequent experiments compared this basic MA with MAs endowed with BE for performing recombination as described in Sect. 3.3 (denoted as MABE). Since the use of BE for recombination has a higher computational cost than a simple blind recombination, and there is no guarantee that recombining two infeasible solutions will result in a feasible solution, we have defined two variants

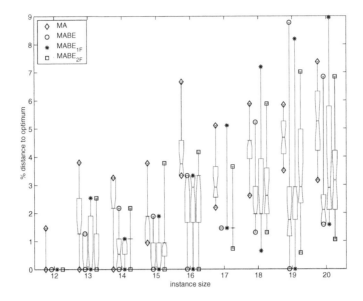

**Fig. 5.** Relative distances to optimum for each algorithm for sizes ranging from 12 up to 20. Each box summarizes 20 runs.

of MABE: in the first one $-MABE_{1F}-$ we require that at least one of the parents is feasible in order to apply BE; otherwise blind recombination is used. In the second one $-MABE_{2F}-$ we require the two parents being feasible, thus being more restrictive in the application of BE. With these two variants, we intend to explore the computational tradeoffs involved in the application of BE as an embedded component of the MA. For these algorithms, mutation was performed prior to recombination in order to better exploit good solutions provided by BE.

Fig. 5 shows the empirical performance of the different algorithms evaluated (relative to the optimum). Results show that MABE improves over MA on average and can find better solutions specially for larger instances. For example, average relative distance to the optimal solution is just 2.39% for $n = 20$. Note that results for $n = 19$ and $n = 20$ were obtained giving to each run of the evolutionary algorithm just 10 and 11 minutes respectively. As a comparison, recall that the approach in [7] respectively requires over 15 hours and over 2 days for these same instances, and that other approaches are unaffordable for $n > 15$. Note also that MABE can find the optimal solution in at least one run for $n < 17$ and $n = 19$ and the distance to the optimum for other instances is less than 1.58%. Results for $MABE_{1F}$ and $MABE_{2F}$ show that these algorithms do not improve over MABE. It seems that the effort saved not recombining unfeasible solutions does not further improve the performance of the algorithm. Fig. 6 extends these results up to size 28. The trend is essentially the same for sizes 21 and 22. Quite interestingly, it seems that for much larger instances the plain MA starts to catch up with MABE. This may be due to the increased computational cost for performing recombination. Recall that we are linearly

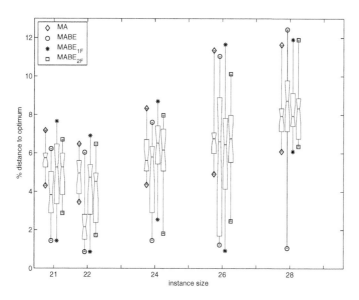

**Fig. 6.** Relative distances to the best known solutions for each algorithm for sizes ranging from 21 up to 28. Each box summarizes 20 runs. Results are only displayed for sizes for which an upper bound is available in the literature.

increasing the allowed computational time with the instance size, whereas the computational complexity of BE is superlinear.

The statistical significance of the results has been evaluated using a nonparametric test, the *Wilcoxon ranksum* test [13]. It has been found that differences are statistically significant (at the standard 5% level) when comparing the plain EA to any other algorithm in all cases. When comparing $MABE_{[*]}$ and MA, differences are significant for all instances except for size 12 (where all algorithms find systematically the optimum in most runs) and size $> 24$ (where the allowed computational time might be not enough for $MABE_{[*]}$ to progress further in the search). Finally, improvements for MABE over $MABE_{1F}$ and $MABE_{2F}$ are only significant in some cases (sizes 20, 22 and 24 for the former, and sizes 14, 15, 19, 20, 21 and 22 for the latter). The fact that MABE is significantly better than $MABE_{2F}$ in more cases than it is for $MABE_{1F}$ correlates well with the fact that BE is used less frequently in the former than in the latter.

## 5   Conclusions and Future Work

We have presented a model for the hybridization of BE, a well-known technique in the domain of constraint programming, with EAs. The experimental results for this model have been very positive, solving to optimality large instances of a hard constrained problem, and outperforming other evolutionary approaches, including a memetic algorithm incorporating tabu search.

There are many interesting extensions to this work. As it was outlined in Sect. 3.3, the proposed optimal recombination operator can be used with more than two parents. Furthermore, the resulting operator is fixed-parameter tractable when the number of parents is taken as parameter. An experimental study of multiparent recombination in this context can provide very interesting results. Work is currently underway in this direction.

Further directions for future work can be found in a more-in-depth exploitation of the problem symmetries [5, 6, 7, 8]: for any stable pattern an equivalent one can be created by rotating (by $90°$, $180°$ or $270°$) or reflecting the board. The presented approach could be adapted to incorporate them in the recombination process, probably boosting the search capabilities. We also plan to analyze this possibility. Finally, in [7], a hybrid algorithm for the MDSLP that combines BE and branch-and-bound search is presented providing excellent results. Hybridizing this algorithm with an EA seems a promising line of research as well.

**Acknowledgements.** This work was partially supported by Spanish MCyT under contracts TIC2002-04498-C05-02 and TIN2004-7943-C04-01.

# References

1. Gardner, M.: The fantastic combinations of John Conway's new solitaire game. Scientific American **223** (1970) 120–123
2. E.R. Berlekamp, J.C., Guy, R.: Winning Ways for your Mathematical Plays. Volume 2 of Games in Particular. Academic Press, London (1982)
3. Gardner, M.: On cellular automata, self-reproduction, the garden of Eden and the game of "life". Scientific American **224** (1971) 112–117
4. Gardner, M.: Wheels, Life, and Other Mathematical Amusements. W.H. Freeman, New York (1983)
5. Bosch, R., Trick, M.: Constraint programming and hybrid formulations for three life designs. In: CP-AI-OR. (2002) 77–91
6. Smith, B.M.: A dual graph translation of a problem in 'life'. In Hentenryck, P.V., ed.: 8th International Conference on Principles and Practice of Constraint Programming - CP'2002. Volume 2470 of Lecture Notes in Computer Science., Ithaca, NY, USA, Springer (2002) 402–414
7. Larrosa, J., Morancho, E., Niso, D.: On the practical use of variable elimination in constraint optimization problems: 'still life' as a case study. Journal of Artificial Intelligence Research **23** (2005) 421–440
8. Larrosa, J., Morancho, E.: Solving 'still life' with soft constraints and bucket elimination. In Rossi, F., ed.: Principles and Practice of Constraint Programming - CP 2003. Volume 2833 of Lecture Notes in Computer Science., Kinsale, Ireland, Springer (2003) 466–479
9. Bistarelli, S., Montanari, U., Rossi, F.: Semiring-based constraint satisfaction and optimization. Journal of the ACM **44** (1997) 201–236
10. Dechter, R.: Bucket elimination: A unifying framework for reasoning. Artificial Intelligence **113** (1999) 41–85

11. Radcliffe, N.: The algebra of genetic algorithms. Annals of Mathematics and Artificial Intelligence **10** (1994) 339–384
12. Downey, R., Fellows, M.: Fixed parameter tractability and completeness I: Basic theory. SIAM Journal of Computing **24** (1995) 873–921
13. Lehmann, E., D'Abrera, H.: Nonparametrics: Statistical Methods Based on Ranks. Prentice-Hall, Englewood Cliffs, NJ (1998)

# Effects of Scale-Free and Small-World Topologies on Binary Coded Self-adaptive CEA

Mario Giacobini[1], Mike Preuss[2], and Marco Tomassini[1]

[1] Information Systems Department, University of Lausanne, Switzerland
mario.giacobini@unil.ch, marco.tomassini@unil.ch
[2] Systems Analysis Group, Computer Science Department,
University of Dortmund, Germany
mike.preuss@uni-dortmund.de

**Abstract.** In this paper we investigate the properties of CEAs with populations structured as Watts–Strogatz small-world graphs and Albert–Barabási scale-free graphs as problem solvers, using several standard discrete optimization problems as a benchmark. The EA variants employed include self-adaptation of mutation rates. Results are compared with the corresponding classical panmictic EA showing that topology together with self-adaptation drastically influences the search.

## 1 Introduction

The standard population structure used in evolutionary algorithms (EAs) is the *panmictic* structure. In panmictic populations, also called mixing, any individual is equally likely to interact with any other individual. This setting is the most straightforward and many theoretical results have been obtained for it. However, since at least two decades, several researchers have suggested that EAs populations might have structures endowed with spatial features, like many natural populations (for recent reviews see [1, 2] and references therein). Empirical results suggest that using structured populations is often beneficial owing to better diversity maintenance, formation of niches, and lower selection pressures in the population favoring the slow spreading of solutions and relieving premature convergence and stagnation. The most popular models are the *island model* and the *cellular model*. In the island model the whole population is subdivided into several subpopulations each of which is panmictic. A standard EA runs in each subpopulation and, from time to time, a fraction of individuals migrate between islands. Although this model may offer some advantages over a single mixing population, it is still rather close to the latter.

Here we shall focus on cellular models instead, which are a more radical departure from the standard setting. What sets them apart is the fact that all the operators act locally, within a small pool of individuals. The customary cellular topology is the regular lattice. Cellular evolutionary algorithms (CEAs) on regular lattices, usually rings and two-dimensional grids, have been often used with good results and some of their theoretical properties are known (see [2]). However, there is no reason why cellular models should be limited to regular lattices. Other graph structures are possible, such as random graphs and *small-world* networks. These small-world networks are not regular nor

J. Gottlieb and G.R. Raidl (Eds.): EvoCOP 2006, LNCS 3906, pp. 86–98, 2006.

completely random, and have recently attracted a lot of attention in many areas because of their surprising topological properties [3, 4]. Random graphs and small-world networks have been recently studied from the point of view of the selection intensity in the population [5]. Random graphs are roughly equivalent to panmictic structures in behavior, at least for not too small probability of having an edge between two arbitrary vertices. The families of small-world graphs are potentially more interesting, as they can induce widely variable global selection pressures, depending on the value of some graph characteristic parameter [5]. A first investigation on the use of such structured populations for optimization problems has been proposed by Preuss and Lasarczyk [6].

In this paper we investigate the properties of CEAs with populations structured as Watts–Strogatz small-world graphs and Albert–Barabási scale-free graphs as problem solvers, using several standard discrete optimization problems as a benchmark. We should like to point out at the outset that it is not our intention to compete with the best heuristics for the problems. We do not use problem information, nor do we include any kind of local or enhanced search. Our goal is simply to compare these irregular population structures with regular lattices CEAs and the panmictic EA using the simplest settings and only few parameters. We are especially interested in answering the following questions:

- What is the influence of different node degree distributions on CEAs when the overall connectivity (number of connections) remains constant?
- Are scale-free topologies worthwhile alternatives to standard small-world ones? If so, for which problem types?
- When —if at all— does self-adaptation of mutation parameters provide an advantage over fixed mutation rates?

When dealing with evolutionary algorithms on binary represented problems, a sporadically suggested [7] and rarely used technique is the self-adaptation of mutation parameters. Although well established for continuous representations [8], its applicability is rather unclear for test problems typically approached with genetic algorithms. It is our hope that self-adaptation proves worthwhile for CEAs, especially in connection with small-world topologies.

## 2   Test Problems

In this section we present the set of problems chosen for this study. The benchmark is representative because it contains many different interesting features in optimization, such as epistasis, multimodality, deceptiveness, and problem generators. These are important ingredients in any work trying to evaluate algorithmic approaches with the objective of getting reliable results, as stated by Whitley et al. in [9].

We experiment with the massively multimodal deceptive problem (MMDP), a modified version of the multimodal problem generator P-PEAKS, error correcting code design (ECC), and the countsat problem (COUNTSAT). The choice of this set of problems is justified by both their difficulty and their application domains (combinatorial optimization, telecommunications, etc.). This gives us a fair level of confidence in the results, although no benchmark will ever be able to assert the superiority of a particular algorithm on all problems and problem instances [10]. The problems selected for this benchmark are briefly presented in the following paragraphs.

***Massively Multimodal Deceptive Problem (MMDP).*** The MMDP is a problem that has been specifically designed to be difficult for an EA [11]. It is made up of $k$ deceptive subproblems ($s_i$) of 6 bits each, whose value depends on the number of ones (*unitation*) a binary string has (see Figure 1). These subfunctions possess two global maxima and a deceptive attractor in the middle point.

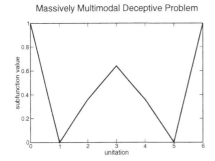

| unitation | subfunction value |
|-----------|-------------------|
| 0 | 1.000000 |
| 1 | 0.000000 |
| 2 | 0.360384 |
| 3 | 0.640576 |
| 4 | 0.360384 |
| 5 | 0.000000 |
| 6 | 1.000000 |

**Fig. 1.** Basic deceptive bipolar function ($s_i$) for MMDP

In MMDP each subproblem $s_i$ contributes to the fitness value according to its *unitation* (Figure 1). The global optimum has a value of $k$ and it is attained when every subproblem is composed of zero or six ones. The number of local optima is quite large ($22^k$), while there are only $2^k$ global solutions. Therefore, the degree of multimodality is regulated by the $k$ parameter. To avoid floor and ceiling effects (none or all EA are able to solve the problem) we use a moderately difficult instance with $k = 20$. Fitness is computed after after Eq. 1, utilizing subfunction $s_i$ as depicted in Figure 1. Note that this problem is separable; its constituents could be optimized individually if its boundaries in the genome were known to the EA.

$$f_{MMDP}(s) = \sum_{i=1}^{k} fitness(s_i) \qquad (1)$$

***Multimodal Problem Generator (wP-PEAKS).*** A problem generator is an easily parameterizable task which has a tunable degree of epistasis, thus permitting to derive instances with growing difficulty at will. With a problem generator we evaluate our algorithms on a high number of random problem instances. Since a different instance is solved each time the algorithm runs, the predictive power of the results for the problem class as a whole is increased.

The idea of P-PEAKS is to generate $P$ random $N$-bit strings that represent the location of $P$ peaks in search space. Using a small/large number of peaks results in weakly/strongly epistatic problems. In the original problem formulation [12], the fitness value of a string was the number of bits it had in common with the nearest peak in that space, divided by $N$. However, each peak represented a global optimum. We modified the problem by adding weights $w_i \in \mathbb{R}_+$ with only $w_1 = 1.0$ and $w_{[2...P]} < 1.0$, thereby requiring the optimization algorithm to find the one peak bearing the global optimum instead of just any peak. It should be noted that doing so for one global and nine local peaks (as utilized in our experiments) —tested empirically— appears to be a lot harder than a standard P-PEAKS with $P = 100$.

$$f_{wP-PEAKS}(\boldsymbol{x}) = \frac{1}{N} \max_{1 \leq i \leq p} \{w_i \cdot N - HammingD(\boldsymbol{x}, Peak_i)\} \tag{2}$$

***Error Correcting Code Design Problem (ECC).*** The ECC problem was presented in [13]. We will consider a three-tuple $(n, M, d)$, where $n$ is the length of each codeword (number of bits), $M$ is the number of codewords, and $d$ is the minimum Hamming distance between any pair of codewords. Our objective will be to find a code which has a value for $d$ as large as possible (reflecting greater tolerance to noise and errors), given previously fixed values for $n$ and $M$. The problem we have studied is a simplified version of that in [13]. In our case we search half of the codewords ($M/2$) that will compose the code, and the other half is made up by the complement of the codewords computed by the algorithm. The fitness function to be maximized is:

$$f_{ECC} = \frac{1}{\displaystyle\sum_{i=1}^{M} \sum_{j=1, i \neq j}^{M} d_{ij}^{-2}}, \tag{3}$$

where $d_{ij}$ represents the Hamming distance between codewords $i$ and $j$ in the code $C$ (made up of $M$ codewords of length $n$). In the present paper, we consider an instance with $M = 24$ and $n = 12$, yielding optimum fitness of 0.0674 [14].

***COUNTSAT Problem.*** The COUNSAT problem has been proposed by Droste *et al.* [15] as an instance of the MAXSAT problem difficult to be solved by Evolutionary Algorithms. In COUNSAT, the solution value is the number of clauses (among all the possible 3-variables Horn clauses) that are satisfied by an $n$-bit input string, where the binary value 0 and 1 are considered as a *false* and a *true* boolean value, respectively. It is easy to check that the optimum value is that of the solution with all the variables assigned to 1. Droste *et al.* have proved that the fitness of a tentative solution $\boldsymbol{x}$ can be easily computed using the following equation:

$$f_{COUNTSAT}(\boldsymbol{x}) = s + n(n-1)(n-2) - 2(n-2)\binom{s}{2} + 6\binom{s}{3}, \tag{4}$$

where $s$ is the unitation of the solution $\boldsymbol{x}$ (i.e. the number of 1 entries in $\boldsymbol{x}$), and $n$ is the length of $\boldsymbol{x}$. In this paper we will study an instance of $n = 20$ variables, with normalized optimum fitness of 1.0.

## 3  Small-World Graph Topologies

It has been shown in recent years that graphs occurring in many social, biological, and man-made systems are often neither completely regular, such as lattices, nor completely random [16]. They have instead what has been called a *small-world* topology, in which nodes are highly clustered yet the path length between them is small. This behavior is due to the presence of *shortcuts* i.e., a few direct links between nodes that would otherwise be far removed. Following Watts' and Strogatz's discovery, Barabasi *et al.* [3] found that several important networks such as the World Wide Web, Internet, author citation networks, and metabolic networks among others, also have the small world property but their degree distribution function differs: they have more nodes of high degree that are likely in a random graph of the same size and edge density. These graphs have been called *scale-free* because the degree probability distribution function follows a power law. In the next sections we briefly describe how small-world and scale-free graphs can be constructed, more details can be found in [3, 4, 16].

## 3.1   The Watts–Strogatz Model

Although this model has been a real breakthrough in the technical sense when it appeared, today it is clear that it is not a good representation of real networks as it retains many features of the random graph model. In spite of this, the Watts–Strogatz model, because of its simplicity of construction and the richness of behavior, is still an interesting topology in artificial systems where there is no "natural" constraint on the type of connectivity.

According to Watts and Strogatz [16], a small-world graph can be constructed starting from a regular ring of nodes in which each node has $k$ neighbors ($k \ll N$) by simply systematically going through successive nodes and "rewiring" a link with a certain probability $\beta$. When the edge is deleted, it is replaced with an edge to a randomly chosen node. If rewiring an edge would lead to a duplicate edge, it is left unchanged. This procedure will create a number of *shortcuts* that join distant parts of the lattice.

Shortcuts are the hallmark of small worlds. While the average path length[1] between nodes scales logarithmically in the number of nodes for a random graph, in Watts-Strogatz graphs it scales approximately linearly for low rewiring probability but goes down very quickly and tends to the random graph limit as $\beta$ increases. This is due to the progressive appearance of shortcut edges between distant parts of the graph, which obviously contract the path lengths between many vertices. However, small worlds typically have a higher clustering coefficient[2] than random graphs. Small-world networks have a degree distribution $P(k)$ close to Poissonian.

## 3.2   The Barabási-Albert Model

Albert and Barabási were the first to realize that real networks grow incrementally and that their evolving topology is determined by the way in which new nodes are added to the network and proposed an extremely simple model based on these ideas [3]. At the beginning one starts with a small clique of $m_0$ nodes. At each successive time step a new node is added such that its $m \leq m_0$ edges link it to $m$ nodes already in the

---

[1] The average path length $L$ of a graph is the average value of all pair shortest paths.

[2] The clustering coefficient $C$ of a node is a measure of the probability that two nodes that are its neighbors are also neighbors among themselves. The average $\langle C \rangle$ is the average of the $C$s of all nodes in the graph.

graph. When choosing the nodes to which the new nodes connects, it is assumed that the probability $\pi$ that a new node will be connected to node $i$ depends on the degree $k_i$ of $i$ such that nodes that have already many links are more likely to be chosen over those that have few. This is called *preferential attachment* and is an effect that can be observed in several real networks. The probability $\pi$ is given by:

$$\pi(k_i) = \frac{k_i}{\sum_j k_j},$$

where the sum is over all nodes already in the graph. The model evolves into a stationary scale-free network with power-law probability distribution for the vertex degree $P(k) \sim k^{-\gamma}$, with $\gamma \sim 3$.

## 4   Experiment

*Focus.* Investigate the effects of varied scale-free and small-world topologies on cellular EA with and without self-adaptation.

*Pre-experimental Planning.* First tests employed the parameter optimization method SPO as recently suggested by Bartz-Beielstein [17]. They revealed that, keeping the population size constant at $400$ and the number of connections at $800$, in most cases no significant performance increase could be gained by varying the number of offspring per generation or the maximum lifespan of an individual (the latter would lead to a $\kappa$-type or comma-type environmental selection/replacement scheme). This also holds for the mutation rate meta-parameter $\tau$ needed for self-adaptation, which has therefore been fixed at $0.5$. Furthermore, the mutation rate default setting $p_m = 1/l$, with $l$ the representation length, could be verified as a good compromise when using a fixed mutation rate for different problems.

A notable exception is the COUNTSAT problem, where self-adaptation together with large birth surplus and comma-type environmental selection performed very well. However, to simplify interpretation of results, we limited experimentation to plus selection, that is, any parent survives as long as it is not outperformed by its offspring. Our tests also showed that choosing a large population size for the panmictic EA is well-founded for the given problem set, at least when striving for high success rates.

For all problems, we determined suitable run lengths in order to measure success rates that approximate the ones for an infinite number of evaluations. The resulting run lengths are given in Table 1. In most cases, the actual average amount of evaluations needed to reach the global optimum is much lower.

When mutation rates are allowed to change, they still must be initialized with meaningful values. Our testing revealed that either starting with $p_m = 1/l$ or $p_m = 0.5$ for all individuals is advantageous, as opposed to initializing $p_m$ uniformly within $]0, 1[$.

*Task.* The character of our experiment is explorative; we want to find evidence that helps to answer the questions posed in the Introduction, namely situations in which small-world/scale-free topology based CEAs and/or self-adaptation appear advantageous over a standard, panmictic EA.

**Table 1.** Problem designs, common (top) and individual (bottom) part. SR stands for success rate, and AES is the average number of evaluations to solution. Each run was stopped at the given maximum number of evaluations as the only termination criterion.

| Initialization | Number of runs | Performance measures |
|---|---|---|
| randomized | 100 | SR/AES |

| Problem | Instance | Bits | Max. eval | Optimum |
|---|---|---|---|---|
| MMDP | 20 blocks of 6 bits | 120 | 120000 | 20.0 |
| wP-PEAKS | 10 peaks, $w_1 = 1.0$, $w_{[2...10]} = 0.99$ | 100 | 200000 | 1.0 |
| ECC | 12 codes à 12 bits, 12 complementary codes | 144 | 400000 | 0.0674 |
| COUNTSAT | 20 bits | 20 | 120000 | 1.0 |

*Setup.* Utilized problem designs, including initialization and termination criterion, are documented in Table 1. The EA variants employed all use bit-flipping mutation with probability $p_m$ and 2-point crossover. Mating selection is done randomly in the neighborhood of each individual, i.e. uniform selection, or among the whole population for the panmictic variant. We set the crossover probability to 1, so that during each generation, every individual produces one offspring. Replacement —or environmental selection— is performed simultaneously (synchronous) for all individuals, taking the better one of the current individual and its offspring each. Population size (400) and number of connections (800) are kept at CEA standard values to allow for comparison with previous studies [18].

Self-adaptation is performed as suggested by Rudolph [19] for discrete variables, differing only in that a mutation event always flips the accordant bit instead of computing its new value from the old one or choosing it randomly from $\{0, 1\}$. We apply it to the mutation probability only, as depicted in Eqn. 5, where $\tau$ is a constant meta-parameter and $N(0, 1)$ stands for a standard normally distributed random variable.

$$p'_{mut} = p_{mut} \cdot \exp\left(\tau \cdot N(0, 1)\right) \tag{5}$$

Thus every individual gets a mutation probability that it bequeaths to approximately half of its children by discrete recombination. We follow the standard scheme of evolution strategies by first applying mutation to the mutation rate, then utilizing the acquired mutation rate for mutating the rest of the genome [8].

Summarizing, four EA variants are run on the test problem set: A panmictic EA, a CEA with fixed mutation rate $p_m = 1/l$, and two CEAs with self-adaptive mutation rates, starting with $p_m = 1/l$ and $p_m = 0.5$, respectively. Except for the panmictic EA, different graphs are tested: For the Watts and Strogatz model topologies tried, we vary the rewiring factor $\beta$ between 0 and 0.2 . Whereas 0 stands for an unmodified ring structure, $\beta > 0.2$ produces networks that rapidly approach random graphs. The scale-free topologies were created for kernel sizes from the minimum 2 to 28, in which almost half of the available connections must be spent for the kernel, so that at least one per remaining node is left for preferential attachment. With the given parameters, actual topologies have been created anew for every single run.

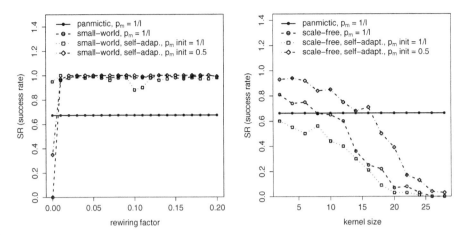

**Fig. 2.** Success rates for small-world (left) and scale-free (right) topology CEAs, compared to a similarly parametrized panmictic EA, on the MMDP. Each point is generated from 100 runs.

***Experimentation/Visualization.*** Due to space limitations, we only depict SR (success rate) results for the four problems, see Figures 2 and 3. Table 2 additionally provides numerical values obtained for AES (average number of evaluations to solution) and SR criteria on the MMDP, tables for the other problems are omitted for the same reason.

***Observations.*** The first thing to note is that the SR performance curves look very different for the four test problems. We therefore decided to describe the obtained results separately.

MMDP: Success rates for all small-world topology CEAs (except when $\beta = 0$) are near 1 and thus much higher than 0.66 of the panmictic variant. At the same time, they are a lot slower than the panmictic EA (see Table 2). Ring topology CEAs, i.e. $\beta = 0$, may have failed to succeed because their time consumption would have been even higher than the given limit. Scale-free CEAs with small kernels perform comparable to small-world CEAs with medium rewiring factor, in success rates as well as in speed. For larger kernels, success rates drop dramatically, even below the ones for the panmictic EA. Simultaneously, the length of successful runs increases. Self-adaptation of mutation rates works well in all small-world CEAs and quite good for scale-free CEAs with small kernels. It remarkably lowers the AES if started with $p_m = 1/l$. Interestingly, it was observed that learned mutation rates, especially when started at 0.5, tend to develop towards both ends of the allowed interval, namely 0 and 1, within the same population.

ECC: For both topology types, the fixed mutation CEA outperforms all other variants with respect to the SR criterion. Rewiring rates and kernel sizes seem to have little influence here. The panmictic EA is slightly faster but achieves much worse success rates. Self-adaptation does not seem to work at all for this problem, it delays the CEAs while also reducing success rates.

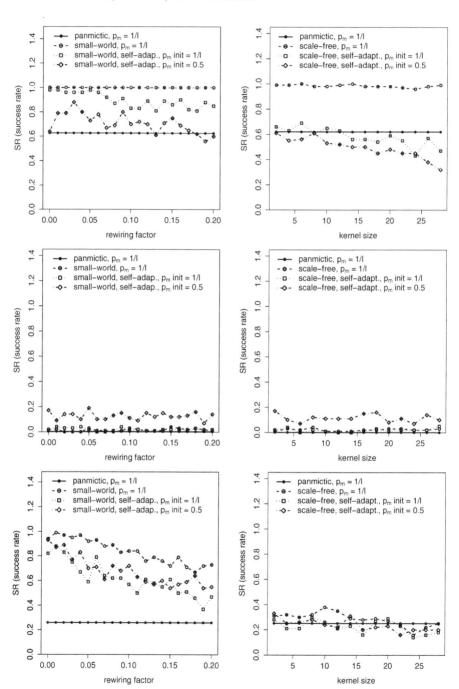

**Fig. 3.** Success rates for small-world (left) and scale-free (right) topology CEAs, compared to a similarly parametrized panmictic EA, on problems (top to bottom) ECC, COUNTSAT and wP-PEAKS. Points are generated from 100 runs each.

**Table 2.** Panmictic versus parametrized scale-free and small-world topology cellular EA on problem MMDP. Performance values for EA variants are ordered into blocks of three rows, giving success rates (SR), average evaluations to solution (AES) and AES standard deviations, respectively. Success rates are averaged from 100 runs, AES values and standard deviations computed from the fraction reaching the global optimum. Characteristic path lengths (cpl) and clustering coefficients (C) are determinedempirically.

| EA variant: panmictic, non-adaptive, initial $p_m = 1/l$ | | | | | | |
|---|---|---|---|---|---|---|
| SR | 0.66 | 0.66 | 0.66 | 0.66 | 0.66 | 0.66 | 0.66 |
| AES | 28894 | 28894 | 28894 | 28894 | 28894 | 28894 | 28894 |
| AES std.dev. | 11727 | 11727 | 11727 | 11727 | 11727 | 11727 | 11727 |
| kernel sizes ⇒ | 2 | 4 | 6 | 10 | 14 | 20 | 28 |
| cpl/C ⇒ | 3.7/0.05 | 3.7/0.06 | 3.6/0.07 | 3.4/0.12 | 3.3/0.20 | 3.1/0.39 | 3.1/0.76 |
| EA variant: scale-free, non-adaptive, initial $p_m = 1/l$ | | | | | | |
| SR | 0.81 | 0.74 | 0.75 | 0.65 | 0.36 | 0.07 | 0.0 |
| AES | 58420 | 59027 | 57307 | 65785 | 65500 | 74571 | — |
| AES std.dev. | 10507 | 12162 | 8016 | 9998 | 9960 | 12816 | — |
| EA variant: scale-free, self-adaptive, initial $p_m = 1/l$ | | | | | | |
| SR | 0.60 | 0.55 | 0.50 | 0.44 | 0.30 | 0.03 | 0.0 |
| AES | 40067 | 38473 | 38480 | 40364 | 41067 | 44000 | — |
| AES std.dev. | 10334 | 2878 | 3407 | 3675 | 3714 | 4320 | — |
| EA variant: scale-free, self-adaptive, initial $p_m = 0.5$ | | | | | | |
| SR | 0.93 | 0.94 | 0.92 | 0.85 | 0.68 | 0.39 | 0.03 |
| AES | 64129 | 63745 | 62348 | 67365 | 71706 | 84051 | 110000 |
| AES std.dev. | 10373 | 10592 | 9239 | 11950 | 14014 | 15416 | 10198 |
| rewiring factor ⇒ | 0.0 | 0.01 | 0.02 | 0.05 | 0.10 | 0.15 | 0.20 |
| cpl/C ⇒ | 50.4/0.5 | 15.5/0.47 | 10.8/0.45 | 7.4/0.37 | 5.9/0.28 | 5.4/0.20 | 5.0/0.16 |
| EA variant: small-world, non-adaptive, initial $p_m = 1/l$ | | | | | | |
| SR | 0.00 | 0.96 | 0.98 | 1.00 | 0.98 | 0.98 | 0.99 |
| AES | — | 100960 | 84850 | 70320 | 61460 | 56570 | 55270 |
| AES std.dev. | — | 9652 | 9376 | 6240 | 5295 | 4989 | 7614 |
| EA variant: small-world, self-adaptive, initial $p_m = 1/l$ | | | | | | |
| SR | 0.95 | 1.00 | 0.98 | 0.99 | 0.88 | 1.00 | 0.98 |
| AES | 94880 | 59220 | 52420 | 45030 | 40200 | 40200 | 37320 |
| AES std.dev. | 1081 | 5871 | 4609 | 3622 | 3790 | 7180 | 2489 |
| EA variant: small-world, self-adaptive, initial $p_m = 0.5$ | | | | | | |
| SR | 0.35 | 0.97 | 1.00 | 1.00 | 1.00 | 1.00 | 0.99 |
| AES | 109540 | 88920 | 81040 | 68680 | 61380 | 58480 | 56040 |
| AES std.dev. | 8842 | 1223 | 1180 | 7854 | 7001 | 6400 | 6780 |

COUNTSAT: Only one of the four algorithms is able to solve the problem with non-significant success rates: The self-adaptive CEA starting with $p_m = 0.5$ . Topology differences seem to have little influence. Unfortunately, we did not try a panmictic EA with self-adaptation to see if topology has an effect at all. Our impression is that this is not the case but success rather depends on high mutation rates.

wP-PEAKS: Here, the small-world CEAs clearly dominate the panmictic EA, with the fixed mutation rate CEA performing best. Self-adaptation only lowers the success rates. Measured AES values for all small-world variants are largely constant and around 2 to 3-times higher than for the panmictic variant, regardless of the rewiring factor. The scale-free CEAs achieve no better success rates than the panmictic EA, but also require 2 to 3-times more evaluations than the panmictic.

***Interpretation.*** At a first glance, it seems hard to perceive a clear trend within the obtained results. The most we can state is that scale-free topologies do not seem to provide a worthwhile alternative to panmictic or Watts-Strogatz small-worlds. Nevertheless, when thinking about the properties of the utilized test problems and linking them to the algorithm properties of the EA variants regarded as most successful (Table 3), we may derive some generalizable conjectures. It seems that problems with a certain degree of separability may profit from localizing operators. However, this also happens for the wP-PEAKS problem which is non-separable.

**Table 3.** Test problem properties next to algorithm properties found successful. The ECC problem is not fully separable but organized in blocks. Solution permutable means that fitness of a solution, or subsolution in case of the MMDP, solely depends on the number of ones, not their location.

| Problem | Separable | Solution permutable | Topology matters | Self-adaptation works |
| --- | --- | --- | --- | --- |
| MMDP | X | X | X | X |
| ECC | partly | – | X | – |
| COUNTSAT | – | X | – | X |
| wP-PEAKS | – | – | X | – |

Concerning self-adaptation, the picture is much clearer. Within our experiments, it worked well for problems with permutable best solutions. That is, several optimal solutions exist that share the number of ones, either in the whole genome as for the COUNTSAT problem, or in the separate building blocks as for the MMDP. Lacking further investigations, we can only speculate why self-adaptation provides an advantage, or at least does not diminish optimization success here. Possibly, the temporary appearance of several different, namely higher mutation probabilities in the course of the optimization process leads to better results.

## 5   Conclusions

The results of this empirical study indicate that small-world topologies allow for a trade-off between robustness and speed of the search; this is in agreement with the results of [5] on selection pressure, especially when Watts–Strogatz networks are used. In terms of success rate, these population topologies behave at least as well, and often better, than the panmictic case. However, their convergence speed is lower. This effect had already been reported in the case of regular lattice population structures for the same class of problems [18].

On the other hand, scale-free topologies do not seem very helpful in their current form, especially for large kernel sizes. Smaller clique sizes work better but, overall, they do not outperform the standard panmictic setting. This confirms that the selection pressure induced by these topologies on the population may be too high, similar to the panmictic, thus causing premature convergence [5]. However, we have only experimented with static scale-free topologies: we feel that playing with highly connected nodes in a graph would open new perspectives in the control of the exploration/exploitation trade-off, and we intend to try out these ideas in the future.

As far as the EA strategies are concerned, self-adaptation helps if the solution/sub-solution is permutable, while fixed mutation performs best overall. In the future we also intend to extend the investigation to continuous problems and to study the dynamics of birth surplus (comma) strategies.

# References

1. E. Cantú-Paz. *Efficient and Accurate Parallel Genetic Algorithms*. Kluwer Academic Press, 2000.
2. M. Tomassini. *Spatially Structured Evolutionary Algorithms: Artificial Evolution in Space and Time*. Springer, Berlin, Heidelberg, New York, 2005.
3. R. Albert and A.-L. Barabasi. Statistical mechanics of complex networks. *Reviews of Modern Physics*, 74:47–97, 2002.
4. M. E. J. Newman. The structure and function of complex networks. *SIAM Review*, 45:167–256, 2003.
5. M. Giacobini, M. Tomassini, and A. Tettamanzi. Takeover time curves in random and small-world structured populations. In H.-G. Beyer at al., editor, *Proceedings of the Genetic and Evolutionary Computation Conference GECCO'05*, pages 1333–1340. ACM Press, 2005.
6. M. Preuss and C. Lasarczyk. On the importance of information speed in structured populations. In Xin Yao et al., editor, *Parallel Problem Solving from Nature, PPSN VIII*, pages 91–100. Lecture Notes in Computer Science Vol. 3242, Springer-Verlag, 2004.
7. T. Bäck. Self-adaptation in genetic algorithms. In F. J. Varela and P. Bourgine, editors, *Toward a Practice of Autonomous Systems – Proc. First European Conf. Artificial Life (ECAL'91)*, pages 263–271, Cambridge MA, 1992. The MIT Press.
8. H.-G. Beyer and H.-P. Schwefel. Evolution strategies: A comprehensive introduction. *Natural Computing*, 1(1):3–52, 2002.
9. D. Whitley, S. Rana, J. Dzubera, and K. E. Mathias. Evaluating evolutionary algorithms. *Artif. Intelligence*, 85:245–276, 1997.
10. D. H. Wolpert and W. G. Macready. No free lunch theorems for optimization. *IEEE Transactions on Evolutionary Computation*, 1(1):67–82, 1997.
11. D. E. Goldberg, K. Deb, and J. Horn. Massively multimodality, deception and genetic algorithms. In R. Männer and B. Manderick, editors, *Parallel Prob. Solving from Nature II*, pages 37–46. North-Holland, 1992.
12. K. A. De Jong, M. A. Potter, and W. M. Spears. Using problem generators to explore the effects of epistasis. In T. Bäck, editor, *Proceedings of the Seventh ICGA*, pages 338–345. Morgan Kaufmann, 1997.
13. F. J. MacWilliams and N. J. A. Sloane. *The Theory of Error-Correcting Codes*. North-Holland, Amsterdam, 1977.
14. H. Chen, N. S. Flann, and D. W. Watson. Parallel genetic simulated annealing: A massively parallel SIMD algorithm. *IEEE Transactions on Parallel and Distributed Systems*, 9(2):126–136, 1998.
15. S. Droste, T. Jansen, and I. Wegener. A natural and simple function which is hard for all evolutionary algorithms. In *IEEE International Conference on Industrial Electronics, Control, and Instrumentation (IECON 2000)*, pages 2704–2709, Piscataway, NJ, 2000. IEEE Press.
16. D. J. Watts and S. H. Strogatz. Collective dynamics of 'small-world' networks. *Nature*, 393:440–442, 1998.

17. T. Bartz-Beielstein. *New Experimentalism Applied to Evolutionary Computation.* PhD thesis, University of Dortmund, April 2005.
18. B. Dorronsoro, E. Alba, M. Giacobini, and M. Tomassini. The influence of grid shape and asynchronicity on cellular evolutionary algorithms. In *2004 Congress on Evolutionary Computation (CEC 2004)*, pages 2152–2158. IEEE Press, Piscataway, NJ, 2004.
19. G. Rudolph. An evolutionary algorithm for integer programming. In Y. Davidor, H.-P. Schwefel, and R. Männer, editors, *Proc. Parallel Problem Solving from Nature – PPSN III, Jerusalem*, pages 139–148, Berlin, 1994. Springer.

# Particle Swarm for the Traveling Salesman Problem

Elizabeth F. Gouvêa Goldbarg, Givanaldo R. de Souza, and Marco César Goldbarg

Department of Informatics and Applied Mathematics,
Federal University of Rio Grande do Norte,
Campus Universitário Lagoa Nova, Natal , Brazil
{beth, gold}@dimap.ufrn.br, givanaldo@yahoo.com.br

**Abstract.** This paper presents a competitive Particle Swarm Optimization algorithm for the Traveling Salesman Problem, where the velocity operator is based upon local search and path-relinking procedures. The paper proposes two versions of the algorithm, each of them utilizing a distinct local search method. The proposed heuristics are compared with other Particle Swarm Optimization algorithms presented previously for the same problem. The results are also compared with three effective algorithms for the TSP. A computational experiment with benchmark instances is reported. The results show that the method proposed in this paper finds high quality solutions and is comparable with the effective approaches presented for the TSP.

## 1 Introduction

The Traveling Salesman is a classical NP-hard combinatorial problem that has been an important test ground for many algorithms. Given a graph G = (N,E), where N = {1,...,$n$} is the set of nodes and E = {1,...,$m$} is the set of edges of G, and costs, $c_{ij}$, associated with each edge linking vertices $i$ and $j$, the problem consists in finding the minimal total length Hamiltonian cycle of G. The length is calculated by the summation of the costs of the edges in a cycle. If for all pairs of nodes {$i,j$}, the costs $c_{ij}$ and $c_{ji}$ are equal then the problem is said to be symmetric, otherwise it is said to be asymmetric. A survey of the TSP is presented by Gutin and Punnen [10].

The term A-Life is utilized to describe researches on systems that simulate essential properties of the real life, with two general lines:

- how computational techniques may help the investigation of natural phenomena, and
- how biological techniques may help to solve computational problems.

Several bio-inspired methods were proposed to solve Combinatorial Optimization problems, such as Genetic Algorithms [12], Memetic Algorithms [17], Cultural Algorithms [24] and Ant Systems [5], among others. Particle swarm optimization, PSO, algorithms belong also to the class of bio-inspired methods [14]. This is a population-based technique introduced by a Psychologist, James Kennedy, and an Electrical Engineer, Russell Eberhart, who based their method upon the behavior of bird flocks. PSO algorithms for the TSP were presented previously by Onwubulu and Clerc [18] and Wang et al. [26]. Hybrid PSO approaches for the same problems were also presented by Machado and Lopes [16] and Pang et al. [19].

J. Gottlieb and G.R. Raidl (Eds.): EvoCOP 2006, LNCS 3906, pp. 99–110, 2006.
© Springer-Verlag Berlin Heidelberg 2006

This paper presents a new PSO algorithm for the Traveling Salesman Problem. The main difference between the approach reported in this paper and the previous ones regards the velocity operators. In this paper local search and path-relinking procedures are proposed as velocity operators for a discrete optimization problem.

Local search is a traditional optimization method that starts with an initial solution and proceeds by searching for a better solution in a neighborhood defined for the starting solution. If a better solution is found, then it is assumed as the new current solution and the process of searching in its neighborhood is re-started. This process continues until no improvement of the current solution is found [1].

Path-relinking is an intensification technique which ideas were originally proposed by Glover [8] in the context of scheduling methods to obtain improved local decision rules for job shop scheduling problems [9]. The strategy consists in generating a path between two solutions creating new solutions. Given an origin, $x_s$, and a target solution, $x_t$, a path from $x_s$ to $x_t$ leads to a sequence $x_s$, $x_s(1)$, $x_s(2)$, ..., $x_s(\mathrm{r}) = x_t$, where $x_s(i+1)$ is obtained from $x_s(i)$ by a move that introduces in $x_s(i+1)$ an attribute that reduces the distance between attributes of the origin and target solutions. The roles of origin and target can be interchangeable. Some strategies for considering such roles are:

- forward: the worst among $x_s$ and $x_t$ is the origin and the other is the target solution;
- backward: the best among $x_s$ and $x_t$ is the origin and the other is the target solution;
- back and forward: two different trajectories are explored, the first using the best among $x_s$ and $x_t$ as the initial solution and the second using the other in this role;
- mixed: two paths are simultaneously explored, the first starting at the best and the second starting at the worst among $x_s$ and $x_t$, until they meet at an intermediary solution equidistant from $x_s$ and $x_t$.

The paper is organized as follows. Particle swarm optimization is described in Section 2. The proposed algorithm and its variants are described in Section 3. A computational experiment is reported in Section 4. The experiment compares the results of the proposed approach with previous PSO algorithms. To date PSO algorithms for the TSP have failed to produce results comparable to competitive techniques. A comparison with three effective techniques proposed for the Traveling Salesman Problem shows that the proposed approach produces high quality solutions. Finally, some concluding remarks are presented in Section 5.

## 2  Particle Swarm Optimization

Particle swarm optimization, PSO, is an evolutionary computation technique inspired in the behavior of bird flocks, fish schools and swarming theory. PSO algorithms were first introduced by Kennedy and Eberhart [14] for optimizing continuous nonlinear functions. The fundamentals of their method lie on researches on computer simulations of the movements of social creatures [11], [21], [23].

Given a population of solutions (the swarm) for a given problem, each solution is seen as a social organism, also called particle. The method attempts to imitate the behavior of the real creatures making the particles "fly" over a solution space, thus balancing the efforts of search intensification and diversification. Each particle has a value associated with it. In general, particles are evaluated with the objective function being optimized. A velocity is also assigned to each particle in order to direct the "flight" through the problem space. The artificial creatures have a tendency to follow the best ones among them. In the classical PSO algorithm, each particle

- has a position and a velocity
- knows its own position and the value associated with it
- knows the best position it has ever achieved, and the value associated with it
- knows its neighbors, their best positions and their values

As pointed by Pomeroy [20], rather than exploration and exploitation what has to be balanced is individuality and sociality. Initially, individualistic moves are preferable to social ones (moves influenced by other individuals), however it is important for an individual to know the best places visited by its neighbors in order to "learn" good moves.

The neighborhood may be physical or social [18]. Physical neighborhood takes distances into account, thus a distance metric has to be established. This approach tends to be time consuming, since each iteration distances must be computed. Social neighborhoods are based upon "relationships" defined at the very beginning of the algorithm.

The move of a particle is a composite of three possible choices:

- To follow in its own way
- To go back to its best previous position
- To go towards its best neighbor's previous position or towards its best neighbor

Equations (1) and (2) are utilized to update the particle's position and velocity at each time step [6].

$$x_{t+1} = x_t + v_t \tag{1}$$

$$V_{t+1} = w.v_t + c_1.rand_1.(pbest_t - x_t) + c_2.rand_2 .(gbest_t - x_t) \tag{2}$$

Where $x_t$ and $v_t$ are the particle's position and velocity at instant t, respectively, $pbest_t$ is the particle's best previous position, $gbest_t$ is the best position that any particle has achieved so far, $w$ is the inertia factor [25], $c_1$ and $c_2$ are social/cognitive coefficients that quantify how much the particle trusts its experience and how much it trusts the best neighbor, $rand_1$ and $rand_2$ are randomly generated numbers.

To apply PSO to discrete problems, one is required to define a representation for the position of a particle and to define velocity operators regarding the movement options allowed for the particles (and ways for, possibly, combining movements).

A general framework of a PSO algorithm for a minimization problem is listed in the following.

```
procedure PSO
{
   Initialize a population of particles
   do
     for each particle p
```

```
    Evaluate particle p
    If the value of p is better than the value of pbest
        Then, update pbest with p
    end_for
    Define gbest as the particle with the best value
    for each particle p do
     Compute p's velocity
     Update p's position
    end_for
  while (a stop criterion is not satisfied)
}
```

## 3   PSO for the Traveling Salesman Problem

Usually, evolutionary algorithms represent TSP solutions as a permutation on the $n$ vertices of a given instance. This representation is also adopted in this work. A social neighborhood containing only the global best particle (representing the best current solution) is defined for each member of the population. A particle has three movement options:

- To follow in its own way
- To go back to its best previous position
- To go towards the global best solution (particle)

The main difference between the proposed algorithm and the previous approaches lies on the way the velocity operators are defined. The first option of a particle, that is to follow its own way, is implemented by means of a local search procedure. Two local search procedures were utilized, each of them defining a version of the algorithm. The first local search procedure is based upon an inversion operator to build neighbor solutions. The inversion operator inverts the elements of a particle between two indices, $a$ and $b$. The difference $b$-$a$ varies between 1 (simulating a 2-swap move) and $n$-1. Thus, when $v_1$ is applied to a particle $p$, the local search procedure starts inverting sequences of two elements, then inverts sequences of three elements, and so on. Figure 1 illustrates an inversion of a sequence of elements defined by the interval $a$=2 and b=5 (a sequence with four elements) of particle $p$ resulting on particle $p$'.

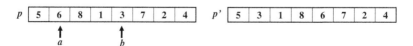

**Fig. 1.** Inversion of elements among indices $a$ and $b$

The second local search procedure is the Lin-Kernighan neighborhood, LK [15]. This is a recognized efficient improvement method for the TSP. The basic LK algorithm has a number of decisions to be made and depending on the strategies adopted by programmers distinct implementations of this algorithm may result on different performances. The literature contains reports of many LK implementations with widely varying behavior [13]. In this work the algorithm utilized the LK implementation of the Concorde solver [3].

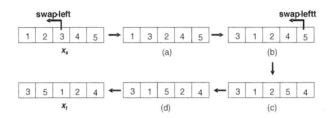

**Fig. 2.** Path-relinking operator

Another velocity operator is considered when a particle has to move from its current position to another (pbest or gbest). In these cases the authors consider that a natural way to accomplish this task is to perform a path-relinking operation between the two solutions.

The velocity utilized to move a particle from an origin to a target position is defined as a path-relinking operation. The relinking operator utilized in this work is illustrated in Figure 2. The operator swaps one element with its right (left) neighbor. The steps of a path-relinking procedure that explores solutions in the path between the origin solution (1 2 3 4 5) and the target solution (3 5 1 2 4) is shown. First, the element 3 is moved to the first position by swapping it with elements 2 (Figure 2(a)) and 1 (Figure 2(b)). At this point, element 5 has to be moved to the second position. It is swapped with element 4 (Figure 2(c)), element 2 (Figure 2(d)) and, finally, it is swapped with element 1, when the target solution is reached. The swap operators lead to $O(n^2)$ procedures.

The path-relinking is applied simultaneously from the origin to the target solution and vice-versa (back and forward). It is also utilized the swap-left and swap-right operations.

Combinations of movements are possible. For instance, to combine the first and third option of movement one can stop the local search procedure after a given number of iterations and then do path-relinking among the solution obtained in the local search and global best solution. Although combinations could be done, in the proposed algorithms, no combination of movements was implemented. Initial probabilities are assigned to each one of the three possible movements. These probabilities change as the algorithm runs. A general framework of the algorithm implemented in this work is listed in the following.

```
procedure PSO_TSP
{
  Define initial probabilities for particles' moves:
      p₁ = x /*to follow its own way*/
      p₂ = y /*to go forward pbest*/
      p₃ = z /*to go forward gbest*/
      /* x+y+z=1 */
  Initialize a population of particles
  do
    for each particle p
     Evaluate particle p
     If the value of p is better than the value of pbest
        Then, update pbest with p
    end_for
    Define gbest as the particle with the best value
    for each particle p do
     Choose p's velocity
     Update p's position
    end_for
    Update probabilities:
       p₁ = p₁×0.95; p₂ = p₂×1.01; p₃ = 100%-(p₁+p₂)
  while (a stop criterion is not satisfied)
}
```

At the beginning, the algorithm defines the probabilities associated with each velocity for the particles, where $p_1$, $p_2$ and $p_3$ correspond, respectively, to the likelihood that the particle follows its own way, goes toward the best previous position and goes toward the best solution found so far. Then, the algorithm proceeds modifying the particle's position according to the velocity operator randomly chosen. Finally, the probabilities are updated.

Initially, a high probability is set to $p_1$, and low values are assigned to $p_2$ and $p_3$. The goal is to allow that individualistic moves occur more frequently in the first iterations. During the execution this situation is being modified and, at the final iterations, $p_3$ has the highest value. The idea is to intensify the search in good regions of the search space in the final iterations.

Particles are initialized with a random adaptive version of the heuristic nearest neighbor [2]. The procedure is similar to the construction phase of a GRASP algorithm [7]. At first, a city is randomly chosen, then, other cities are added to the solution in each step. A restricted candidate list is built with the 5% cities closest to the last one inserted in the solution. A city is chosen at random from this list and is added to the solution. This step is repeated until a TSP solution is completed.

# 4 Computational Experiments

The proposed algorithms were implemented in C++ on a Pentium IV (3.0 GHz and 512 Mb of RAM) running Linux. The algorithms were applied to symmetric instances of the benchmark TSPLIB [22] with sizes ranging from 51 to 7397. The stop criteria were:

- To find the optimum
- To reach a maximum number of iterations = 2000
- To reach a maximum number of iterations with no improvement of the best current solution = 20
- To reach a maximum processing time = 60 seconds for instances with $n < 1000$, 300 seconds when $1000 \leq n < 5000$, 1000 seconds when $5000 \leq n < 7000$ and 2000 seconds for $n \geq 7000$.

The population had a fixed size with 20 particles.

A first experiment compared the two proposed algorithms in 11 symmetric instances with sizes ranging from 51 to 2103. Twenty independent runs of each algorithm were performed. The results are showed in Table 1 in terms of percent deviation from the optimal solution. This gap is computed with equation (3), where *Sol* and *Opt* denote, respectively, the (best or average) solution obtained by the algorithm and the optimal solution.

$$(Sol - Opt) \times 100 / Opt \tag{3}$$

The columns show the name of the TSPLIB instances, the best solution (Min) and the average solution of the two versions of the proposed PSO algorithms denoted by PSO-INV and PSO-LK, the versions with the inversion and LK local search procedures, respectively.

Not surprisingly, PSO-LK exhibits a better performance than PSO-INV, since the local search procedure embedded in the former version is more powerful than the local search procedure of the latter version. However, a comparison of the results of the weakest version of the proposed algorithm with the ones reported in the work of Pang et al. [19] reveals that the former finds the best results. Once Wang et al. [26] reported results just for one asymmetric TSP instance, the authors implemented their algorithm in order to compare its performance with the proposed algorithm. Table 2 shows the best results (percent deviations) of the weakest version of the proposed algorithm, PSO-INV, and of the algorithms PSO-P [19] and PSO-W [26] for the three TSP symmetric instances with $n > 50$ of the computational experiment reported by Pang et al. [19].

**Table 1.** Comparison of the two versions of the proposed algorithms

| Instances | PSO-INV | | PSO-LK | |
|---|---|---|---|---|
| | Min | Av | Min | Av |
| eil51 | 0.2347 | 1.9836 | 0 | 0 |
| berlin52 | 0 | 2.0041 | 0 | 0 |
| eil76 | 2.4164 | 4.5167 | 0 | 0 |
| rat195 | 5.8114 | 8.7581 | 0 | 0 |
| pr299 | 5.8476 | 7.9952 | 0 | 0 |
| pr439 | 4.4200 | 8.0111 | 0 | 0 |
| d657 | 6.9656 | 9.6157 | 0 | 0 |
| pr1002 | 9.8574 | 11.1900 | 0 | 0 |
| d1291 | 13.2104 | 15.5505 | 0 | 0.0113 |
| rl1304 | 10.4432 | 11.9942 | 0 | 0 |
| d2103 | 16.7383 | 18.4180 | 0.0087 | 0.0267 |

**Table 2.** Comparison of two PSO heuristics with PSO-INV

| Instance | PSO-W | PSO-P | PSO-INV |
|---|---|---|---|
| eil51 | 114 | 2.54 | 0.2347 |
| berlin52 | 115 | 2.12 | 0 |
| eil76 | 162 | 4.75 | 2.4164 |

The results showed in Table 2 correspond to the best solution found in 20 independent runs. The proposed algorithm PSO-INV was also applied to the asymmetric instance br17 reported by Wang et al. [26]. The results are showed in Table 3, where columns show the algorithm, the best and average percent deviation from the optimal solution, and the average runtime in seconds.

**Table 3.** Comparison of PSO-INV and PSO-W for the TSPLIB instance br17

| Algorithm | Min | Average | T (s) |
|---|---|---|---|
| PSO-INV | 0 | 0 | < 0.01 |
| PSO-W | 0 | 16.66 | 60.04 |

A computational experiment investigated the differences between the results obtained by the LK procedure [2] and the PSO-LK algorithm. This experiment aimed at finding out if the proposed PSO approach was able to improve the LK [2] results. Table 4 shows the percent difference from the optimum, for 30 TSPLIB symmetric instances with $n$ ranging from 439 to 7397. Bold elements mark when an improvement occurred. Twenty independent runs of each algorithm were performed. The columns T show the average time of each algorithm in seconds. From the thirty instances of the experiment, the PSO approach was able to improve twenty-three minimal results and all the averages. A statistical analysis shows that, the means of columns Min and Average of LK are 0.1115, 0.3431, respectively. The means of PSO-LK are 0.0022 and 0.0154. Although the runtimes of PSO-LK are higher than the LK, in average, improvements of 98% and 95% were achieved on the best and mean results, respectively.

**Table 4.** Comparison of LK and PSO-LK results

| Instance | LK | | | PSO-LK | | |
|---|---|---|---|---|---|---|
| | Min | Average | T (s) | Min | Average | T (s) |
| pr439 | 0.0000 | 0.0463 | 0.88 | 0.0000 | **0.0000** | 0.78 |
| pcb442 | 0.0000 | 0.1119 | 0.67 | 0.0000 | **0.0000** | 0.80 |
| d493 | 0.0029 | 0.1216 | 2.21 | **0.0000** | **0.0000** | 19.38 |
| rat575 | 0.0295 | 0.1277 | 0.62 | **0.0000** | **0.0000** | 6.47 |
| p654 | 0.0000 | 0.0078 | 3.72 | 0.0000 | **0.0000** | 1.90 |
| d657 | 0.0020 | 0.1500 | 1.70 | **0.0000** | **0.0000** | 12.42 |
| rat783 | 0.0000 | 0.0704 | 0.90 | 0.0000 | **0.0000** | 5.25 |
| dsj1000 | 0.0731 | 0.2973 | 5.89 | **0.0027** | **0.0031** | 178.48 |
| pr1002 | 0.0000 | 0.1318 | 1.96 | 0.0000 | **0.0000** | 9.50 |
| u1060 | 0.0085 | 0.1786 | 3.12 | **0.0000** | **0.0000** | 38.18 |
| vm1084 | 0.0017 | 0.0669 | 2.59 | **0.0000** | **0.0010** | 34.74 |
| pcb1173 | 0.0000 | 0.1814 | 1.78 | 0.0000 | **0.0001** | 48.18 |
| d1291 | 0.0039 | 0.4333 | 1.79 | **0.0000** | **0.0000** | 29.86 |
| rl1304 | 0.0202 | 0.3984 | 2.67 | **0.0000** | **0.0000** | 21.62 |
| rl1323 | 0.0463 | 0.2300 | 2.34 | **0.0000** | **0.0092** | 225.32 |
| nrw1379 | 0.0547 | 0.1354 | 2.45 | **0.0017** | **0.0085** | 417.80 |
| fl1400 | 0.0000 | 0.1215 | 14.87 | 0.0000 | **0.0000** | 15.42 |
| fl1577 | 0.7371 | 2.2974 | 6.30 | **0.0000** | **0.0135** | 461.99 |
| vm1748 | 0.0903 | 0.1311 | 4.66 | **0.0000** | **0.0018** | 854.17 |
| u1817 | 0.1976 | 0.5938 | 1.88 | **0.0000** | **0.0863** | 789.18 |
| rl1889 | 0.1836 | 0.3844 | 4.58 | **0.0000** | **0.0073** | 894.43 |
| d2103 | 0.0597 | 0.3085 | 2.82 | **0.0000** | **0.0043** | 1137.56 |
| u2152 | 0.2381 | 0.5548 | 2.16 | **0.0000** | **0.0717** | 1415.32 |
| pr2392 | 0.0775 | 0.3904 | 3.78 | **0.0000** | **0.0021** | 577.78 |
| pcb3038 | 0.1598 | 0.2568 | 5.42 | **0.0101** | **0.0396** | 323.94 |
| fl3795 | 0.5665 | 1.0920 | 15.60 | **0.0000** | **0.0142** | 621.63 |
| fnl4461 | 0.0882 | 0.1717 | 9.08 | **0.0296** | **0.0462** | 583.78 |
| rl5915 | 0.3528 | 0.5343 | 10.08 | **0.0122** | **0.0633** | 1359.25 |
| rl5934 | 0.2221 | 0.4761 | 10.65 | **0.0012** | **0.0650** | 983.04 |
| pla7397 | 0.1278 | 0.2912 | 21.85 | **0.0075** | **0.0253** | 1563.22 |

Finally, Table 5 shows a comparison of the results obtained by PSO-LK and three effective heuristics: Tourmerge [4], NYYY iterated Lin-Kermighan variant (reported at http://www.research.att.com/~dsj/chtsp/) and JM iterated Lin-Kernighan variant. The results of these heuristics were obtained in the DIMACS Challenge page: http://www.research.att.com/~dsj/chtsp/results.html.

The columns related to PSO-LK show the best and average tours found in twenty independent runs. The columns related to the Tourmerge algorithm show the best and average tours obtained in five independent runs. Results are not reported for instances fnl446 and pla7397. The columns related to the NYYY and JM iterated Lin-Kernighan variants show the best tours obtained in ten $n$ iterations runs.

**Table 5.** Comparison of heuristics for TSP symmetric instances

| Instance | PSO-LK | | Tourmerge | | ILKNYYY | ILKJM |
|----------|--------|---------|-----------|---------|---------|--------|
|          | Min    | Average | Min       | Average | Nb10    | Nb10   |
| dsj1000  | 0.0027 | 0.0031  | 0.0027    | 0.0478  | 0       | 0.0063 |
| pr1002   | 0      | 0       | 0         | 0.0197  | 0       | 0.1482 |
| u1060    | 0      | 0       | 0         | 0.0049  | 0.0085  | 0.0210 |
| vm1084   | 0      | 0.0010  | 0         | 0.0013  | 0.0217  | 0.0217 |
| pcb1173  | 0      | 0.0001  | 0         | 0.0018  | 0       | 0.0088 |
| d1291    | 0      | 0       | 0         | 0.0492  | 0       | 0      |
| rl1304   | 0      | 0       | 0         | 0.1150  | 0       | 0      |
| rl1323   | 0      | 0.0092  | 0.01      | 0.0411  | 0.01    | 0      |
| nrw1379  | 0.0017 | 0.0085  | 0         | 0.0071  | 0.0247  | 0.0018 |
| fl1400   | 0      | 0       | 0         | 0       | 0       | 0      |
| fl1577   | 0      | 0.0135  | 0         | 0.0225  | 0       | 0      |
| vm1748   | 0      | 0.0018  | 0         | 0       | 0       | 0      |
| u1817    | 0      | 0.0863  | 0.0332    | 0.0804  | 0.1643  | 0.2657 |
| rl1889   | 0      | 0.0073  | 0.0082    | 0.0682  | 0.0082  | 0.0041 |
| d2103    | 0      | 0.0043  | 0.0199    | 0.3170  | 0.0559  | 0      |
| u2152    | 0      | 0.0717  | 0         | 0.0794  | 0       | 0.1743 |
| pr2392   | 0      | 0.0021  | 0         | 0.0019  | 0.0050  | 0.1495 |
| pcb3038  | 0.0101 | 0.0396  | 0.0036    | 0.0327  | 0.0247  | 0.1213 |
| fl3795   | 0      | 0.0142  | 0         | 0.0556  | 0       | 0.0104 |
| fnl4461  | 0.0296 | 0.0462  | ---       | ---     | 0.0449  | 0.1358 |
| rl5915   | 0.0122 | 0.0633  | 0.0057    | 0.0237  | 0.0580  | 0.0168 |
| rl5934   | 0.0012 | 0.0650  | 0.0023    | 0.0104  | 0.0115  | 0.1723 |
| pla7397  | 0.0075 | 0.0253  | ---       | ---     | 0.0209  | 0.0497 |

From the twenty-one instances for which Tourmerge presented results, PSO-LK finds five best minimal results, Tourmerge finds three best minimal results and both algorithms find the same quality tours for thirteen instances. Regarding average results Tourmerge finds the best values on six instances and PSO-LK finds the best values on thirteen instances. The mean values of the "Min" and "Average" columns for the twenty-one instances are 0.0013 and 0.0186 for PSO-LK and 0.0041 and 0.0467 for Tourmerge. These results show that, in average, PSO-LK has a better performance than Tourmerge regarding minimal and average results.

Comparing the proposed algorithm with the NYYY iterated Lin-Kernighan variant, one can observe that from the twenty-three instances of the experiment, PSO-LK finds the best tours for thirteen instances, ILK-NYYY finds the best results for one instance and same results are found for nine instances. The averages of the best results of these two algorithms for the twenty-three instances are 0.0028 and 0.0199, PSO-LK and ILK-NYYY, respectively.

The comparison with the JM iterated Lin-Kernighan version shows that the PSO-LK algorithm finds the best results for sixteen instances and both algorithms find the same tours quality for seven instances. The averages for PSO-LK and ILK-JM are 0.0028 and 0.0569, respectively.

# 5  Conclusions

This paper presented a PSO algorithm for the TSP where the concept of velocity is differentiated when the particle has to follow its own way and when it goes forward someone's position. Local search procedures were applied in the former and path-relinking operations were applied in the latter case.

Although some works where PSO algorithms were applied to the TSP were presented previously, those algorithms did not reported results which could be compared with the ones obtained by effective heuristics. In this work, an effective PSO approach was presented for the TSP which results were compared with three high quality heuristics for this problem. The comparison showed that the proposed algorithm outperforms the three heuristics regarding best tours. The comparison among the average results of PSO-LK and the Tourmerge also shows that the first heuristics finds better results.

# References

1. Aarts, E., Lenstra, J.K.: Local Search in Combinatorial Optimization. John Wiley & Sons, Chichester, England (1997)
2. Bellmore, M., Nemhauser, G.L.: The Traveling Salesman Problem: A Survey. Operations Research Vol. 16 (1968) 538–582
3. Concorde TSP Solver: http://www.tsp.gatech.edu/concorde.html. Last access on 01/18/2005
4. Cook, W.J., Seymour, P.: Tour Merging via Branch-decomposition. INFORMS Journal on Computing Vol. 15 (2003) 233-248
5. Dorigo, M., Gambardella, L.M.: Ant Colony System: A Cooperative Learning Approach to the Traveling Salesman Problem. IEEE Transactions on Evolutionary Computation vol. 1, N. 1 (1997) 53-66
6. Eberhart, R.C., Shi, Y.:Comparing Inertia Weights and Constriction Factors in Particle Swarm Optimization. In: Proceedings of the 2000 Congress on Evolutionary Computation Vol. 1 (2000) 84-88
7. Feo, T.A., Resende, M.G.C.: A Probabilistic Heuristic for a Computationally Difficult Set Covering Problem. Operations Research Letters Vol. 8 (1989) 67–71
8. Glover, F.: Parametric Combinations of Local Job Shop Rules. Chapter IV, ONR Research Memorandum N. 117, GSIA, Carnegie Mellon University, Pittsburgh, PA (1963)
9. Glover, F., Laguna, M., Martí, R.: Fundamentals of Scatter Search and Path Relinking. Control and Cybernetics Vol. 29, N. 3 (2000) 653-684
10. Gutin, G., Punnen, A. P. (Ed.): Traveling Salesman Problem and Its Variations. Kluwer Academic Publishers (2002)
11. Heppner, F., Grenander, U.: A Stochastic Nonlinear Model for Coordinated Bird Flocks. In: Krasner, S. (eds.), The Ubiquity of Caos, AAAS Publications, Washington, DC (1990)
12. Holland, J. H.: Adaptation in Natural and Artificial Systems, University of Michigan Press, Ann Arbor, MI (1975)
13. Johnson, D. S., McGeoh, L.A.: Experimental Analysis of Heuristics for the STSP. In: Guttin, G., Punnen, A.P. (eds.): Traveling Salesman Problem and Its Variations, Kluwer Academic (2002)

14. Kennedy, J., Eberhart, R.: Particle Swarm Optimization, Proceedings of the IEEE International Conference on Neural Networks Vol. 4 (1995) 1942-1948
15. Lin, S., Kernighan, B.: An Effective Heuristic Algorithm for the Traveling-salesman Problem. Operations Research Vol. 21 (1973) 498-516
16. Machado, T.R., Lopes, H.S.: A Hybrid Particle Swarm Optimization Model for the Traveling Salesman Problem. In: Ribeiro, H., Albrecht, R.F., Dobnikar, A. (eds.): Natural Computing Algorithms, Wien: SpringerWienNewYork (2005) 255-258
17. Moscato, P.: On Evolution, Search, Optimization, Genetic Algorithms and Martial Arts: Towards Memetic Algorithms, Caltech Concurrent Computation Program, C3P Report 826, 1989
18. Onwubulu, G.C., Clerc, M.: Optimal Path for Automated Drilling Operations by a New Heuristic Approach Using Particle Swarm Optimization. International Journal of Production Research Vol. 42, N. 3 (2004) 473-491.
19. Pang,W., Wang, K.-P., Zhou, C.-G., Dong, L.-J., Liu, M., Zhang, H.-Y., Wang, J.-Y.: Modified Particle Swarm Optimization Based on Space Transformation for Solving Traveling Salesman Problem. Proceedings of the Third International Conference on Machine Learning and Cybernetics (2004) 2342-2346
20. Pomeroy, P.: An Introduction to Particle Swarm Optimization. Electronic document available at www.adaptiveview.com/ipsop1.html
21. Reeves, W.T.: Particle Systems Technique for Modeling a Class of Fuzzy Objects. Computer Graphics Vol. 17, N. 3 (1983) 359-376
22. Reinelt, G.: TSPLIB, 1995. Available: http://www.iwr.uni-heidelberg.de/iwr/comopt/ software/TSPLIB95/
23. Reynolds, C. W.: Flocks, Herds and Schools: a Distributed Behavioral Model, Computer Graphics Vol. 21, N. 4 (1987) 24-34
24. Reynolds, R. G.: An Introduction to Cultural Algorithms. In: Proceedings of Evolutionary Programming, EP94, World Scientific, River Edge, NJ (1994) 131-139
25. Shi, Y., Eberhart, R.C.: Parameter Selection in Particle Swarm Optimization. Evolutionary Progamming VII, Proceedings of EP98, New York (1998) 591-600
26. Wang, K.-P., Huang, L., Zhou, C.-G., Pang, W.: Particle Swarm Optimization for Traveling Salesman Problem. Proceedings of the Second International Conference on Machine Learning and Cybernetics (2003) 1583-1585

# Hierarchical Cellular Genetic Algorithm

Stefan Janson[1], Enrique Alba[2],
Bernabé Dorronsoro[2], and Martin Middendorf[1]

[1] Department of Computer Science, University of Leipzig, Germany
{janson, middendorf}@informatik.uni-leipzig.de
[2] Department of Computer Science, University of Málaga, Spain
{eat, bernabe}@lcc.uma.es

**Abstract.** Cellular Genetic Algorithms (cGA) are spatially distributed Genetic Algorithms that, because of their high level of diversity, are superior to regular GAs on several optimization functions. Also, since these distributed algorithms only require communication between few closely arranged individuals, they are very suitable for a parallel implementation. We propose a new kind of cGA, called hierarchical cGA (H-cGA), where the population structure is augmented with a hierarchy according to the current fitness of the individuals. Better individuals are moved towards the center of the grid, so that high quality solutions are exploited quickly, while at the same time new solutions are provided by individuals at the outside that keep exploring the search space. This algorithmic variant is expected to increase the convergence speed of the cGA algorithm and maintain the diversity given by the distributed layout. We examine the effect of the introduced hierarchy by observing the variable takeover rates at different hierarchy levels and we compare the H-cGA to the cGA algorithm on a set of benchmark problems and show that the new approach performs promising.

## 1 Introduction

The cellular Genetic Algorithm (cGA) [1] is a kind of decentralized GA in which the population is arranged in a toroidal grid, usually of dimension 2. The characteristic feature of cGAs is a neighbourhood for each individual that is restricted to a certain subset of individuals in the immediate vicinity of its position. Individuals are allowed to interact only with other individuals belonging to their neighbourhood. This endows the cGA with useful properties for the optimization [2, 3] and also facilitates a parallel implementation because of its inherent parallel design. The cGA has already been successfully implemented on parallel platforms [1, 4] and in sequential computers [5].

While the distributed arrangement of the population of the cGA preserves a high level of diversity, it can delay the convergence speed of the algorithm because the cooperative search by the entire population is restricted. In order to increase the convergence speed of the cGA algorithm, we introduce a hierarchy into the grid population that orders the individuals according to their current

J. Gottlieb and G.R. Raidl (Eds.): EvoCOP 2006, LNCS 3906, pp. 111–122, 2006.

fitness. Such a population hierarchy has already been successfully applied to particle swarm optimization algorithms [6]. Here the convergence speed of the PSO algorithm was accelerated after using a hierarchical population when solving several continuous optimization problems.

In this work we examine the effect of introducing a hierarchy into a cellular Genetic Algorithm. Our new algorithm is called *Hierarchical Cellular Genetic Algorithm* (H-cGA). In the H-cGA a hierarchy is established within the population of a regular cGA by arranging individuals according to their fitness: the worse the fitness value of an individual is, the farther it is located from the center of the population. Thus, in this hierarchical population, the best individuals are placed in the center of the grid where they can mate with other high quality individuals. This way we hope to reach better solutions faster, while still preserving the diversity provided by a locally restricted parent selection and keeping the opportunity of an efficient parallel implementation.

The paper is structured as follows. In Section 2 we present our newly proposed algorithm, H-cGA, as well as a new selection operator we have designed for this algorithm. We provide a closer examination of the takeover behaviour of the algorithm in Section 3. Section 4 contains experiments on a benchmark of functions, where the performance of the H-cGA algorithm is compared with that of a canonical cGA. Finally, we give our conclusions and further research directions for our new approach.

## 2   The H-cGA Algorithm

In this section we present the hierarchical cGA algorithm. First we will briefly outline the procedure of a cellular GA. Then we describe the hierarchical ordering of the population that is introduced for the H-cGA algorithm and how this ordering is obtained. After that we introduce a new selection operator for this algorithm.

In a Cellular GA [1] the population is mapped onto a 2-dimensional toroidal grid of size $x \times y$, where an individual can only mate with individuals within a locally restricted neighbourhood. In each generation, an offspring is obtained for every individual, where one of the parents is the current individual and the other one is selected from its neighbourhood. Each offspring is mutated with a given probability, and replaces the current individual, if it has a better fitness value. The offspring (or the current individual, if better) is stored in a temporary auxiliary population, and this population replaces the whole current one after each generation. In the H-cGA algorithm the population is re-arranged after every generation with the hierarchical swap operation, as described in the following section. In Algorithm 1. we give a pseudocode of H-cGA.

### 2.1   Hierarchy

The hierarchy is imposed on the cellular population of the GA by defining a center at position $(x/2, y/2)$ and assigning hierarchy levels according to the

**Algorithm 1.** Pseudocode of H-cGA

```
 1: proc Steps_Up(hcga)      //Algorithm parameters in 'hcga'
 2: while not TerminationCondition() do
 3:   for x ← 1 to WIDTH do
 4:     for y ← 1 to HEIGHT do
 5:       n_list←Compute_Neigh(cga,position(x,y));
 6:       parent1←Individual_At(cga,position(x,y));
 7:       parent2←Local_Select(n_list);
 8:       Recombination(cga.Pc,n_list[parent1],n_list[parent2],aux_ind.chrom);
 9:       Mutation(cga.Pm,aux_ind.chrom);
10:       aux_ind.fit←cga.Fit(Decode(aux_ind.chrom));
11:       Insert_New_Ind(position(x,y),aux_ind,cga,aux_pop);
12:     end for
13:   end for
14:   cga.pop←aux_pop;
15:   Swap_Operation(cga.pop);
16:   Update_Statistics(cga);
17: end while
18: end_proc Steps_Up;
```

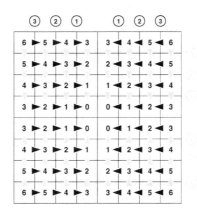

**Fig. 1.** The H-cGA and its different hierarchy levels

distance from the center. The center has level 0 and the level increases with increasing distance to the center (compare Fig. 1). The hierarchy is updated after each iteration of the cGA and individuals with high fitness are moved towards the center. Note, that the population topology is still toroidal when selecting parents.

In Fig. 1 we show how this swap operation is performed. It is applied between cells indicated by the arrows, in the order denoted by the numbers outside of the grid. The update of the hierarchy is performed alternatingly horizontally (black) and vertically (grey) by the swap operation. We assume an even number for the population dimensions $x$ and $y$, so that the population can be uniquely divided into left (upper) and right (lower) half, for the horizontal (vertical) swap. Note, that this implies that there are 4 individuals in the center of the population, i.e. on the highest level of the hierarchy.

In the following we describe the horizontal swap operation, the vertical swap is done accordingly. Each individual $(i, j)$ in the left half compares itself with its left neighbor $(i - 1, j)$ and if this one is better they swap their positions. These comparisons are performed starting from the center of the grid towards the outside, see Fig. 1. Thus, at first individuals in columns $(i, \frac{x}{2} - 1)$ and $(i, \frac{x}{2} - 2)$, for $i = 0, \ldots, y$, are compared. If the fitness value at position $(i, \frac{x}{2} - 2)$ is better they swap positions. These pairwise comparisons are then continued towards the outside of the grid. Hence, an individual can advance only one level at a time but can drop several levels within one iteration.

## 2.2 Dissimilarity Selection

The proposed hierarchy promotes the recombination of good individuals within the population. In this respect, H-cGA is similar to a panmictic GA with a fitness-biased selection. In our H-cGA algorithm this selective recombination of the elite individuals of the population is already included in the hierarchy. Therefore, we examined a new selection operator that is not based on the relative fitness of the neighbouring individuals but instead considers the difference between the respective solution strings. As for the Binary Tournament (BT) selection, two neighbours are selected randomly, but in contrast to BT, where the better one is selected, the one that is more different from the focal individual is selected. All the considered problems are binary encoded, hence we use the Hamming-Distance for determining the dissimilarity.

The overall optimization progress of the algorithm is ensured by only replacing an individual if the newly generated individual, by crossover and mutation, is better than the previous one.

## 3   First Theoretical Results: Takeover Times

We are providing a closer examination of the properties of the proposed algorithm by studying the *takeover* time of the algorithm and comparing it to that of a canonical cGA. The takeover time is the time required until a single best individual has filled the entire population with copies of itself under the effects of selection only. First we are looking at a deterministic takeover process, where the best individual within the neighbourhood is always selected. Later we also consider BT selection and the newly proposed Dissimilarity selection. Initially all individuals get assigned random fitness values from [0:4094] and one individual gets the maximum fitness value of 4095. Then the selection-only algorithm is executed and the proportion of the entire population that holds the maximum fitness value at each iteration is recorded. The considered grid is of size $64 \times 64$ and hence the population consists of 4096 individuals.

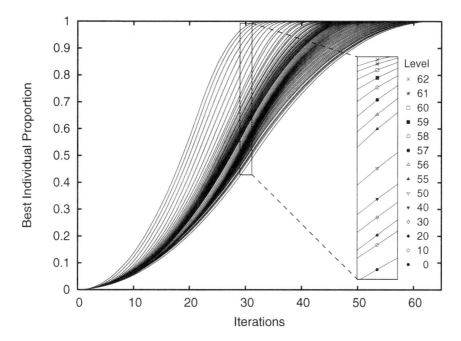

**Fig. 2.** Takeover curve for initially placing the best individual at different levels

In the H-cGA the different levels of the hierarchy influence the time required for takeover. In order to accurately determine this influence, we use the deterministic selection operator. The best individual is initially placed on each possible position on the grid and the takeover time for these 4096 different setups is measured. Then the results for all positions on a specific level of the hierarchy are averaged to obtain the takeover rate for introducing the best individual at this particular level. In Fig. 2 the obtained takeover curves are shown for introducing the best individual at different levels of the hierarchy. The slowest takeover rate is achieved when placing the best value at the center of the grid on level 0. This takeover rate is identical to the deterministic takeover for the regular cGA. The hierarchy level 62 consists of the 4 cells on the corners of the grid, where the fastest takeover rate is obtained. This increasing takeover speed with increasing hierarchy level is very regular, as can be seen in the detail display of iteration 30. The reason why having the best individual near the outside of the hierarchy accelerates takeover is, that, since for selection the topology is still toroidal, adjacent individuals on the opposite end of the grid also adopt the highest fitness value at the beginning of the run. Then the hierarchy swap operation moves the maximal value towards the center from several sides and therefore the actual takeover speed is increased.

We also measured the time required for takeover with BT and Dissimilarity selections. The best individual was placed at a random position and the experiments were repeated 100 times. For the experiments with Dissimilarity selection, the individuals are using a binary string that corresponds to the 12 bit representation of their current fitness value. Our results are shown in Table 1 (average number of iterations, standard deviation, minimum and maximum). As can be seen, initially placing the best individual at the center (H-cGA level 0) slows down the takeover compared to the regular cGA algorithm. This is because once the maximal value spreads it will be delayed until all of the central cells hold the best fitness value, otherwise it will be swapped towards the center again. In the two algorithms, the use of the BT selection induces a higher selection pressure than in the case of using the Dissimilarity selection. After applying the Kruskal-Wallis test (in some cases the data were not normally distributed) to the results in Table 1, we obtained that there exist statistically significant differences at a 95% confidence level.

**Table 1.** Takeover times for the algorithms with BT and Dissimilarity selections

| Algorithm | Avg | Stddev | Min–Max |
|---|---|---|---|
| cGA BT | 75.2 | ± 1.5 | 72.0–80.0 |
| cGA Dis | 78.7 | ± 1.7 | 75.0–83.0 |
| H-cGA BT | 71.0 | ± 6.3 | 48.0–81.0 |
| H-cGA Dis | 79.6 | ± 3.8 | 71.0–89.0 |
| H-cGA BT level 0 | 81.5 | ± 1.5 | 78.0–87.0 |
| H-cGA BT level 62 | 46.3 | ± 2.1 | 42.0–57.0 |

### 3.1   Fitting of the Takeover Curves

The takeover time decreases as the level at which the best individual is introduced increases, as shown in Fig. 2. In order to observe this distinction more precisely, we fitted the obtained takeover curves with a parameterized function, so that we can simply compare the respective fitting parameter.

The takeover time in Genetic Algorithms is usually fitted by a logistic growth curve (1) that models bounded population growth at a rate of $a$ [7]. The size of the population at time $t$ is given by $N(t)$ and the number of individuals at the beginning is $N(0)$. This growth curve is not applicable to the cGA, because instead it exhibits quadratic growth [8]. We used a simple formulation for the quadratic growth in the first half and the symmetrical saturation phase in the second half of the takeover (2). This quadratic growth curve is also controlled by a single parameter $a$.

$$N(t) = \frac{1}{1 + (\frac{1}{N(0)} - 1)\, e^{-at}} \tag{1}$$

$$N(t) = \begin{cases} a\, t^2 & \text{if } t < T/2, \\ -a\, (t - T)^2 + 1 & \text{otherwise.} \end{cases} \tag{2}$$

We fitted the parameter $a$ to the takeover curves obtained for different levels of the hierarchy. This is done by minimizing the mean squared error (MSE)

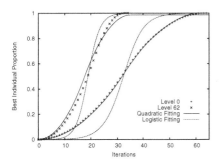

**Fig. 3.** Takeover curves fitted

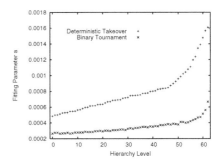

**Fig. 4.** The parameter $a$ for different levels

between the observed data points and the parameterized curve. In Fig. 3 the deterministic takeover curves for introducing the best individual at level 0 and level 62 and the respective fitted curves are reported.

Similar to the takeover curves at different levels, the parameters for fitting these curves are also well ordered. In Fig. 4 we depict the parameter $a$ returned by the fitting for the different levels. This is done for deterministic takeover and takeover with BT selection. The growth rates for the deterministic takeover are fully ordered and, with few exceptions, also for the BT takeover $a$ increases as the level increases.

# 4    Computational Experiments

In this section, we present the results of our tests on a selected benchmark of problems. We briefly describe the problems composing the benchmark used in our testing in Section 4.1, and test the algorithms on those problems in Section 4.2.

## 4.1    Test Problems

For testing our algorithms, we have selected a benchmark of problems with many different features, such as multimodality, deceptiveness, use of constraints, or problem generators. The proposed problems are Onemax, the Massively Multimodal Deceptive Problem (MMDP), P-PEAKS, and the Minimum Tardy Task Problem (MTTP). We show in Table 2, for each of these problems, the fitness function to optimize, the chromosome length (size of the problem), and the value of this optimum.

**Onemax.** The objective of the Onemax [9] problem is to maximize the number of 1s in the binary string. We are considering strings of length 500 bits.

**MMDP.** The Massively Multimodal Deceptive Problem (MMDP) is a problem specifically designed for being difficult to solve by evolutionary algorithms [10]. It consists of $k$ deceptive subproblems each of size 6 bits. These substrings are

**Table 2.** Benchmark of problems

| Problem | Fitness function | $n$ | Optimum |
|---------|-----------------|-----|---------|
| **Onemax** | $f(\boldsymbol{x}) = \sum_{i=1}^{n} x_i$ | 500 | 500 |
| **MMDP** | $f(\boldsymbol{s}) = \sum_{i=1}^{k} \text{fitness}(\boldsymbol{s}_i); \ k = n/6$ | 240 | 40 |
| **P-PEAKS** | $f(\boldsymbol{x}) = \frac{1}{n} \max_{1 \leq i \leq p} \{n - \text{HamDist}(\boldsymbol{x}, \text{Peak}_i)\}$ | 100 | 1.0 |
| **MTTP** | $f(\boldsymbol{x}) = \sum_{i=1}^{n} x_i \cdot w_i$ | 20 | 0.02429 |
| | | 100 | 0.005 |
| | | 200 | 0.0025 |

evaluated according to the number of ones in the substring as given in Fig. 5. It is easy to see that these subfunctions have two global obtima and a deceptive attractor in the middle point. We are using an instance of size $k = 40$ with optimum value 40.0. The number of local optima for this instance is very large $(22^k)$, while there are only $2^k$ global solutions.

| Unitation | Subfunction value |
|-----------|-------------------|
| 0 | 1.000000 |
| 1 | 0.000000 |
| 2 | 0.360384 |
| 3 | 0.640576 |
| 4 | 0.360384 |
| 5 | 0.000000 |
| 6 | 1.000000 |

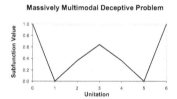

**Fig. 5.** Basic deceptive bipolar function ($\boldsymbol{s}_i$) for MMDP

**P-PEAKS.** In this multimodal problem generator [11] $P$ peaks are generated and the fitness value is calculated as the distance to the nearest peak divided by the string length, having a maximum value of 1.0. We are using a test instance with 100 peaks of length 100.

**MTTP.** The Minimum Tardy Task Problem (MTTP) [12] is a task scheduling problem. Each task has a length (the time it takes), a deadline (before which it has to be finished), and a weight $w_i$. The objective is to maximize the weights of the scheduled tasks. We are considering here three instances of sizes 20, 100 and 200, taken from [13].

### 4.2 Results

In this section we present and analyze the results of our experiments. We compare the behaviour of H-cGA to the cGA algorithm. The same parametrization is used for the two algorithms (see Table 3), and both BT and Dissimilarity selection have been tested. We used a population of size 400 arranged in a grid of size

**Table 3.** Parametrization used in our algorithms

| | |
|---|---|
| *Population* | $20 \times 20$ individuals |
| *Selection of Parents* | itself + (BT or Dissimilarity) |
| *Recombination* | DPX, $p_c = 1.0$ |
| *Bit Mutation* | Bit-flip, $p_m = 1/\#bits$ |
| *Replacement* | Replace if Better |
| *Stopping Criterion* | Find Optimum or Reach 2500 Generations |

**Table 4.** Results for the different test functions. Given are the success rate, average number of required steps to reach the optimum, standard deviation, maximum and minimum, and statistical significance of the results.

| Problem | Algorithm | Success | Avg | Stddev | Min– Max | $p$-value |
|---|---|---|---|---|---|---|
| Onemax | cGA BT | 100% | 129.4 | $\pm$ 7.3 | 111.2–145.2 | |
| | cGA Dis | 100% | 140.7 | $\pm$ 8.1 | 121.6–161.2 | + |
| | H-cGA BT | 100% | 94.1 | $\pm$ 5.0 | 83.2–106.4 | |
| | H-cGA Dis | 100% | 103.1 | $\pm$ 5.6 | 90.4–116.8 | |
| MMDP | cGA BT | 67% | 202.4 | $\pm$ 154.7 | 120.8–859.2 | |
| | cGA Dis | 97% | 179.8 | $\pm$ 106.3 | 116.8–846.0 | + |
| | H-cGA BT | 55% | 102.6 | $\pm$ 76.1 | 68.8–652.8 | |
| | H-cGA Dis | 92% | 122.3 | $\pm$ 111.7 | 73.2–837.6 | |
| P-Peaks | cGA BT | 100% | 41.9 | $\pm$ 3.0 | 32.0– 48.4 | |
| | cGA Dis | 100% | 52.9 | $\pm$ 5.2 | 38.4– 66.0 | + |
| | H-cGA BT | 100% | 47.2 | $\pm$ 8.6 | 30.8– 71.2 | |
| | H-cGA Dis | 100% | 81.1 | $\pm$ 17.1 | 45.2–130.8 | |
| MTTP-20 | cGA BT | 100% | 5.1 | $\pm$ 1.2 | 1.6– 8.0 | |
| | cGA Dis | 100% | 6.0 | $\pm$ 1.3 | 2.0– 9.2 | + |
| | H-cGA BT | 100% | 4.7 | $\pm$ 1.1 | 1.6– 7.2 | |
| | H-cGA Dis | 100% | 5.5 | $\pm$ 1.2 | 2.8– 8.0 | |
| MTTP-100 | cGA BT | 100% | 162.2 | $\pm$ 29.3 | 101.6–241.6 | |
| | cGA Dis | 100% | 174.6 | $\pm$ 26.3 | 96.4–238.8 | + |
| | H-cGA BT | 100% | 138.3 | $\pm$ 35.4 | 62.0–245.6 | |
| | H-cGA Dis | 100% | 132.4 | $\pm$ 26.2 | 64.0–186.8 | |
| MTTP-200 | cGA BT | 100% | 483.1 | $\pm$ 55.3 | 341.6–632.4 | |
| | cGA Dis | 100% | 481.0 | $\pm$ 71.6 | 258.8–634.8 | + |
| | H-cGA BT | 100% | 436.2 | $\pm$ 79.7 | 270.4–631.2 | |
| | H-cGA Dis | 100% | 395.3 | $\pm$ 72.6 | 257.6–578.8 | |

$20 \times 20$ with a Linear5 neighbourhood (the cell itself and its North, East, South and West neighbours are considered). In all our experiments, one parent is the center individual itself and the other parent is selected either by BT or Dissimilarity selection (it is ensured that the two parents are different). An individual is replaced only if the newly generated fitness value is better. The recombination method used is the two point crossover (DPX), and the selected offspring is the one having the largest part of the best parent. The mutation and crossover probabilities are 1.0 and bit mutation is performed with probability $1/\#bits$ for genome string of length $\#bits$. In order to have statistical confidence, all the presented results are average over 100 runs, and the analysis of variance –ANOVA– statistical test (or Kruskal-Wallis if the data is not normally distributed) is applied to the results. We consider in this paper a 95% confidence level.

In Table 4 we present the results we have obtained for all the test problems. Specifically, we show the success rate (number of runs in which the optimum was found), and some measures on the number of evaluations made to find the optimum, such as the average value, the standard deviation, and the maximum and minimum values. The results of our statistical tests are in column $p$-values, where symbol '+' means that there exists statistically significant differences. The evaluated algorithms are cGA and H-cGA both with BT and Dissimilarity selections.

As can be seen in Table 4, both the cGA and the H-cGA algorithms were able to find the optimum value in each run for all the problems, with the exception of MMDP. For this problem, the cGA algorithm achieves slightly better success rates than H-cGA. Regarding the average number of evaluations required to reach the optimum, the hierarchical algorithm always outperforms cGA, except for the P-peaks problem. Hence, the use of a hierarchical population allows us to accelerate the convergence speed of the algorithm to the optimum, while it retains the interesting diversity management of the canonical cGA.

If we now compare the results of the algorithms when using the two different studied selection schemes, we notice that with the Dissimilarity selection the success rate for the MMDP problem can be increased for both the cGA and the H-cGA algorithm.

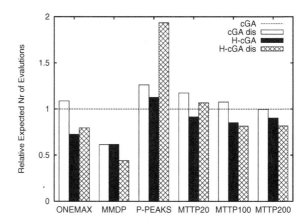

**Fig. 6.** Expected number of evaluations required to reach the optimum, relative to the steps required by cGA, for the different benchmark problems

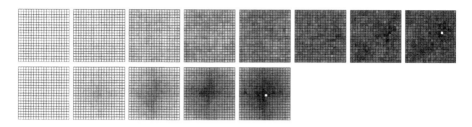

**Fig. 7.** Evolution of the population for the cGA (top) and the H-cGA (bottom)

In Fig. 6 we plot the expected number of evaluations, defined as the average number of evaluations divided by the success rate, required to find the optimum value for each problem. The displayed results are relative to the expected number of evaluations for the cGA.

The expected number of evaluations is increased when using the Dissimilarity selection compared to the equivalent algorithms with BT. For the cGA the Dissimilarity selection was able to reduce the number of required evaluations only for the MMDP problem. But for the H-cGA it proved to be useful also for the two large MTTP instances. In general, the expected number of evaluations is lower for the two studied versions of H-cGA.

Finally, in order to illustrate the effects of using a hierarchical population in the cGA, we show a sample run of the cGA (top) and the H-cGA (bottom) algorithm in Fig. 7. The pictures are snapshots of the population taken every 50 iterations for the MDDP problem until the optimum is found (iteration 383 for cGA and 233 for H-cGA). The darker an individual is coloured the higher its fitness is; the white cell in the last image contains the maximum fitness. As can be seen, the H-cGA algorithm quickly focuses on promising solutions, while at the same time different solutions of lower quality are kept at the outside of the hierarchy.

# 5   Conclusions and Future Work

In this work we have presented a new algorithm called hierarchical cGA, or H-cGA. We included the idea of establishing a hierarchy among the individuals of the population of a canonical cGA. With this hierarchical model we achieve different levels of the exploration/exploitation tradeoff of the algorithm in distinct zones of the population simultaneously. We studied these specific behaviours at different hierarchy levels by examining the respective takeover rates.

We have compared the H-cGA with two different selection methods to the equivalent cGAs and the hierarchical algorithm performed better on almost all test functions. The newly proposed Dissimilarity selection was not useful in all the scenarios, but for the H-cGA algorithm it improved the performance for both the MMDP and the MMTP functions. This selection promotes more diversity into the population but, as a consequence, the convergence is usually slower.

Future work will concentrate on other forms of hierarchies, with respect to the shape or to the criterion that allows ascending in levels. Furthermore, we will consider position-dependant algorithm changes, to emphasize different search strategies on different levels of the hierarchy.

# References

1. Manderick, B., Spiessens, P.: Fine-grained parallel genetic algorithm. In Schaffer, J., ed.: Proceedings of the Third International Conference on Genetic Algorithms, Morgan-Kaufmann (1989) 428–433
2. Alba, E., Tomassini, M.: Parallelism and Evolutionary Algorithms. IEEE Trans. on Evolutionary Computation **6**(5) (2002) 443–462
3. Cant-Paz, E.: Efficient and Accurate Parallel Genetic Algorithms. 2nd edn. Volume 1 of Book Series on Genetic Algorithms and Evolutionary Computation. Kluwer Academic Publishers (2000)
4. Alba, E.: Parallel Metaheuristics: A New Class of Algorithms. Wiley (2005)
5. Alba, E., Dorronsoro, B.: The exploration/exploitation tradeoff in dynamic cellular evolutionary algorithms. IEEE Trans. on Evolutionary Computation **9**(2) (2005) 126–142
6. Janson, S., Middendorf, M.: A hierarchical particle swarm optimizer and its adaptive variant. IEEE Systems, Man and Cybernetics - Part B **35**(6) (2005) 1272–1282
7. Goldberg, D., Deb., K. Foundations of Genetic Algorithms. In: A comparative analysis of selection scheme used in genetic algorithms. Morgan Kaufmann Publishers (1991) 69–93
8. Giacobini, M., Tomassini, M., Tettamanzi, A., Alba, E.: Synchronous and asynchronous cellular evolutionary algorithms for regular lattices. IEEE Transactions on Evolutionary Computation **9**(5) (2005) 489–505
9. Schaffer, J., Eshelman, L.: On crossover as an evolutionary viable strategy. In: 4th ICGA, Morgan Kaufmann (1991) 61–68
10. Goldberg, D., Deb, K., Horn, J.: Massively multimodality, deception and genetic algorithms. In: Proc. of the PPSN-2, North-Holland (1992) 37–46
11. Jong, K.D., Potter, M., Spears, W.: Using problem generators to explore the effects of epistasis. In: 7th ICGA, Morgan Kaufman (1997) 338–345
12. Stinson, D.: An Introduction to the Design and Analysis of Algorithms. The Charles Babbage Research Center, Winnipeg, Manitoba, Canada (1985 (second edition, 1987))
13. S.K., Bäck, T., Heitkötter, J.: An evolutionary approach to combinatorial optimization problems. In: Proceedings of the 22nd annual ACM computer science conference (CSC '94), New York, NY, USA, ACM Press (1994) 66–73

# Improving Graph Colouring Algorithms and Heuristics Using a Novel Representation

István Juhos[1] and Jano I. van Hemert[2]

[1] Department of Computer Algorithms and Artificial Intelligence,
University of Szeged, Hungary, P. O. Box 652., Hungary
juhos@inf.u-szeged.hu
[2] Institute of Computer Graphics and Algorithms,
Vienna University of Technology, Vienna, Austria
jano@vanhemert.co.uk

**Abstract.** We introduce a novel representation for the graph colouring problem, called the Integer Merge Model, which aims to reduce the time complexity of an algorithm. Moreover, our model provides useful information for guiding heuristics as well as a compact description for algorithms. To verify the potential of the model, we use it in DSATUR, in an evolutionary algorithm, and in the same evolutionary algorithm extended with heuristics. An empiricial investigation is performed to show an increase in efficiency on two problem suites , a set of practical problem instances and a set of hard problem instances from the phase transition.

**Keywords:** graph colouring, representation, heuristics, merge model.

## 1 Introduction

The Graph Colouring Problem (GCP) plays an important role in graph theory. It arises in a number of applications—for example in time tabeling and scheduling, register allocation, and printed circuit board testing (see [1–3]). GCP deals with the assigment of colours to the vertices of an undirected graph such that adjacent vertices are not assigned the same colour. The primary objective is to minimize the number of colours used. The minimum number of colours necessary to colour the vertices of a graph is called the chromatic number. Finding it is an NP-hard problem, but deciding whether a graph is $k$-colourable or not is NP-complete [4]. Thus one often relies on heuristics to compute a solution or an approximation.

Graph colouring algorithms make use of adjacency checking during colouring, which plays a key role in the overall performance (see [5–7]). The number of checks depends on the problem representation and the algorithm that uses it. The Integer Merge Model (IMM) introduced here directly addresses the issues mentioned above. Generally, there are two main data structures used to represent graphs: the adjacency matrix and the adjacency list. In [6] a novel graph representation for the colouring problem called the Binary Merge Model (BMM) is introduced. IMM is a generalization of BMM, which is a useful and efficient representation of the GCP ([6,7]). IMM preserves BMM's beneficial feature of improving upon efficiency. Moreover, it provides useful information about the

J. Gottlieb and G.R. Raidl (Eds.): EvoCOP 2006, LNCS 3906, pp. 123–134, 2006.

graph structure during the colouring process, which enables one to define more sophisticated colouring algorithms and heuristics with a compact description. To demonstrate its potential, IMM is embedded in the DSATUR algorithm [8]—a standard and effective heuristic GCP solver—and in a meta-heuristic environment driven by an evolutionary meta-heuristic. On standard problem sets, we compare the effectiveness and efficiency of these three algorithms, with and without the use of IMM.

# 2    Representing the Graph k-Colouring Problem

The problem class known as the graph $k$-colouring problem is defined as follows. Given a graph $G(V, E)$ which is a structure of nodes and edges, where $V = \{v_1, ..., v_n\}$ is a set of nodes and $E = \{(v_i, v_j)|v_i \in V \land v_j \in V \land i \neq j\}$ is a set of edges, the edges define the relation between the nodes $(V \times V \to E)$. The graph $k$-colouring problem is to colour every node in $V$ with one of $k$ colours such that no two nodes connected with an edge in $E$ have the same colour. The smallest such $k$ is called the chromatic number, which will be denoted here by $\chi$.

Graph colouring algorithms make use of adjacency checking during the colouring process, which has a large influence on the performance. Generally, when assigning a colour to a node, all adjacent or coloured nodes must be scanned to check for equal colouring, so constraint checks need to be performed. The number of constraint checks performed lies between two bounds, the current number of coloured neighbours and $|V| - 1$. With the IMM approach the number of checks is greater than zero and less than the number of colours used up to this point. These bounds arise from the model-induced hyper-graph structure and they guarantee that the algorithms will perform better under the same search.

## 2.1    Integer Merge Model

The *Integer Merge Model* (IMM) implicitly uses hyper-nodes and hyper-edges (see Figure 1). A hyper-node is a set of nodes that have the same colour. A hyper-edge connects a hyper-node with other nodes, regardless of whether it is normal or hyper. A hyper-node and a normal-node or hyper-node are connected by a hyper-edge if and only if they are connected by at least two normal edges. IMM concentrates on the operations between hyper-nodes and normal nodes. We try to merge the normal nodes with another node, and when the latter is a hyper-node, a reduction in adjacency checks is possible. These checks can be performed along hyper-edges instead of normal edges, whereby we can introduce significant savings. This is because the initial set of normal edges is folded into hyper-edges. The colouring data is stored in an Integer Merge Table (IMT) (see Figure 2). Every cell $(i, j)$ in this table has non-negative integer values. The columns refer to the nodes and the rows refer to the colours. A value in cell $(i, j)$ is greater than zero if and only if node $j$ cannot be assigned a colour $i$ because of the edges in the original graph $\langle V, E \rangle$. The initial IMT is the adjacency matrix of the graph, hence a unique colour is assigned to each of the nodes. If the graph is not a complete graph, then it might be possible to reduce the number of necessary colours. This corresponds to the reduction of rows in the IMT. To reduce the rows we introduce an Integer Merge Operation, which attempts to merge two rows. When this is possible, the number of colours is decreased by one. When it is not, the number of colours remains the same. It is achievable only when two nodes are not connected by a normal edge or a hyper-edge. An example of both cases is found in Figures 1 and 2.

**Definition 1.** *The Integer Merge Operation $\cup$ merges an initial row $r_i$ into an arbitrary (initial or merged) row $r_j$ if and only if $(j, i) = 0$ (i.e., the hyper-node $x_j$ is not connected to the node $x_i$) in the IMT. If rows $r_i$ and $r_j$ can be merged then the result is the union of them.*

Formally, let $I$ be the set of initial rows of the IMT and $R$ be the set of all possible $|V|$ size integer-valued rows (vectors). Then an integer merge operation is defined as

$$\cup : R \times I \to R$$
$$r'_j := r_j \cup r_i, \quad r'_j, r_j \in R, \quad r_i \in I, \text{ or by components}$$
$$r'_j(l) := r_j(l) + r_i(l), \quad l = 1, 2, \ldots, |V|$$

A merge can be associated with an assignment of a colour to a node, because two nodes are merged if they have the same colour. Hence, we need as many merge operation as the number of the nodes in a valid colouring of the graph, apart from the nodes which are coloured initially and then never merged, i.e., a colour is used only for one node. If $k$ number of rows left in the IMT (i.e., the number of colours used) then the number of integer merge operations was $|V| - k$, where $k \in \{\chi, \ldots, |V|\}$.

With regard to the time complexity of a merge operation, we can say that it uses as many integer additions as the size of the operands. In fact, we just need to increment the value in the row $r_j$, where the corresponding element in the row $r_i$ is non-zero (i.e., has a value of one), that is $d(x_i)$ number of operations. The number of all operations are at most $\sum_i d(x_i) = 2|E|$ for a valid colouring. This occurs when a list based representation of the rows is applied in an implementation. Using special hardware instructions available on modern computers, merge operations can be reduced to one computer instruction. For instance, a merge operation can be performed as one VDOT operation on a vector machine [9].

When solving a graph colouring while using the original graph representation for checking violations, approximately $|V|^2$ constraint checks are required to get to a valid colouring and the IMM supported scheme uses at most $|V| \cdot \kappa$ ($\kappa \approx \chi$ the number of colors in the result coloring) number of checks, because each node ($|V|$ items) has to be compared at most the number of existing hyper-nodes/colours, which is not more than $\kappa$ and $\chi$ if a solution is found. Hence, their quotient is the improvement of an IMM supported colouring, which is proportional to the $|V|/\kappa$ ratio.

## 2.2    Permutation Integer Merge Model

The result of colouring a graph after two or more integer merge operations
depends on the order in which these operations were performed. Consider the
hexagon in Figure 1(a) and its corresponding IMT in Figure 2. Now let the
sequence $P_1 = 1, 4, 2, 5, 3, 6$ be the order in which the rows are considered for
the integer merge operations, i.e., for the colouring.

This sequence of merge operations results in a 4-colouring of the graph depicted in Figure 1(c). However, if we use the sequence $P_2 = 1, 4, 2, 6, 3, 5$ then
the result will be only a 3-colouring, as shown in Figure 1(e) with the merges
$1 \cup 4$, $2 \cup 6$ and $3 \cup 4$. The defined merge is greedy, i.e., it takes a row and tries to
find the first row from the top of the table that it can merge. The row remains
unaltered if there is no suitable row. After performing the sequence $P$ of merge
operations, we call the resulting IMT the *merged* IMT.

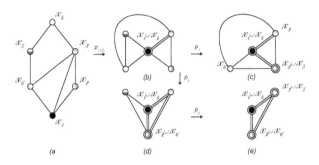

**Fig. 1.** Examples of the result of two different merge orders $P_1 = 1, 4, 2, 5, 3, 6$ and
$P_2 = 1, 4, 2, 6, 3, 5$. The double-lined edges are hyper-edges and double-lined nodes are
hyper-nodes. The $P_1$ order yields a 4-colouring (c), but with the $P_2$ order we get a
3-colouring (e).

Finding a minimal colouring for a graph $k$-colouring problem using the IMT
representation and integer merge operations comes down to finding the sequence
of merge operations that leads to that colouring. This can be represented as
a sequence of candidate reduction steps using the greedy approach described
above. The permutations of this representation form the Permutation Integer
Merge Model (PIMM). It is easy to see that these operations and the colouring
are equivalent.

## 2.3    Extracting Useful Information: Co-structures

The IMM can be incorporated into any colouring algorithm that relies on a construction based form of search. The hyper-graph structure introduced can save
considerable computational effort as we have to make only one constraint check
along a hyper-edge instead of checking all the edges it contains. Besides this
favourable property, the model gives incremental insight into the graph structure

| (a) | $x_1$ | $x_2$ | $x_3$ | $x_4$ | $x_5$ | $x_6$ |
|---|---|---|---|---|---|---|
| $r_1$ | 0 | 1 | 1 | 0 | 0 | 1 |
| $r_2$ | 1 | 0 | 1 | 0 | 0 | 0 |
| $r_3$ | 1 | 1 | 0 | 1 | 0 | 1 |
| $r_4$ | 0 | 0 | 1 | 0 | 1 | 0 |
| $r_5$ | 0 | 0 | 0 | 1 | 0 | 1 |
| $r_6$ | 1 | 0 | 1 | 0 | 1 | 0 |

| (b) | $x_1$ | $x_2$ | $x_3$ | $x_4$ | $x_5$ | $x_6$ |
|---|---|---|---|---|---|---|
| $r_1 \cup r_4$ | 0 | 1 | 2 | 0 | 1 | 1 |
| $r_2$ | 1 | 0 | 1 | 0 | 0 | 0 |
| $r_3$ | 1 | 1 | 0 | 1 | 0 | 1 |
| $r_5$ | 0 | 0 | 0 | 1 | 0 | 1 |
| $r_6$ | 1 | 0 | 1 | 0 | 1 | 0 |

| (c) | $x_1$ | $x_2$ | $x_3$ | $x_4$ | $x_5$ | $x_6$ |
|---|---|---|---|---|---|---|
| $r_1 \cup r_4$ | 0 | 1 | 2 | 0 | 1 | 1 |
| $r_2 \cup r_5$ | 1 | 0 | 1 | 1 | 0 | 1 |
| $r_3$ | 1 | 1 | 0 | 1 | 0 | 1 |
| $r_6$ | 1 | 0 | 1 | 0 | 1 | 0 |

| (d) | $x_1$ | $x_2$ | $x_3$ | $x_4$ | $x_5$ | $x_6$ |
|---|---|---|---|---|---|---|
| $r_1 \cup r_4$ | 0 | 1 | 2 | 0 | 1 | 1 |
| $r_2 \cup r_6$ | 2 | 0 | 2 | 0 | 1 | 0 |
| $r_3$ | 1 | 1 | 0 | 1 | 0 | 1 |
| $r_5$ | 0 | 0 | 0 | 1 | 0 | 1 |

| (e) | $x_1$ | $x_2$ | $x_3$ | $x_4$ | $x_5$ | $x_6$ |
|---|---|---|---|---|---|---|
| $r_1 \cup r_4$ | 0 | 1 | 2 | 0 | 1 | 1 |
| $r_2 \cup r_6$ | 2 | 0 | 2 | 0 | 1 | 0 |
| $r_3 \cup r_5$ | 2 | 1 | 0 | 2 | 0 | 2 |

**Fig. 2.** Integer Merge Tables corresponding to the graphs in Figure 1

with the progress of the merging steps. This information can be used in a beneficial way, e.g., for defining colouring heuristics.

In this section, the co-structures are defined. These structures contain information about some useful graph properties obtained during the merging process. How this information is used precisely is explained in Sections 3 and 4, where we describe the two algorithms in which we have embedded the Integer Merge Model.

In practice the initial graphs are uncoloured, the colouring being performed by colouring the nodes in steps. Here, we deal with the sub-graphs of the original graphs defined by the colouring steps. The related merge tables contain partial information about the original one. For example, let the original graph with its initial IMT be defined by Figure 2.3(a) on which the colouring will be performed. Taking the $x_1, x_4, x_2, x_6, x_3, x_5$ order of the nodes into account for colouring $G$, then $P_1 = 1, 4, 2, 6, 3, 5$ ordered merges of the IMT rows will be performed. After the greedy colouring of the $x_1, x_4, x_2$ nodes there is a related partial or sub-IMT along with the (sub-)hyper-graph. These are depicted in Figure 2.3(b). The 1st and the 4th rows are merged together, but the 2nd cannot be merged with the $1 \cup 4$ merged row, thus the 2nd row remains unaltered in the related sub-IMT. The left, top, right and bottom bars are defined around the sub-IMT to store the four co-structures (see Figure 2.3(b)).

*The left and top co-structures* are associated with the original graph and contain the sum of the rows and the columns of the current IMT, respectively. The sum of the cell values of a row is equal to the sum of the degree of the nodes associated with the row (merged or initial), while the sum of the elements of the $j$-th column provides the coloured degree of the node $x_j$, i.e., the number of coloured neighbours.

*The right and the bottom co-structures* supply information about the hyper-graph represented by the sub-IMT. They are calculated by counting the number of non-zero values in the rows and columns in the order described. The bottom bar value is the colour degree, i.e., the number of adjacent colours of a node.

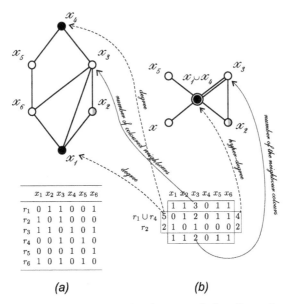

**Fig. 3.** The left side shows the partial colouring of the $G$ graph according to the $x_1, x_4, x_2$ greedy order and the adjacency matrix of the graph. The right one shows the partial or sub-IMT related to this colouring with its co-structures and sub-IMT induced hyper-graph.

The right bar gives the hyper-degree value of the nodes, which is especially interesting in case of hyper-nodes. The hyper-degree tells us how many different normal nodes are connected to the hyper-node being examined. This counts a node once even though it is connected to the hyper-node in question by more than one normal edge folded in a hyper-edge.

By extending the IMT we are able to describe efficient heuristics in a compact manner. To demonstrate this we will formulate two effective heuristic using the Integer Merge Model to get novel colouring algorithms. Two kinds of implementations of the two heuristic algorithms are considered during the experiments, when IMM is used and when it is not used.

## 3   The DSATUR Heuristic

This algorithm of Brélaz's [8] uses a heuristic to dynamically change the ordering of the nodes and it then applies the greedy method to colour them. It works as follows. One node with the highest saturation degree (i.e., number of adjacent colours) is selected from the uncoloured subgraph and is assigned the lowest indexed colour that still yields a valid colouring (first-order heuristic). If there exist several such nodes, the algorithm chooses a node with the highest degree (second-order heuristic). The result can also be a set of nodes. If this is the case, we choose the first node in a certain order (third-order 'heuristic'). The top and bottom co-structures are used to define the DSATUR heuristic (see Figure 4). Let us denote the top co-structure by $\tau_t$ (i.e., the number of coloured neighbours) and the bottom co-structure by $\tau_b$ (i.e., saturation degree). In our terminology the highest saturated node is the node which has the largest $\tau_b$ value. Here, $\tau_t$ is used in the second order heuristic.

**Procedure DSATUR**$_{IMM}$

   1. *Find those uncoloured nodes which have the highest saturated value*
      *$S = \{v | \tau_b(v) = \max_u (\tau_b(u)), v, u \in V\}$*
   2. *Choose those nodes from $S$ that have the highest uncoloured-degree*
      *$N = \arg \max_v (d(v) - \tau_t(v))$*
   3. *Choose the first row/node from the set $N$*
   4. *Merge it with the first non-neighbor hyper-node*
   5. *If there exists an uncoloured node then continue with Step 2*

**Fig. 4.** The DSATUR heuristic is defined by the IMM top $(\tau_t)$ and bottom $(\tau_b)$ co-structures. Here, $d$ is the degree of a node.

A backtracking algorithm is used to discover a valid colouring [10]. It achieves either an optimal solution or a near optimal solution when the maximum number of constraint checks is reached. For comparison purposes, two algorithms were implemented using this heuristic. The first one, DSATUR$_{IMM}$ is based on the IMM structures, while the second one DSATUR$_{pure}$, uses the traditional colouring scheme, where we only make use of the adjacency matrix.

# 4   Evolutionary Algorithm to Guide the Models

We have two goals with this meta-heuristic. The first is to find a successful order of the nodes (see Section 2.2) and the second is to find a successful order for assigning colours. This approach differs from DSATUR, where a greedy color assignment is used. For the first goal, we must search the permutation search space of the model described in Section 2.2, which is of size $n!$. Here, we use an evolutionary algorithm to search through the space of permutations. The genotype consists of the permutations of the nodes, i.e., the rows of the IMT. The phenotype is a valid colouring of the graph after using a colour assignment strategy on the permutation to select the order of the integer merge operations. The colour assignement strategy is a generalization of the one introduced in [7]. We say that the c-th vector of the sub-IMT $r'(c)$ is the most suitable candidate for merging with $r_{p_i}$ if they share the most constraints. The dot product of two vectors provides the number of shared constraints. Thus, by reverse sorting all the sub-IMT vectors on their dot product with $r_{p_i}$, we can reduce the number of colours by merging $r_{p_i}$ with the most suitable match. Here, the dot product operates on integer vectors instead of binary ones, thus generalize that.

    An intuitive way of measuring the quality of an individual $p$ in the population is by counting the number of rows remaining in the final BMT. This equals to the number of colours $k(p)$ used in the colouring of the graph, which needs to be minimised. When we know the optimal colouring is $\chi$ then we may normalise this fitness function to $g(p) = k(p) - \chi$. This function gives a rather low diversity of fitnesses of the individuals in a population because it cannot distinguish between two individuals that use an equal number of colours. This problem is called the fitness granularity problem. We modify the fitness function introduced in [7] so that to use Integer Merge Model structures instead of the appropriate binary one. This fitness relies on the heuristic that one generally wants to avoid highly constraint nodes and rows in order to have a higher chance of successful merges at a later stage, commonly called a succeed-first strategy. It works as follows. After the final merge the resulting IMT defines the colour groups. There are $k(p) - \chi$ over-coloured nodes, i.e., merged rows. Generally, we use the indices of the over-coloured nodes to calculate the number of nodes that need to be minimised (see $g(p)$ above). But these nodes are not necessarily responsible for the over-coloured graph. Therefore, we choose to count the hyper-nodes that violates the least constraints in the final hyper-graph. To cope better with the fitness granularity problem we should modify the $g(p)$ according to the constraints of the over-coloured nodes discussed previously. The final fitness function is then defined as follows. Let $\zeta(p)$ denote the number of constraints, i.e., non-zero elements, in the rows of the final IMT that belong to the over-coloured nodes, i.e., the sum of the smallest $k(p) - \chi$ values of the right co-structure. The fitness function becomes $f(p) = g(p)\zeta(p)$. Here, the cardinality of the problem is known, and used as a stopping criterium ($f(p) = 0$) to determine the efficiency of the algorithm. If $\chi$ is unknown, we can use the worst approximation which is $\chi' = 0$. We must modify the stop condition to, reaching a time limit or to fitness $\leq 0$ due to under-approximation ($\chi' \leq \chi$) or over-approximation ($\chi' > \chi$). Alternatively, the normalisation step can be left out, but this might seriously effect the quality of the evolutionary algorithm in a negative way.

**Procedure EA**$_{IMM}$

```
1. population = generate initial permutations randomly
2. while stop condition allows
   − evaluate each p permutation {
     −− merge p_j − th uncoloured node into c − th hyper-node by c = max_j ⟨r'_j, r_{p_i}⟩
     −− calculate f(p) = (k(p) − χ)ζ(p)   }
   − population_{xover} = xover(population, prob_{xover})
   − population_{mut} = mutate(population_{xover}, prob_{mut}))
   − population = select_{2−tour}(population ∪ population_{xover} ∪ population_{mut})
3. end while
```

**Fig. 5.** The EA$_{imm}$ meta-heuristic uses directly the IMM structure

We use a generational model with 2-tournament selection and replacement, where we employ elitism of size one. This setting is used in all experiments. The initial population is created with 100 random individuals. Two variation operators are used to provide offsprings. First, the 2-point order-based crossover (OX2) [11, in Section C3.3.3.1] is applied. Second, the other variation operator is a simple swap mutation operator, which selects at random two different items in the permutation and then swaps. The probability of using OX2 is set to 0.3 and the probability for using the simple swap mutation is set to 0.8. These parameter settings are taken from the experiments in [7].

# 5    Experiments

The goal of these experiments are twofold. First, to show the improvement in efficiency possible when adding the Integer Merge Model to an existing technique. Second, to show further improvement possible in the evolutionary algorithm by adding heuristics that are based on the additional bookkeeping in the form of the co-structures.

## 5.1    Methods of Comparison

How well an algorithm works depends on its effectiveness and efficiency in solving a problem instance. The first is measured by determining the ratio of runs where the optimum is found, this ratio is called the success ratio; it is one if the optimum, i.e., the chromatic number of the graph, is found in all runs. The second is measured by counting the number of constraint checks an algorithm requires to find the optimum. A *constraint check* is defined equally for each algorithm as checking whether the colouring of two nodes is allowed or not. This measurement is independent of the hardware used and is known to grow exponentially with the problem size for the worst-case.

## 5.2    Definition of the Problem Suites

The first test suite consists of problem instances taken from "The Second DIMACS Challenge" [12] and Michael Trick's graph colouring repository [12]. These graphs originate from real world problems, with some additional artificial ones.

The second test suite is generated using the well known graph $k$-colouring generator of Culberson [13]. It consists of 3-colourable graphs with 200 nodes. The edge density of the graphs is varied in a region called the phase transition. This is where hard to solve problem instances are generally found, which is shown using the typical easy-hard-easy pattern. The graphs are all equipartite, which means that in a solution each colour is used approximately as much as any other. The suite consists of nineteen groups where each group has five instances, one each instance we perform ten runs and calculate averages over these 50 runs. The connectivity is changed from 0.010 to 0.100 by steps of 0.005 over the groups. To characterize better the area of the phase transition, a simplification technique is used introduced by Cheeseman et al in [14]. This three steps node reduction removes the 0.010–0.020 groups, and simplify the graphs in the other groups to get the core of the problems.

## 5.3    Results

In this section, the results of the three kinds of algorithms are presented with and without using the Integer Merge Model, i.e., DSATUR, EA which uses the introduced fitness $f$ and colour choosing heuristics and $EA_{noheur}$ which does not apply these heuristics, it uses a greedy colouring with the fitness $g$. Each algorithm was stopped when they reached an optimal solution or $150\,000\,000$ number of constraint checks. DSatur with backtracking is an exact solver, it tries to explore the search space systematically by its heuristics. Thus, only one run is enough to get its result. Because of the stochastic nature of EAs, we use ten independent runs.

We can summarise the results on test suite one found in Table 1 as follows,

- The performance of an algorithm improves significantly if it employs the IMM framework.
- The evolutionary algorithms perform better than DSATUR, even after improving the efficiency of the latter with IMM.
- Adding heuristics to the evolutionary algorithms is useful to improve upon the efficiency for harder problem instances.
- All algorithms find a solution for almost every problem within the maximum number of constraint checks, except for the extremely hard queen8_8 and r75_5g_8 problems.

**Table 1.** Number of constraint checks required for test suite one using DSATUR and EA with and without IMM (latter is denoted by pure). Ten runs are averaged with different random seeds for EA-s. Prefix-indices show the success ratios if it is not one.

| GRAPH | $|V|$ | $|E|$ | $\chi$ | DSATUR$_{imm}$ | DSATUR$_{pure}$ | EA$_{imm}$ | EA$^{noheur}_{imm}$ | EA$_{pure}$ | EA$^{noheur}_{pure}$ |
|---|---|---|---|---|---|---|---|---|---|
| fpsol2.i.2 | 451 | 8691 | 30 | 3059091 | 40527833 | 3414 | 4541 | 42022 | 56027 |
| fpsol2.i.3 | 425 | 8688 | 30 | 2660498 | 32683629 | 3174 | 4988 | 39151 | 61015 |
| homer | 561 | 1629 | 13 | 2085004 | 75198957 | 2455 | 3672 | 57586 | 171641 |
| inithx.i.1 | 864 | 18707 | 54 | 22305812 | 345876238 | 4328 | 5456 | 120348 | 142315 |
| inithx.i.2 | 645 | 13979 | 31 | 6030391 | 95778467 | 2606 | 3680 | 84603 | 112000 |
| inithx.i.3 | 621 | 13969 | 31 | 5762200 | 86482594 | 2480 | 3804 | 79458 | 124508 |
| miles500 | 128 | 1170 | 20 | 147922 | 1046162 | 9066 | 46276 | 10366 | 75445 |
| miles750 | 128 | 2113 | 31 | 204871 | 1121864 | 120051 | 693403 | 145459 | 5103811 |
| miles1000 | 128 | 3216 | 42 | 244886 | 1249001 | 57934 | 559636 | 116054 | 1120068 |
| miles1500 | 128 | 5198 | 73 | 329361 | 1500956 | 5436 | 14584 | 7032 | 19550 |
| mulsol.i.5 | 186 | 3973 | 31 | 472872 | 2750261 | 1221 | 1370 | 7916 | 8905 |
| myciel6 | 95 | 755 | 7 | 27807 | 624340 | 283 | 331 | 1499 | 2146 |
| myciel7 | 191 | 2360 | 8 | 134956 | 4810974 | 901 | 1350 | 5602 | 11163 |
| queen5_5 | 25 | 160 | 5 | 1665 | 12408 | 678 | 1777 | 906 | 2488 |
| queen7_7 | 49 | 476 | 7 | 1176441 | 9106599 | 1092455 | 6675813 | 2793682 | 25332278 |
| queen8_8 | 64 | 728 | 9 | – | – | $_{0.6}$87482316 | $_{0.4}$102517235 | $_{0.2}$125298157 | – |
| r75_5g_8 | 75 | 1407 | 13 | 35693383 | – | 18668080 | $_{0.2}$122257875 | $_{0.9}$29609833 | $_{0.2}$129031499 |

Figure 6 shows the performance measured by success ratio and by average constraint checks performed of the algorithms on test suite two where 50 independent runs are used for every setting of the density. Both evolutionary algorithms show a sharp dip in the success ratio in the phase transition (see Figure 6), which is accompanied with a rise in the average number of constraint checks. IMM has significant influence on the performance, the improvement lies in between 6 and 48 times on average (see Figure 6). DSatur provides good results on the whole suite. Both the low target colour and the sparsity of the graphs are favourable terms for the heuristics it employs. Furthermore, the order of the graphs does not imply combinatorial difficulties for the backtracking algorithm. Beside these facts, the suite is appropriate to get valuable information about the behaviour of the algorithms. Even if the DSATURS perfom well on the problem sets, the EA, using the IMM abilities, can outperform the pure version of DSATUR in the critical region. In the phase transition it is 50% better on average. In practice, increasing the size of the graph leads to better performance of the EAs as opposed to the two exact DSATUR algorithms. By employing EA heuristics, i.e., the fitness function $f$ and the colour choosing strategy, we clearly notice an improvement in both efficiency and effectiveness over the simple greedy colouring strategy with the simple fitness $g$. Furthermore, the confidence intervals for this range are small and non-overlapping. These two approaches give a much robust algorithm for solving graph $k$-colouring.

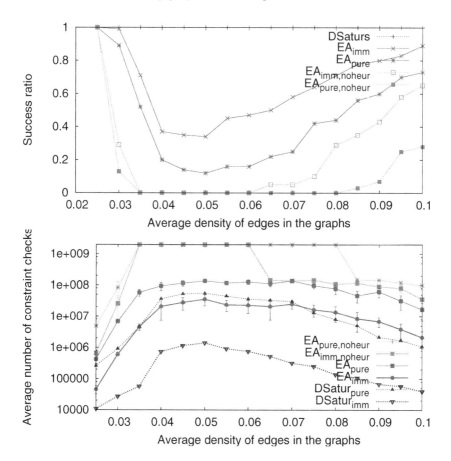

**Fig. 6.** Success ratio (DSATUR is always one) and average constraint checks to a solution for the DSATUR variants, the EAs with and without heuristics (with 95% confidence intervals)

## 6   Conclusions

In this paper, we introduced the Integer Merge Model for representing graph colouring problems. It forms a good basis for developing efficient graph colouring algorithms because of its three beneficial properties, a significant reduction in constraint checks, the availability of useful information for guiding heuristics, and the compact description possible for algorithms.

We showed how the popular DSATUR can be described in terms of the Integer Merge Model and we empirically investigated how much it can benefit from the reduction in constraint checks. Similarly, we showed how an evolutionary algorithm can be made more effective by adding heuristics that rely on the Integer Merge Model. Here we have shown a significant increase in both effectiveness and effectivity.

Further studies may include incorporating the Integer Merge Model in other algorithms, including more heuristics. Also, other constraint problems may be considered.

# References

1. de Werra, D.: An introduction to timetabling. European Journal of Operations Research **19** (1985) 151–162
2. Briggs, P.: Register allocation via graph coloring. Technical Report TR92-183 (1998)
3. Garey, M., Johnson, D., So, H.: An application of graph colouring to printed circuit testing. IEEE Transaction on Circuits and Systems **CAS-23** (1976) 591–599
4. Garey, M., Johnson, D.: Computers and Instractability: A Guide to the Theory of NP-Completeness. W.H. Freeman, San Francisco, CA (1979)
5. Craenen, B., Eiben, A., van Hemert, J.: Comparing evolutionary algorithms on binary constraint satisfaction problems. IEEE Transactions on Evolutionary Computation **7** (2003) 424–444
6. Juhos, I., Tóth, A., van Hemert, J.: Binary merge model representation of the graph colouring problem. In Raidl, G., Gottlieb, J., eds.: Evolutionary Computation in Combinatorial Optimization. Volume 3004 of LNCS., Springer (2004) 124–134
7. Juhos, I., Tóth, A., van Hemert, J.: Heuristic colour assignment strategies for merge models in graph colouring. In: Evolutionary Computation in Combinatorial Optimization. Volume 3448 of LNCS., Springer (2005) 132–143
8. Brélaz, D.: New methods to color the vertices of a graph. Communications of the ACM **22** (1979) 251–256
9. CNET: Xbox specs revealed (http://cnet.com/xbox+specs+revealed/2100-1043_3-5705372.html) (2005) Accessed: Nov. 10, 2005.
10. Golomb, S., Baumert, L.: Backtrack programming. ACM **12** (1965) 516–524
11. Bäck, T., Fogel, D., Michalewicz, Z., eds.: Handbook of Evolutionary Computation. Institute of Physics Publishing Ltd, Oxford University Press (1997)
12. Johnson, D., Trick, M.: Cliques, Coloring, and Satisfiability. American Mathematical Society, DIMACS (1996)
13. Culberson, J.: Iterated greedy graph coloring and the difficulty landscape. Technical Report TR 92-07, University of Alberta, Dept. of Computing Science (1992)
14. Cheeseman, P., Kenefsky, B., Taylor, W.: Where the really hard problems are. In: Proceedings of the IJCAI'91. (1991) 331–337

# Minimizing Makespan on a Single Batch Processing Machine with Non-identical Job Sizes: A Hybrid Genetic Approach

Ali Husseinzadeh Kashan[1], Behrooz Karimi[1,*], and Fariborz Jolai[2]

[1] Department of Industrial Engineering, Amirkabir University of Technology,
P.O.Box: 15875-4413, Tehran, Iran
{a.kashani, B.Karimi}@aut.ac.ir
[2] Department of Industrial Engineering, Faculty of Engineering, University of Tehran,
P.O.Box: 11365-4563, Tehran, Iran
fjolai@ut.ac.ir

**Abstract.** This paper addresses minimizing makespan by genetic algorithm (GA) for scheduling jobs with non-identical sizes on a single batch processing machine. We propose two different genetic algorithms based on different encoding schemes. The first one is a sequence based GA (SGA) that generates random sequences of jobs and applies the batch first fit (BFF) heuristic to group the jobs. The second one is a batch based hybrid GA (BHGA) that generates random batches of jobs and ensures feasibility through using knowledge of the problem. A pairwise swapping heuristic (PSH) based on the problem characteristics is hybridized with BHGA that has the ability of steering efficiently the search toward the optimal or near optimal schedules. Computational results show that BHGA performs considerably well compared with a modified lower bound and significantly outperforms the SGA and a simulated annealing (SA) approach addressed in literature. In comparison with a constructive heuristic named FFLPT, BHGA also shows its superiority.

## 1 Introduction

Batch processing machines (BPM) are encountered in many different environments. This research is motivated by burn-in operations in semiconductor manufacturing [1]. The purpose of burn-in operations is to test the integrated circuits by subjecting them to thermal stress for an extended period. Since the burn-in operation constitutes a bottleneck in the final testing operation, efficient scheduling of this operation to maximize throughput is of great concern to management. In scheduling problems, makespan ($C_{max}$) is equivalent to the completion time of the last job leaving the system. The minimum $C_{max}$ usually implies a high utilization. The utilization for bottleneck station is closely related to throughput rate of the system. Therefore, reducing the $C_{max}$ should also lead to a higher throughput rate. Te following assumptions are considered to our problem:

---

* Corresponding author.

J. Gottlieb and G.R. Raidl (Eds.): EvoCOP 2006, LNCS 3906, pp. 135 – 146, 2006.

(1) There are $n$ jobs to be processed. The minimum required processing time of job $i$ is denoted by $t_i$.

(2) Each machine has a capacity $B$ and each job $i$ has a size $s_i$. The sum of the sizes of jobs in a batch must be less than or equal to $B$. We assume that no job has a size exceeding the machine capacity.

(3) Once processing of a batch is initiated, it cannot be interrupted and other jobs cannot be introduced into the machine until processing is completed. The processing time of batch $j$, $T_j$, is given by the longest processing time of the jobs in the batch.

(4) The objective is minimizing the makespan, $C_{max}$.

Based on the cited assumptions we can get the following optimization model related to our problem:

$$Minimize \quad C_{max} = \sum_{j=1}^{k} T_j \tag{1}$$

$$St: \quad \sum_{j=1}^{k} X_{ij} = 1 \quad i = 1, \cdots, n \tag{2}$$

$$\sum_{i=1}^{n} s_i X_{ij} \leq B \quad j = 1, \cdots, k \tag{3}$$

$$T_j \geq t_i X_{ij} \quad i = 1, \cdots, n \quad j = 1, \cdots, k \tag{4}$$

$$X_{ij} \in \{0,1\} \quad i = 1, \cdots, n \quad j = 1, \cdots, k \tag{5}$$

$$T_j \geq 0, \tag{6}$$

$$\left\lceil \sum_{i=1}^{n} s_i / B \right\rceil \leq k \leq n, \quad k : \text{integer} \tag{7}$$

The objective is minimizing the makespan. Constraint set (2) ensures the assignment of each job $i$ to only one batch $j$. Constraint set (3) indicates that sum of the sizes of jobs in a batch must be less than or equal to machine capacity. Constraint set (4) gives processing time of batch $j$. Constraint sets (5) and (6) denote the binary restriction of the variables $X_{ij}$ and the non-negativity restriction for variables $T_j$, respectively. The last constraint denotes the minimum and the maximum number of required batches to group all jobs.

The arrangement of the paper is as follows. In Section 2, we review works related to batch processing models. Section 3, reviews the batch first fit heuristic. In this section we also develop a heuristic approach for the considered problem. GAs implementation and parameters setting are described in Section 4. In Section 5, the performance of proposed GAs is evaluated. We conclude the paper with a summary in Section 6.

## 2 Literature Review

Batch machine scheduling problems, where the batch processing time is given by the processing time of the longest job in the batch have been addressed extensively.

Given different size for jobs, Uzsoy [1] gave complexity results for the minimizing makespan ($C_{max}$) and total completion time ($\Sigma C_i$) criteria and provided some heuristics and a branch and bound algorithm. Dupont and Jolai [2] proposed heuristics to minimize $C_{max}$. In another research Jolai and Dupont [3] considered $\Sigma C_i$ criterion for the same problem. A branch-and-bound procedure for minimizing the $C_{max}$ was also developed by Dupont and Dhaenens-Flipo [4]. Melouk, et al., [5] developed a simulated annealing (SA) approach for minimizing $C_{max}$ to schedule a single BPM with different job sizes. We use their proposed SA as a base algorithm in our computational experiments. Shuguang et al. [6] recently presented an approximation algorithm with worst-case ratio 2+ε, for minimizing the makespan.

# 3   Heuristics

Uzsoy [1] constructed a set of heuristics based on the First-Fit procedure developed for the bin-packing problem. The batch first fit (BFF) heuristic adapted to the scheduling problem is as follows:

*Step 1*. Arrange the jobs in some arbitrary order.
*Step 2*. Select the job at the head of the list and place it in the first batch with enough space to accommodate it. If it fits in no existing batch, create a new batch. Repeat step 2 until all jobs have been assigned to a batch.
*Step 3*. Sequence the batches on the machine in any arbitrary order.

It has been shown that if in the BFF the sorting of jobs is in LPT order at the first step, then the algorithm will be superior in both the average and the worst-case performance comparing to non-LPT based BFF algorithms [1]. In this case it is called FFLPT algorithm.

   Based on our encoding scheme used for the BHGA, we develop a heuristic procedure named random batches procedure (RBP), which gives a feasible batching scheme by considering both size and processing time of jobs. The heuristic is as follows:

*Step1*. Assign all jobs to $L$ batches, randomly without considering the capacity restriction. If the derived batching scheme is feasible, stop. Otherwise go to step 2.
*Step2*. Choose a batch with capacity violation and the longest batch processing time. Select the job with the longest processing time in it. Put the selected job in the first feasible batch having the batch processing time longer than processing time of the selected job with the smallest residual capacity, then go to step 3. If there is no such a feasible batch with longer processing time, put the job in the feasible batch with longest processing time and go to step 3. If the selected job fits in no existing batches, create a new batch and assign the job to it.
*Step3*. Repeat step 2 until getting a feasible batching scheme.

The above procedure tries to group jobs with longer processing time in the same batches and simultaneously minimizing the residual capacity of batches.

   Uzsoy [1] proposed a lower bound on the optimal makespan by relaxing the problem to allow jobs to be split and processed in different batches. When there are some jobs that cannot be grouped with any other jobs in a same batch, we modify the lower bound as follows:

1) Put the jobs satisfying following relation in the set $J$, and remove them from the set of whole jobs.

$$J = \{k \mid B - s_k < \min_{i \in \{1,...,n\}} \{s_i\}\}. \tag{8}$$

2) For the reduced problem, construct an instance of $C_{max}$ ($C_{max}^{LB}$) with unit job sizes where each job $m$ is replaced by $s_m$ jobs of unit size and processing time $t_m$. This can be solved by ordering jobs in decreasing order of processing times, successively grouping the $B$ jobs with longest processing times into the same batch.

3) The modified lower bound, $C_{max}^{MLB}$, can be obtained as: $C_{max}^{MLB} = \sum_{i \in J} t_i + C_{max}^{LB}$.

To show the computation of the lower bound solution, we use a test problem whose data are shown in Table 1.

$B = 10$;

**Table 1.** Data for the test problem

| Job $i$ | 1 | 2 | 3 | 4 | 5 | 6 | 7 |
|---------|----|----|----|----|----|----|----|
| $t_i$   | 10 | 3  | 5  | 2  | 6  | 3  | 1  |
| $s_i$   | 7  | 5  | 3  | 9  | 6  | 4  | 8  |

1) Since the minimum size is equal to 3, then jobs 4 and 7 cannot be grouped by any other job in a same batch. So we set:
   $J = \{4,7\}$, batch $1 = \{4\}$, batch $2 = \{7\}$.
2) For the remain jobs, sorting the jobs in decreasing order of their processing times and successively grouping the jobs which are allowed to be split across the batches, we get the following batching scheme:
   batch $3 = \{1,5\}$, batch $4 = \{5,3,2\}$, batch $5 = \{2,6\}$;
   $T_3 = \max\{10,6\} = 10$, $T_4 = \max\{6,5,3\} = 6$, $T_5 = \max\{3,3\} = 3$;
3) The lower bound solution is: $C_{max}^{MLB} = 2+1+10+6+3 = 22$.

## 4   Proposed GAs and Their Implementation

Genetic algorithms are powerful and broadly applicable stochastic search and optimization techniques based on principles of evolution theory. In the past few years, genetic algorithms have received considerable attention for solving difficult combinatorial optimization problems.

   This paper proposes two different GAs based on different chromosome representations for scheduling a single batch processing machine. The first one is SGA that searches the solutions space via generating random sequences of jobs to obtain sequences that yield the better $C_{max}$. For each generated sequence, BFF heuristic is applied to batch the jobs in an order similar to the corresponding sequence. In the second GA entitled BHGA, searching the solutions space is done by generating random feasible batches of jobs. A pairwise swapping heuristic is used in BHGA to improve its performance. The following steps describe in detail how the GAs are implemented in this research.

## 4.1  Coding

In our coding schemes, each gene corresponds to one of $n$ jobs. In SGA, each chromosome is related to a sequence of jobs. To generate the chromosomes we have used random key representation of Bean [7], which represents order of jobs with a sequence of uniformly distributed random numbers. A simple example of using random key representation for a batch-scheduling problem can be found in Wang and Uzsoy [8]. In BHGA, a solution for the problem of assigning jobs to batches is an array whose length is equal to the number of jobs to be scheduled. Each gene represents the batch to which the job is assigned. Figures 1 and 2 show the chromosomes related to GA heuristics for a batch-scheduling problem with 5 jobs.

| $J_3$ | $J_5$ | $J_4$ | $J_2$ | $J_1$ |
|-------|-------|-------|-------|-------|
| 0.461 | 0.106 | 0.281 | 0.514 | 0.515 |

**Fig. 1.** Chromosome representation for SGA

| $J_1$ | $J_2$ | $J_3$ | $J_4$ | $J_5$ |
|-------|-------|-------|-------|-------|
| 1 | 4 | 2 | 3 | 4 |

**Fig. 2.** Chromosome representation for BHGA

## 4.2. Initial Population

The initial population for SGA is constructed randomly, where the initial population for BHGA is generated by RBP. The performance of RBP is directly dependent on the number of initial batches. This implies that starting RBP with large number of initial batches would be undesirable. However, starting RBP with small number of initial batches to form all chromosomes of BHGA's initial population may cause to trap in locality. To come up with poor quality of initial seeds and to avoid trapping in locality, we use a truncated geometric distribution to generate randomly the number of initial batches used in RBP to form a feasible initial population of chromosomes. Using a geometric distribution to simulate the number of initial batches ensures that the probability of starting RBP with large number of batches would be small, against the high probability for starting with less number of batches. The following relation gives the random number of initial batches to start RBP for constructing the chromosomes of initial population of BHGA.

$$L = \left\lceil \frac{\ln(1-(1-(1-p)^{l-1})R)}{\ln(1-p)} \right\rceil + 1 \quad : L \in \{2,3,\cdots,l\}. \tag{9}$$

where $L$ is a random number of initial batches for starting RBP, distributed by a truncated geometric distribution, $R$ is a uniformly distributed variable belonging to [0,1], $p$ is the probability of success, and $l$, is the minimum number of required batches, $l = \left\lceil \sum_{i=1}^{n} s_i \middle/ B \right\rceil$. As stated before, scheduling jobs with RBP, starting from one batch is almost equivalent to scheduling jobs with FFLPT heuristic; so for fair comparison between SGA and BHGA, we start RBP with more than one batch.

The above approach for generating the initial population in BHGA has some advantages, since it uses efficiently the knowledge of the problem to group the jobs with longer processing times in the same batches as long as possible. However this is not the case with SGA through using the random key representation. Also one can adjust the quality of initial population by choosing a proper value of $p$. Setting a high value for $p$, seems to be effective on accelerating the convergence rate trough a good quality initial seeds but may cause trapping in local optimum. Low value of $p$ causes to less-accelerate converging due to the relatively low quality of initial population. So a tuned value of $p$ would be drastic to exploit advantages of such an intelligent mechanism of generating initial population.

### 4.3  Selection

For selecting chromosomes, the roulette wheel technique is followed. The fitness of the chromosome is the makespan value. The probability of selecting a certain chromosome is proportional to its fitness.

### 4.4  Crossover

A parameterized uniform crossover of Bean [7] is used in which, a bias coin is used to determine which one of the two selected parents should pass its gene. Using parameterized uniform crossover ensures the feasibility of the generated offspring to be still a sequence and allows us to bias the search more strongly toward components of better solutions and accelerate convergence. A simple example of using parameterized uniform crossover for a batch-scheduling problem can be found in Wang and Uzsoy [8]. It should be noted that in BHGA we don't use random key representation because in this case the chromosomes are not represented as sequences. For a fair comparison between BHGA and SGA, we extend the use of parameterized uniform crossover to BHGA. Our computational results report the good performance of applying parameterized uniform crossover on BHGA. In crossover stage of BHGA it may happen to lose some of batches in the generated offspring. In this case we rename the batches of the offspring in consecutive order. For any infeasible generated offspring, the RBP is applied to ensure the feasibility. In this case each generated offspring indicates an initial batching plan, which is infeasible. To ensure the feasibility we apply steps 2 and 3 of RBP to get a feasible batching scheme.

### 4.5  Mutation

The swapping mutation is used as mutation operator for both SGA and BHGA. It selects two random genes of a selected chromosome and swaps them with each other. The feasibility of the mutated schedule is kept through using steps 2 and 3 of RBP.

## 4.6  Pairwise Swapping Heuristic (PSH)

For BHGA, we developed an iterative pairwise swapping heuristic (PSH), which reduces $C_{max}$ value of a given schedule by consecutively swapping jobs between batches as much as possible. The PSH adapted to BHGA is as follows:

For any selected schedule while there is not any improvement in $C_{max}$ do:

*Step1.* List the batches that have only one job with maximum processing time. If the list is empty, consider the current schedule as PSH offspring and stop. Otherwise, consider the first batch in the list (call it $a$) and find a batch with batch processing time equal or longer than the processing time of the considered batch (call it $b$). Swap the job with longest processing time in $a$, ( $j^a$ ), with a job in $b$, ( $j^b$ ), which has the processing time less than processing time of job $j^a$, with respect to capacity violation constraint. If there is not any batch $b$, or there is not any job $j^b$, consider the current schedule as PSH offspring corresponding to the batch $a$. Repeat step 1 for all batches in the list.

*Step2.* Choose a schedule with better $C_{max}$ among the schedules of step 1 and return to step 1.

In each generation, for all problem categories (after deleting repeated chromosomes) the PSH is applied on the best chromosomes selected based on a defined probability, i.e. $P_{PSH.}$

In each generation of both SGA and BHGA, the new population is formed by offspring generated by GA operators. The rest of the population is filled with the best chromosomes of the former generation.

## 4.7  Parameters Setting

The performance of a GA is generally sensitive to the setting of the parameters that influence the search behavior and the quality of convergence. It is highly desirable to set these parameters to levels that produce high quality solutions.

For both SGA and BHGA, we consider the population size equal to $n$ and the number of simulated generations equal to 200. Also both GAs are stopped if there is no improvement in the best solution obtained in last 100 generations. Worth to mention that to avoid useless computations, we set an extra stopping criterion, which is getting to the modified lower bound solution. For BHGA, based on our primary experiments we found 0.2 to be appropriate value for the truncated geometric distribution parameter (probability of success), for all problem instances.

For tuning crossover rate, mutation rate and head probability, different values are considered; 0.5 and 0.7 for crossover rate, 0.05, 0.1 and 0.15 for mutation rate and 0.7 and 0.9 for head probability. Table 2 shows the results of tuning. Moreover in BHGA the value for $P_{PSH}$ is considered equal to 0.25 for all problem categories. We found this value by our primary experiments.

**Table 2.** Parameter setting for GA's operators

| Parameters | SMOGA | BHMOGA |
|---|---|---|
| Crossover rate | 0.70 | 0.50 |
| Mutation rate | 0.15 | 0.15 |
| Head probability | 0.90 | 0.90 |
| $P_{PSH}$ | - | 0.25 |
| Truncated geometric parameter | - | 0.20 |

## 5  Experimentations and Results

For testing effectiveness of proposed GAs, randomly test problem instances are generated in a manner similar to Melouk et al. [5]. For covering various types of problems, three factors are identified: number of jobs, variation in job processing times, and variation in job sizes. In general 6 categories of problems are generated by combining three levels of job sizes (small, large and combination of small and large (mixed)) and two levels of processing times. Processing times and job sizes are generated from discrete uniform distribution. The factors, their levels and ranges, are shown in Table 3. For example a problem category with 10 jobs, processing times generated from [1, 20] and job sizes generated from [4, 8] is shown by J1t2s3. The machine capacity is assumed to be 10 in all instances.

All the algorithms, SGA, BHGA, SA, FFLPT and the modified lower bound, are coded in MATLAB 6.5.1 and all test problem instances are solved 15 times by GAs and SA on a Pentium III, 800MGz computer with 128 MB RAM.

Table 4 represents the results for 10 jobs instances. Columns 2 and 4, report the best and worst $C_{max}$ among the 15 runs of BHGA. Column 3 and 5 report the average $C_{max}$ and the standard deviation among the solutions gained by 15 runs of each problem category by BHGA, respectively. Column 6 reports the average time taken by BHGA. These values are reported in columns 7, 8, 9, 10 and 11 for the SGA and 12, 13, 14, 15 and 16 for the SA, respectively. Column 17 reports the performance of the FFLPT. Tables 5, 6 and 7 can be interpreted in a same manner. For each algorithm, the best case, average and the worst case performance are computed by the following relation:

$$\{ [C_{\max} - C_{max}^{MLB} ] / C_{\max}^{MLB} \} * 100 . \tag{10}$$

**Table 3.** Factors and levels

| Factors | Levels |
|---|---|
| $|J|$ | 10, 20, 50 and 100 jobs (J1-J4) |
| $t_i$ | t1: Discrete uniform [1, 10] |
|  | t2: Discrete uniform [1, 20] |
| $s_i$ | s1: Combination of small and large jobs (Discrete uniform [1, 10]) |
|  | s2: Small jobs (Discrete uniform [2, 4]) |
|  | s3: Large jobs (Discrete uniform [4,8]) |

**Table 4.** Results for 10 job instances

| Run code | BHGA | | | | | SGA | | | | | SA | | | | | FFLPT |
|---|---|---|---|---|---|---|---|---|---|---|---|---|---|---|---|---|
| | Best | Avg | Worst | S.D | Time | Best | Avg | Worst | S.D | Time | Best | Avg | Worst | S.D | Time | |
| J1t1s1 | 0.00 | 0.00 | 0.00 | 0.00 | 0.15 | 0.00 | 1.74 | 7.40 | 0.63 | 1.64 | 0.00 | 10.85 | 18.51 | 1.94 | 2.72 | 0.00 |
| J1t1s2 | 0.00 | 0.00 | 0.00 | 0.00 | 0.29 | 0.00 | 3.47 | 13.04 | 0.86 | 2.23 | 0.00 | 2.60 | 8.69 | 0.91 | 0.72 | 0.00 |
| J1t1s3 | 0.00 | 0.00 | 0.00 | 0.00 | 0.76 | 0.00 | 0.42 | 2.12 | 0.41 | 0.94 | 0.00 | 1.12 | 6.38 | 0.83 | 2.29 | 2.12 |
| J1t2s1 | 0.00 | 0.00 | 0.00 | 0.00 | 0.92 | 0.00 | 1.15 | 4.34 | 0.63 | 1.74 | 0.00 | 5.50 | 10.86 | 2.29 | 2.28 | 4.34 |
| J1t2s2 | 0.00 | 0.00 | 0.00 | 0.00 | 0.24 | 0.00 | 3.80 | 11.42 | 1.95 | 1.42 | 0.00 | 7.42 | 20.00 | 2.32 | 0.86 | 0.00 |
| J1t2s3 | 0.00 | 0.00 | 0.00 | 0.00 | 0.15 | 0.00 | 0.00 | 0.00 | 0.00 | 0.24 | 0.00 | 1.05 | 5.26 | 1.64 | 0.84 | 0.00 |

**Table 5.** Results for 20 job instances

| Run code | BHGA | | | | | SGA | | | | | SA | | | | | FFLPT |
|---|---|---|---|---|---|---|---|---|---|---|---|---|---|---|---|---|
| | Best | Avg | Worst | S.D | Time | Best | Avg | Worst | S.D | Time | Best | Avg | Worst | S.D | Time | |
| J2t1s1 | 1.56 | 1.56 | 1.56 | 0.00 | 13.40 | 1.56 | 3.75 | 7.81 | 1.05 | 11.61 | 1.56 | 8.23 | 17.18 | 2.85 | 16.97 | 7.81 |
| J2t1s2 | 0.00 | 1.23 | 4.65 | 0.91 | 3.62 | 0.00 | 7.27 | 9.30 | 1.40 | 14.90 | 0.00 | 7.74 | 13.95 | 1.54 | 9.02 | 4.65 |
| J2t1s3 | 0.00 | 0.00 | 0.00 | 0.00 | 0.89 | 0.00 | 1.11 | 4.16 | 0.94 | 8.71 | 0.00 | 3.88 | 8.33 | 2.14 | 19.61 | 5.56 |
| J2t2s1 | 1.86 | 1.86 | 1.86 | 0.00 | 12.97 | 1.86 | 4.11 | 9.34 | 2.79 | 14.15 | 2.80 | 5.14 | 7.47 | 1.30 | 19.31 | 1.86 |
| J2t2s2 | 0.00 | 0.80 | 1.51 | 0.51 | 6.61 | 0.00 | 5.15 | 9.09 | 2.29 | 18.23 | 0.00 | 6.86 | 15.15 | 2.47 | 8.39 | 1.51 |
| J2t2s3 | 0.00 | 0.00 | 0.00 | 0.00 | 2.06 | 0.00 | 1.20 | 3.47 | 1.51 | 11.041 | 0.00 | 1.94 | 3.47 | 1.82 | 19.41 | 8.33 |

**Table 6.** Results for 50 job instances

| Run code | BHGA | | | | | SGA | | | | | SA | | | | | FFLPT |
|---|---|---|---|---|---|---|---|---|---|---|---|---|---|---|---|---|
| | Best | Avg | Worst | S.D | Time | Best | Avg | Worst | S.D | Time | Best | Avg | Worst | S.D | Time | |
| J3t1s1 | 2.02 | 2.09 | 2.70 | 0.25 | 128 | 5.40 | 7.43 | 11.48 | 2.44 | 165.21 | 8.10 | 10.74 | 13.51 | 2.25 | 253.5 | 5.40 |
| J3t1s2 | 0.00 | 0.23 | 1.17 | 0.41 | 27.68 | 7.05 | 11.21 | 16.47 | 2.53 | 148.79 | 15.29 | 17.17 | 22.35 | 1.45 | 62.41 | 0.00 |
| J3t1s3 | 0.55 | 0.55 | 0.55 | 0.00 | 102.9 | 1.67 | 2.75 | 4.46 | 1.57 | 178.56 | 4.46 | 6.20 | 8.93 | 2.32 | 178.5 | 1.67 |
| J3t2s1 | 2.60 | 3.02 | 3.25 | 0.96 | 158.9 | 5.86 | 7.38 | 11.07 | 4.15 | 156.4 | 7.49 | 10.26 | 13.68 | 5.18 | 226.3 | 3.58 |
| J3t2s2 | 0.00 | 0.36 | 1.82 | 1.05 | 30.47 | 10.36 | 12.62 | 17.68 | 3.15 | 138.8 | 13.41 | 17.31 | 21.95 | 4.01 | 64.3 | 1.82 |
| J3t2s3 | 1.66 | 1.69 | 2.22 | 0.51 | 138.51 | 2.50 | 4.36 | 6.11 | 4.18 | 186.57 | 3.05 | 6.08 | 11.38 | 7.41 | 212.4 | 3.33 |

**Table 7.** Results for 100 job instances

| Run code | BHGA | | | | | SGA | | | | | SA | | | | | FFLPT |
|---|---|---|---|---|---|---|---|---|---|---|---|---|---|---|---|---|
| | Best | Avg | Worst | S.D | Time | Best | Avg | Worst | S.D | Time | Best | Avg | Worst | S.D | Time | |
| J4t1s1 | 1.17 | 1.72 | 2.35 | 0.73 | 1182.3 | 7.45 | 9.69 | 13.33 | 4.28 | 935.5 | 10.98 | 13.56 | 16.47 | 4.68 | 1303.2 | 4.31 |
| J4t1s2 | 0.00 | 0.12 | 0.60 | 0.41 | 137.9 | 13.85 | 16.44 | 18.67 | 2.55 | 771.3 | 19.87 | 23.25 | 25.90 | 2.74 | 320.3 | 2.41 |
| J4t1s3 | 1.04 | 1.04 | 1.04 | 0.00 | 568.3 | 2.62 | 3.28 | 3.93 | 1.55 | 1235.5 | 5.77 | 6.69 | 9.18 | 3.94 | 1087.3 | 2.62 |
| J4t2s1 | 1.23 | 1.74 | 2.29 | 1.88 | 1709.4 | 5.65 | 7.46 | 9.54 | 6.65 | 968.4 | 9.54 | 11.25 | 12.19 | 4.31 | 1339.8 | 3.35 |
| J4t2s2 | 0.00 | 0.09 | 0.32 | 0.48 | 151.9 | 14.46 | 17.91 | 20.90 | 5.02 | 745.4 | 19.93 | 23.47 | 25.08 | 4.62 | 291.2 | 1.28 |
| J4t2s3 | 0.96 | 1.04 | 1.24 | 0.91 | 912.7 | 2.89 | 4.15 | 5.10 | 4.80 | 1188.9 | 6.89 | 8.62 | 11.31 | 9.26 | 1066.3 | 1.51 |

**Table 8.** Comparisons for 10 job instances

| Run code | BHGA-SGA | BHGA-SA | BHGA-FFLPT |
|---|---|---|---|
| J1t1s1 | 1.71 | 9.78 | 0.00 |
| J1t1s2 | 3.36 | 2.54 | 0.00 |
| J1t1s3 | 0.42 | 1.11 | 2.08 |
| J1t2s1 | 1.13 | 5.21 | 4.16 |
| J1t2s2 | 3.66 | 6.91 | 0.00 |
| J1t2s3 | 0.00 | 1.04 | 0.00 |
| Average | 1.71 | 4.43 | 1.04 |

**Table 9.** Comparisons for 20 job instances

| Run code | BHGA-SGA | BHGA-SA | BHGA-FFLPT |
|---|---|---|---|
| J2t1s1 | 2.10 | 6.16 | 5.79 |
| J2t1s2 | 5.63 | 6.04 | 3.26 |
| J2t1s3 | 1.09 | 3.74 | 5.26 |
| J2t2s1 | 2.15 | 3.11 | 0.00 |
| J2t2s2 | 4.13 | 5.67 | 0.70 |
| J2t2s3 | 1.18 | 1.90 | 7.69 |
| Average | 2.71 | 4.43 | 3.78 |

**Table 10.** Comparisons for 50 job instances

| Run code | BHGA-SGA | BHGA-SA | BHGA-FFLPT |
|---|---|---|---|
| J3t1s1 | 4.96 | 7.80 | 3.14 |
| J3t1s2 | 9.86 | 14.45 | -0.23 |
| J3t1s3 | 2.13 | 5.31 | 1.09 |
| J3t2s1 | 4.05 | 6.55 | 0.53 |
| J3t2s2 | 10.88 | 14.44 | 1.43 |
| J3t2s3 | 2.56 | 4.13 | 1.58 |
| Average | 5.74 | 8.78 | 1.25 |

**Table 11.** Comparisons for 100 job instances

| Run code | BHGA-SGA | BHGA-SA | BHGA-FFLPT |
|---|---|---|---|
| J4t1s1 | 7.26 | 10.42 | 2.48 |
| J4t1s2 | 14.01 | 18.76 | 2.23 |
| J4t1s3 | 2.16 | 5.28 | 1.53 |
| J4t2s1 | 5.32 | 8.54 | 1.55 |
| J4t2s2 | 15.11 | 18.93 | 1.17 |
| J4t2s3 | 2.98 | 6.97 | 0.46 |
| Average | 7.80 | 11.48 | 1.57 |

where the term $C_{max}$ denotes the performance for each algorithm. Tables 8, 9, 10 and 11 compare BHGA versus SGA, SA and FFLPT in term of $C_{max}$ value differences for the problems with 10, 20, 50 and 100 jobs, respectively. The differences are given by:

$$\{[C_{max}(A)-C_{max}(BHGA)]/\max\{C_{max}(A), C_{max}(BHGA)\}\}*100 .  \quad (11)$$

where, $C_{max}(A)$ denotes the $C_{max}$ for the SGA or SA or FFLPT comparators. The positive differences imply that BHGA works better than the others.

Computational analysis shows that in all categories of test problem instances, BHGA performs significantly better than SGA and SA, especially for the problems with large number of jobs. Concerning to increasing trend in the reported objective value differences, it is promising for BHGA to outperform SGA and SA as the problem size increase.

The resulting tables show that, for all problems with small jobs (J.t.s2), BHGA achieved the optimal makespan. As was mentioned before, for these problems, jobs swapping can be more easily handled by the PSH, so converging to the optimal makespan may be best achieved through the hybridizing PSH with BHGA. However for problems with small jobs both SGA and SA report the worst performance. For the problems with mixed size jobs (J.t.s1) results show the better performance for BHGA compared to SGA and SA. In this case, although the objective value differences are not as large as the case of problems with small jobs, they are still meaningfully large, especially for larger problems. In this case because of the diversity in the job sizes, effective use of the batch capacity in assigning the jobs to batches is an important factor that is satisfied by RBP for the BHGA. The smallest objective value difference exists for the problems with large size jobs (J.t.s3). It is expected that for these problems, 40% of individual jobs lie in exactly one batch (the jobs with sizes 7 and 8) and yet the 60% of jobs need to be assigned efficiently to batches to decrease the makespan. This caused to reduction in the objective value differences among the BHGA and both SGA and SA. The increment in the difference values is more likely as the problem size increases. As it is clear from the results, by reporting smaller standard deviations, BHGA is also more reliable than both SGA and SA.

Compared to FFLPT, results show that BHGA performs better. However in this case the differences are not as large as the case with SGA and SA.

Comparing the required running time for the algorithms, results show that in most of test problem instances the average time needed for BHGA is less than SGA and SA, especially for problems with small and large jobs. This can be related to the high rate of convergence in BHGA for the relatively good quality of initial population and the effective performance of PSH. In BHGA using RBP to get feasibility for a schedule, generally it takes less time than batching a sequence of jobs using BFF heuristic in SGA. This means superiority for BHGA compared to SGA. In SGA for each generated offspring, BFF is applied to assign all $n$ jobs to batches, while in BHGA when an offspring with $m$ batches is formed, to make it feasible, $n$-$m$ jobs in worst case must be reassigned to batches by RBP; this is because at least $m$ jobs do not need to be reassigned in batches.

## 6  Conclusions and Future Research Directions

In this paper we considered the problem of minimizing makespan on a single batch processing machine with non-identical job sizes. The processing time of a batch is given by the longest processing time of the jobs in the batch. We proposed two different genetic algorithms based on different chromosome representations. Our computational results show that our second GA entitled BHGA outperforms the first GA named SGA and also outperforms the simulated annealing approach taken from the literature as a base algorithm, especially for large-scale problems. In most cases this algorithm has the ability to find optimal or near optimal solution(s) in reasonable CPU times. Also its superiority compared to the constructive FFLPT algorithm and its very good performance comparing to a modified lower bound was proved. Some characteristics such as using a robust mechanism for generating initial population and using an efficient pairwise swapping heuristic, which has the ability of steering quickly the search toward the optimal solution, cause BHGA to dominate the others.

For future research, the extension of our approach could include due date related performances and also consider dynamic job arrivals and incompatible families.

## References

1. Uzsoy, R.: A single batch processing machine with non-identical job sizes. International Journal of Production Research. 32 (1994) 1615-1635
2. Dupont, L., Jolai Ghazvini, F., Minimizing makespan on a single batch processing machine with non identical job sizes. European journal of Automation Systems, 1998, 32, 431-440.
3. Jolai Ghazvini, F., Dupont, L.: Minimizing mean flow time on a single batch processing machine with non- identical job size. International Journal of Production Economics. 55 (1998) 273-280.
4. Dupont, L., Dhaenens-Flipo, C.: Minimizing the makespan on a batch machine with non-identical job sizes: an exact procedure. Computers & Operations Research. 29 (2002) 807-819.
5. Melouk, S., Damodaran, P., Chang, P.Y.: Minimizing makespan for single machine batch processing with non-identical job sizes using simulated annealing. International Journal of Production Economics. 87 (2004) 141-147.
6. Shuguang, L., Guojun, L., Xiaoli, W., Qiming, L.: Minimizing makespan on a single batching machine with release times and non-identical job sizes. Operations Research Letter. 33 (2005) 157-164.
7. Bean, J.C.: Genetic algorithms and random keys for sequencing and optimization. ORSA Journal of Computing. 6 (1994) 154-160.
8. Wang, C., Uzsoy, R.: A genetic algorithm to minimize maximum lateness on a batch processing machine. Computers & Operations Research. 29 (2002) 1621-1640.

# A Relation-Algebraic View on Evolutionary Algorithms for Some Graph Problems

Britta Kehden and Frank Neumann

Inst. für Informatik und Prakt. Mathematik,
Christian-Albrechts-Univ. zu Kiel, 24098 Kiel, Germany
{bk, fne}@informatik.uni-kiel.de

**Abstract.** We take a relation-algebraic view on the formulation of evolutionary algorithms in discrete search spaces. First, we show how individuals and populations can be represented as relations and formulate some standard mutation and crossover operators for this representation using relation-algebra. Evaluating a population with respect to their constraints seems to be the most costly step in one generation for many important problems. We show that the evaluation process for a given population can be sped up by using relation-algebraic expressions in the process. This is done by examining the evaluation of possible solutions for three of the best-known NP-hard combinatorial optimization problems on graphs, namely the vertex cover problem, the computation of maximum cliques, and the determination of a maximum independent set. Extending the evaluation process for a given population to the evaluation of the whole search space we get exact methods for the considered problems, which allow to evaluate the quality of solutions obtained by evolutionary algorithms.

## 1 Introduction

Evolutionary algorithms (EAs) have become quite popular in solving real-world problems as well as solving problems from combinatorial optimization. Representing possible solutions for a given problem has been widely discussed. There are for example different representations for the well-known traveling salesman problem (see e.g. Michalewicz (2004)) or NP-hard spanning tree problems ((e.g. Raidl and Julstrom (2003) and Soak, Corne, and Ahn (2004)). Each of these representations leads to a different neighborhood of a particular solution and variation operators such as mutation and crossover have to be adjusted to the considered representation.

In this paper we focus on a different issue of representation. We study the question whether representations using relations can be useful. The issue of hybridization has become quite popular in the area of evolutionary computation. Here one combines evolutionary algorithms with other approaches in order to get better results. We think that it may be useful to combine evolutionary algorithms with the techniques used in the community of relational methods, and focus on how to formulate the different modules of an EA using relational algebra. A first step into this direction has been made by Kehden, Neumann, and

J. Gottlieb and G.R. Raidl (Eds.): EvoCOP 2006, LNCS 3906, pp. 147–158, 2006.

Berghammer (2005) who have given a relational implementation of a $(1+\lambda)$ EA. It may be also worth examining which parts of the search process can be carried out using relational algebraic expressions. We think it is worth building a bridge between relational algebra and evolutionary computation. In particular it may be useful to consider relation-algebraic expressions in some parts of an evolutionary algorithm, which can speed up some computations.

We focus on evolutionary algorithms for the search space $\{0,1\}^n$ and examine how the most important modules of an evolutionary algorithm can be implemented on the basis of relational operations. Relational algebra has been widely used in computer science. Especially in the case of NP-hard combinatorial optimization problems on graphs, a lot of algorithms have been developed. Relational algebra has a small, but efficiently to implement, set of operations. On the other hand it allows a formal development of algorithms and expressions starting usually with a predicate logic description of the problem.

Because of the small set of operations used in relational algebra that can be implemented efficiently we get a common environment for evolutionary algorithms working in the mentioned discrete search space. In addition it has been shown that relations can be implemented efficiently using Ordered Binary Decision Diagrams (OBDDs). In particular in the case that a considered relation has a special structure we can hope to get a much more compact representation of the relation than the standard representation as a matrix. Due to the properties of OBDDs the operations used in relational algebra can be implemented efficiently. Software systems for relational algebra using this kind of representation are for example RelView (e.g. Berghammer and Neumann (2005)) or Croco-Pat (Beyer, Noack, and Lewerentz (2005)). Both systems are able to carry out relation-algebraic expressions and programs based on OBDD operations.

We represent a population which is a set of solutions as one single relation and evaluate this population using relational algebra. It turns out that this approach can be implemented in a way that mainly relies on the relation-algebraic formulation of the specific modules. Considering the evaluation of a given population we show that this process can be made more efficient using relational algebra. We consider three well-known NP-hard combinatorial optimization problems, namely minimum vertex covers, maximum cliques, and maximum independent sets, and show how the whole population can be evaluated with respect to the corresponding constraints of the given problem using relational algebra. It turns out that using this approach can reduce the runtime from $\Theta(n^3)$ to $O(n^{2.376})$ for a population of size $n$ compared with a standard approach.

In the special case where we evaluate all search points of the given search space we are dealing with an exact method for the considered problem. We show how the evaluation process for a population can be turned into a process for evaluating the whole search space and therefore computing an exact solution. Due to the compact representation of relations by OBDDs this approach is often quite successful. Comparing the solutions obtained by an evolutionary approach with the optimal solution we are able to evaluate the quality of the solutions obtained by evolutionary algorithms.

In Section 2 we show how populations are represented as relations and re-call preliminaries of relational algebra. In Section 3 we consider the evaluation process for three important graph problems based on relational algebra, and for-mulate important modules of an evolutionary algorithm using relational algebra in Section 4. The evaluation processes of populations for the given problems are extended to get exact methods in Section 5. Based on this, we investigate in Section 6 the quality of solutions obtained by a standard evolutionary approach for the vertex cover problem with respect to the optimal solutions using some experiments. Finally, we finish with conclusions.

## 2   Representing Populations as Relations

We consider evolutionary algorithms working in the search space $\{0,1\}^n$, and possible solutions are therefore bitstrings of length $n$. For simplicity we also assume that a population has $n$ individuals although our ideas may be adapted to populations of different sizes. We want to represent the whole population as a relation $P$ where each individual of $P$ is stored in one single column. As we want to show how to use relational algebra in an evolutionary algorithm we have to start with some basic definitions. Relational algebra has a surprisingly small set of operations that can be implemented efficiently. In the following we give the basic notations that we will use throughout this paper. In addition we give upper bounds on the runtime needed to execute the different operations. For a more detailed description of relational algebra we refer to Schmidt and Ströhlein (1993).

We write $R : X \leftrightarrow Y$ if $R$ is a relation with domain $X$ and range $Y$, i.e. a subset of $X \times Y$. In the case of finite carrier sets, we may consider a relation as a Boolean matrix. Since this Boolean matrix interpretation is well suited for many purposes, we often use matrix terminology and matrix notation in the fol-lowing. Especially, we speak of the rows, columns and entries of $R$ and write $R_{xy}$ instead of $(x, y) \in R$. The basic operations on relations are $R^\top$ (transposition), $\overline{R}$ (negation), $R \cup S$ (union), $R \cap S$ (intersection), $RS$ (composition), the special relations $\mathsf{O}$ (empty relation), $\mathsf{L}$ (universal relation), and $\mathsf{I}$ (identity relation). A relation $v : X \leftrightarrow \mathbf{1}$ is called vector, where $\mathbf{1} = \{\bot\}$ is a specific singleton set. We omit in such cases the second subscript, i.e. write $v_i$ instead of $v_{i\bot}$. Such a vector can be considered as a Boolean matrix with exactly one column and describes the subset $\{x \in X : v_x\}$ of $X$. Note, that one search point of the considered search space can be represented as a vector of length $n$. A set of $k$ subsets of $X$ can be represented as a relation $P : X \leftrightarrow [1..k]$ with $k$ columns. For $i \in [1..k]$ let $P^{(i)}$ be the $i$-th column of $P$. More formally, every column $P^{(i)}$ is a vector of the type $X \leftrightarrow \mathbf{1}$ with $P_x^{(i)} \Longleftrightarrow P_{xi}$. We use relational expressions to evaluate populations, i.e. to decide which individuals fulfill certain properties by computing vectors of the type $[1..k] \leftrightarrow \mathbf{1}$.

We assume that we are always working with populations that have exactly $n$ individuals which implies $k = n$. Therefore, the relation $P$ representing this population has exactly $n$ rows and $n$ columns. Under the assumption that we work with $n \times n$ relations, the operations transposition, negation, union and intersection can be implemented in time $O(n^2)$. The standard implementation for the composition needs time $\Theta(n^3)$. Using the algorithm proposed by Coppersmith and Winograd (1990) for the multiplication of two $n \times n$ matrices we can reduce the runtime for the composition to $O(n^{2.376})$. In the case that we work with random relations $R$ and $S$, where each entry is set with probability $p$ to 1 and 0 otherwise, the runtime for the composition can be reduced to almost $O(n^2 \log n)$ as shown by Schnorr and Subramanian (1998). This also holds if one of the relations is arbitrary but fixed and the entries of the other one are chosen with probability $p$.

Another advantage of the relation-algebraic approach is that relations can be represented efficiently using Ordered Binary Decision Diagrams (OBDDs) as shown by Berghammer, Leoniuk, and Milanese (2002). OBDDs are implicit representations for boolean functions (see e.g. Wegener (2000)). In the case that the relations have a regular structure we often get a compact representation of relations. The mentioned operations for relations can be carried out efficiently using standard operations on OBDDs. For practical applications this leads sometimes to faster algorithms for a given task. For example Beyer, Noack, and Lewerentz (2005) have shown for practical instances that the transitive closure of a given graph can be computed much faster and with significantly less memory than using standard approaches.

# 3    Testing Properties of Solutions for Some Graph Problems

Assume that we have a relation $P$ that represents a population such that each column is identified with an individual. One important issue is to test whether the individuals of the population fulfill given constraints which means that they are feasible solutions. Often this is the most costly step in one generation of an evolutionary algorithm. In Section 4 we will show that all the other important modules which we consider like crossover, mutation, and selection methods of an evolutionary algorithm can be carried out in time $O(n^2)$ for one generation.

Often even the test whether an individual fulfills the constraints of a given problem needs time $\Omega(n^2)$. In the case that our population consists of $n$ individuals we get a runtime of $\Omega(n^3)$ for evaluating the whole population using a standard approach. In the following we show how the evaluation of a population can be done for some well-known graph problems using relation-algebraic expressions.

Given a graph $G = (V, E)$ with $n$ vertices and $m$ edges represented as an adjacency relation $R$ we want to test each individual to fulfill a given property. We concentrate on the constraints for some of the best-known combinatorial optimization problems for graphs, namely minimum vertex covers, maximum cliques, and maximum independent sets. A vertex cover of a given graph is a set of vertices $V' \subseteq V$ such that $e \cap V' \neq \emptyset$ holds for each $e \in E$. For a clique $C \subseteq V$ the property that $R_{uv}$ for all $u, v \in C$ with $u \neq v$ has to be fulfilled and in an independent set $I \subseteq V$, $\overline{R}_{uv}$ has to hold for all $u, v \in I$ with $u \neq v$. It is well known that computing a vertex cover of minimum cardinality or cliques and independent sets of maximal cadinaliy are NP-hard optimization problems (see e.g. Garey and Johnson (1979)).

Consider one individual $x$ of the population. If we want to test whether this individual fulfills one of the stated properties we need a runtime of $\Theta(n^2)$ to do this in a standard approach. In the case of the vertex cover problem we have to consider whether each edge contains a vertex of $V'$ which needs time $\Theta(n^2)$ for graphs with $m = \Theta(n^2)$. In the case of cliques we have to consider whether $\{u, v\} \in E$ holds for each pair of two distinct vertices $u, v \in C$. If $|C| = \Theta(n)$, $\Theta(n^2)$ edges have to be considered. Similar for the independent set problem and $|I| = \Theta(n)$ we have to test for $\Theta(n^2)$ pairs of vertices if there is no edge between these vertices. Working with a population of size $\Theta(n)$ this means that we need time $\Theta(n^3)$ for evaluating each of these properties.

We want to show that the runtime for evaluating a population that is represented as a relation can be substantially smaller using relation-algebraic expressions. Note, that the size of the solutions, which means the number of ones in the associated column, can be determined for the whole population $P$ in time $O(n^2)$ by examining each entry of the relation at most once. Therefore, the most costly part of the evaluation process for the three mentioned problems seems to be the test whether the given constraints are fulfilled.

Given the two relations $R$ and $P$ we can compute a vector that marks all solutions of the population that are vertex covers.

$$
\begin{aligned}
P^{(i)} \text{ is a vertex cover of } R \iff & \forall u, v : R_{uv} \rightarrow (P_u^{(i)} \vee P_v^{(i)}) \\
\iff & \neg \exists u, v : R_{uv} \wedge \overline{P}_{ui} \wedge \overline{P}_{vi} \\
\iff & \neg \exists u : (R\overline{P})_{ui} \wedge \overline{P}_{ui} \\
\iff & \neg \exists u : (R\overline{P} \cap \overline{P})_{ui} \\
\iff & \neg \exists u : \mathsf{L}_{\perp u} \wedge (R\overline{P} \cap \overline{P})_{ui} \\
\iff & \neg(\mathsf{L}(R\overline{P} \cap \overline{P}))_{\perp i} \\
\iff & \overline{\mathsf{L}(R\overline{P} \cap \overline{P})}_i^{\mathsf{T}} .
\end{aligned}
$$

To obtain the expression for the case of independent sets we can use the fact that $P^{(i)}$ is an independent set iff $\overline{P}^{(i)}$ is a vertex cover.

$$
\begin{aligned}
P^{(i)} \text{ is an independent set of } R \iff & \overline{P}^{(i)} \text{ is a vertex cover of } R \\
\iff & \overline{\mathsf{L}(RP \cap P)}_i^{\mathsf{T}}
\end{aligned}
$$

Since a set of vertices is a clique of $G$ if and only if it is an independent set of the complement graph with adjacency relation $\overline{R} \cap \overline{\mathsf{I}}$, we can conclude

$$
\begin{aligned}
P^{(i)} \text{ is a clique of } R \iff & P^{(i)} \text{ is an independent set of } \overline{R} \cap \overline{\mathsf{I}} \\
\iff & \overline{\mathsf{L}((\overline{\mathsf{I}} \cap \overline{R})P \cap P)}_i^{\mathsf{T}} \\
\iff & \overline{\mathsf{L}((\overline{\mathsf{I}} \cup R)P \cap P)}_i^{\mathsf{T}}
\end{aligned}
$$

Considering the different expressions, the most costly operation that has to be performed is the composition of two $n \times n$ relations. Therefore the evaluation process for a given population $P$ and a relation $R$ can be implemented in time $O(n^{2.376})$ by adapting the algorithm of Coppersmith and Winograd (1990) for the multiplication of two $n \times n$ matrices to relations, which beats the lower bound of $\Omega(n^3)$ for the standard implementation. Note, that also larger populations can be handled saving a factor of $n^{0.624}$ compared with the standard implementation by partitioning the large population into different subpopulations of size $n$ and executing relation-algebraic expressions for these subpopulations.

# 4    Relation-Algebraic Formulation of Important Modules

Variation operators are important to construct new solutions for a given problem. We assume that the current population is represented by a relation $P$ and present a relation-algebraic formulation for some well-known variation-operators. In addition we formulate an important selection method based on relational algebra. It turns out that the runtimes for our general framework are of the same magnitude as in a standard approach.

## 4.1    Mutation

An evolutionary algorithm that uses only mutation as variation operator usually flips each bit of each individual with a certain probability $p$. The approach for our population represented as a relation is straightforward and can be implemented in time $O(n^2)$ as we have to consider each entry of $P$ at most once. To integrate the mutation operator into the general framework we assume that we have constructed a relation $M$ that gives the mask which entries are flipped in the next step. In this case each entry of $M$ is set to 1 with probability $p$. Then we can construct the relation $C$ for the children of $P$ using the symmetric difference of $P$ and $M$. Hence, we can compute the $n$ children of $P$ by

$$C = (P \cap \overline{M}) \cup (\overline{P} \cap M).$$

## 4.2    Crossover

A crossover operator for the current population $P$ takes two individuals of $P$ to produce one child. To create the population of children $C$ by this process, we assume that we have in addition created a relation $P'$ by permuting the columns of $P$. Then we can decide which entry to use for the relation $C$ by using a mask $M$.

$$C = (M \cap P) \cup (\overline{M} \cap P')$$

To implement different crossover operators we have to use different masks in this expression.

In the case of uniform crossover each entry is chosen from $P$ or $P'$ with probability $1/2$. Hence, we can use the mask $M$ where each entry is set to 1 with probability $1/2$ and 0 otherwise.

In the case of 1-point crossover, we choose for each column $j$ a position $k_i$ such that $M_{ij}$ holds for $1 \leq i \leq k_i$ and $\overline{M}_{ij}$ otherwise. In a similar way the masks for the other crossover operators can be constructed.

Often one does not want to apply crossover to each pair of individuals given by $P$ and $P'$. Normally each pair of individuals is used for crossover with a certain probability $p_c$. To specify these pairs of individuals, we construct a vector $m$ of length $n$ where each entry is set to 1 with probability $p_c$. Then we obtain a new mask

$$M' = M \cap (m\mathsf{L})^\top,$$

with $\mathsf{L} : \mathbf{1} \leftrightarrow [1..n]$ which can be used to compute the relation $C$. In a similar way we can mutate individuals of $P$ only with a certain probability $p_m$ by constructing a vector $m$ where each entry is set to 1 with probability $p_m$.

The construction of the masks for crossover or mutation can be carried out in time $O(n^2)$ and each operation for the computation of $C$ needs time $O(n^2)$. Hence, each of the mentioned crossover or mutation operators can be implemented in time $O(n^2)$ using relational algebra.

## 4.3   Selection

We focus on one of the most important selection methods used in evolutionary computation, namely tournament selection, and assume that we have a parent population $P$ and a child population $C$ both of size $n$. To establish $n$ tournaments of size 2 we use a random bijective mapping that assigns each individual of $P$ to an individual of $C$. This can be done by permuting the columns of $C$ randomly. Due to the evaluation process we assume that we have a decision vector $d$ that tells us to take the individual of $P$ or the individual of $C$ for the newly created population $N$.

Let $P, C : X \leftrightarrow [1..n]$ and $d : [1..n] \leftrightarrow \mathbf{1}$, where we assume that the columns of $C$ have already been permuted randomly. We want to construct a new population $N$, such that for each $i \in [1..n]$ either $P^{(i)}$ or $C^{(i)}$ is the $i$-th column of $N$. The vector $d$ specifies which columns should be adopted in the new population $N$.

$$d_i \iff P^{(i)} \text{ should be adopted}$$

and

$$\overline{d}_i \iff C^{(i)} \text{ should be adopted.}$$

More formally, the new relation $N : X \leftrightarrow [1..n]$ is defined by

$$N_{xi} \iff (P_{xi} \wedge d_i) \vee (C_{xi} \wedge \overline{d}_i).$$

This can easily be transformed into a relational expression:

$$N_{xi} \Longleftrightarrow (P_{xi} \wedge d_i) \vee (C_{xi} \wedge \overline{d}_i)$$
$$\Longleftrightarrow (P_{xi} \wedge \mathsf{L}d^\top_{xi}) \vee (C_{xi} \wedge \overline{\mathsf{L}d^\top}_{xi})$$
$$\Longleftrightarrow ((P \cap \mathsf{L}d^\top) \cup (C \cap \overline{\mathsf{L}d^\top}))_{xi}$$

Hence, the new population $N$ is determined by

$$N = (P \cap \mathsf{L}d^\top) \cup (C \cap \overline{\mathsf{L}d^\top}).$$

The expression $\mathsf{L}d^\top$ is the composition of two vectors of length $n$ which can be carried out in time $O(n^2)$. The whole determination of $N$ can be done in time $O(n^2)$ as all other operations used in the expression for $N$ can also be executed in time $O(n^2)$ as mentioned in Section 2.

## 5   Computing Exact Solutions

Using the expressions derived in Section 3 we give exact approaches for the three considered problems. Here we work with a relation M, called membership relation, instead of the relation $P$.

The membership relation has been shown to be very useful for the computation of exact solutions using relational algebra.

$$\mathsf{M} : X \leftrightarrow 2^X \qquad \mathsf{M}_{xY} :\Longleftrightarrow x \in Y .$$

The membership relation has size $n \times 2^n$. This means that even for small dimensions it is not possible to store this relation. Note, that the relation M lists all search points of the search space $\{0,1\}^n$ columnwise and that, in general, it is impossible to evaluate all search points in the search space sequentially in a reasonable amount of time.

In Section 2 we have mentioned that relations can be represented in a compact form using OBDDs. This is indeed the case if we consider the relation M. It can be shown that this relation can be represented by an OBDD using $O(n)$ nodes. The main argument for that is that the underlying function is similar to the direct storage access function (see e.g. Wegener (2000)). For a complete proof on the stated upper bound see Leoniuk (2001). Nevertheless, the stated approach has its limitations as the OBDDs that are constructed during the computation process may be exponentially large in $n$. As described in Section 3, the vector

$$v = \overline{\mathsf{L}(R\overline{\mathsf{M}} \cap \overline{\mathsf{M}})}^\top$$

of type $2^X \leftrightarrow \mathbf{1}$ specifies all vertex covers of $R$. To select the vertex covers with the minimal number of vertices, we need the size-comparison relation

$$\mathsf{S} : 2^X \leftrightarrow 2^X \qquad \text{defined by} \qquad \mathsf{S}_{AB} :\Longleftrightarrow |A| \leq |B|.$$

The relation $S$ has size $4^n$ but can be represented by an OBDD of size $O(n^2)$ as shown by Milanese (2003). We use the ordering $S$ to develop a relational expression which specifies the minimal subsets in $v$.

$$
\begin{aligned}
y \text{ is a smallest element of } v \text{ with respect to } S \iff\ & v_y \wedge \forall z : v_z \to S_{yz} \\
\iff\ & v_y \wedge \neg \exists z : v_z \wedge \overline{S}_{yz} \\
\iff\ & v_y \wedge \overline{\overline{S}v}_y \\
\iff\ & (v \cap \overline{\overline{S}v})_y
\end{aligned}
$$

The relational expression

$$
m = v \cap \overline{\overline{S}v}
$$

describes the set of the smallest elements in $v$ with respect to the order $S$, hence we obtain a vector of the same type as $v$, specifying the smallest vertex covers of $R$. By choosing a point $p \subseteq m$ and computing the product $Mp$ we achieve a vector of the type $X \leftrightarrow \mathbf{1}$ that represents a minimal vertex cover of $R$. Similarly, we can compute all independent sets of $R$ with

$$
v = \overline{\mathsf{L}(RM \cap M)}^{\mathsf{T}}
$$

and all cliques of $R$ with

$$
v = \overline{\mathsf{L}((\overline{\mathsf{I} \cup R})M \cap M)}^{\mathsf{T}}
$$

using the expressions introduced in Section 3. In this cases we use the relational expression

$$
v \cap \overline{\overline{S}^{\mathsf{T}}v}
$$

to select the largest elements of $v$ with respect to the order $S$. Hence, we achieve the maximal independent sets respectively the maximal cliques of $R$.

## 6    The Quality of a Standard Evolutionary Approach

We have shown in the previous sections how the modules of an evolutionary algorithm can be formulated using relational algebra. In addition we have given formulations for the exact computation of the considered problems. In the following we present some experimental results. We consider the vertex cover problem and want to evaluate the quality obtained by a standard evolutionary approach after a small number of generations, namely 100. He, Yao, and Li (2005) have studied the behavior of different evolutionary algorithms for the vertex cover problem. To examine the quality of the algorithms they have considered different graph classes, namely a bipartite graph and a graph with one large cycle, where the value of an optimal solution is known. Here we consider random graphs, where it is hard to obtain optimal solutions. Using the exact method proposed in Section 5 we are able to compare the results obtained by a standard evolutionary algorithm with the value of an optimal solution.

We have carried out all these computations using the RelView-System that allows the evaluation of relation-algebraic terms and programs. Our computations were executed on a Sun-Fire 880 running Solaris 9 at 750 MHz.

The general framework used in this work allows to carry out all the operations using RelView and shows the quality of the results obtained by a standard evolutionary approach with respect to the optimal solutions. But it should be mentioned that RelView is designed as a general tool for relational algebra and it is not tuned with respect to the considered problems. Especially it is not designed in a special way for evolutionary computation. The runtimes for the exact approach were often less then 10 seconds, but quite high, around 120 seconds, in the case of the evolutionary approach. This is due to the fact that that RelView has not been designed for the implementation of randomized algorithms constructing many random relations. To get fast evolutionary algorithms based on relational algebra, one has to implement the modules of an EA in a special way as shown in Sections 2–4. The reason for using RelView is the general framework and the fast computation of the exact solutions.

In our experiments we consider a simple evolutionary algorithm and random graphs with $n = 50$ vertices. The initial population is chosen by applying $n$ times a random version of the well-known approximation algorithm due to Gavril and Yannakakis (see e.g. Cormen, Leiserson, Rivest, and Stein (2001)). Starting with the given graph, in each step an edge $e = \{u, v\}$ is chosen uniformly at random and $u, v$ are included in the solution. After that all edges incident to $u$ or $v$ are deleted from the graph. The process is iterated until the graph is empty. It is well known that solutions constructed in this way have size at most twice the size of an optimal solution.

The population has size $n$, we use uniform-crossover and after that mutation where each bit is flipped with probability $1/n$. The probability for a crossover to take place is $p_c = 0.8$. After $n$ children have been created in this way, we apply tournament selection with $n$ tournaments of size 2 where one participant is chosen from the parents and one from the children as explained in Section 4. Note, that only individuals being vertex covers are accepted for the next generations as in a tournament a feasible solution always wins against an unfeasible one. In the case that both individuals of a tournament are vertex covers an individual with the smallest number of vertices is chosen for the next generation. The algorithm is terminated after 100 generations. Only the initial population is chosen in a way that is special for the vertex cover problem. The other parts of the algorithm are general and are often used in the scenario of black-box optimization. Therefore our aim is not to beat specific algorithms for the vertex cover problem, but to examine the quality of solutions obtained by a standard evolutionary approach with respect to the value of an optimal one.

In our experiments we considered random graphs, where each edge is chosen with a certain probability $p$. We examine for each $p \in \{0.05, 0.1, 0.15, 0.2\}$ ten random graphs. The results obtained by the evolutionary algorithm are compared with the value of an optimal solution computed by using the exact method of Section 5, and the best solution included in the initial population. Note, that

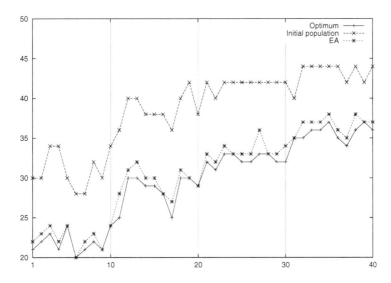

**Fig. 1.** Results on random graphs with 50 vertices

the initial population is constructed by applying $n$ times a random version of the approximation algorithm due to Gavril and Yannakakis.

Figure 1 shows our results. Here the 40 instances are listed at the $x$-axis, where the first ten instances correspond to the value $p = 0.05$, the next ten instances to $p = 0.1$ and so on. On the $y$-axis we list the value of an optimal solution, the results obtained by the initialization of the population, and the results achieved by the evolutionary algorithm. Figure 1 shows that the initial population is always far from optimal, but even within this small number of generations highly improved by the standard evolutionary approach such that optimal or nearly optimal solutions are constructed.

## 7    Conclusions

We have taken a relation-algebraic view on evolutionary algorithms for some graph problems. It turns out that the evaluation of a population can be sped up by using relation-algebraic expressions to test whether the solutions of the population fulfill given constraints. In the case of the three considered graph problems the computation time for one generation can be reduced from $\Theta(n^3)$ to $O(n^{2.376})$. We have also shown how the evaluation process of a population can be extended to the evaluation of the whole search space. This leads to exact methods for the considered problems which allows to evaluate the results obtained by evolutionary algorithms with respect to the optimal solutions. Although we have considered EAs that are not tuned in a special way for the problem considered in our experiments, namely the vertex cover problem, the comparison shows that the solutions obtained by a standard evolutionary algorithm are, even within a small number of generations, not far from optimal.

# References

1. Berghammer, R., Leoniuk, B., and Milanese, U. (2002). Implementation of Relational Algebra Using Binary Decision Diagrams. In: de Swart H.C.M. (ed.): Proc. 6th International Conference RelMiCS 2001, LNCS 2561, Springer, Berlin, Germany, 241–257.
2. Berghammer, R., and Neumann, F. (2005). RELVIEW - An OBDD-based Computer Algebra System for Relations. In: Gansha et al. (Eds.): Proc. of the 8th International Workshop on Computer Algebra in Scientific Computing - CASC 2005, LNCS 3718, Springer, Berlin, Germany, 40–51.
3. Beyer, D., Noack, A., and Lewerentz, C. (2005). Efficient Relational Calculation for Software Analysis. IEEE Transactions on Software Engineering 31(2): 137–149.
4. Coppersmith, D., and Winograd, S. (1990). Matrix multiplication via arithmetic progressions. Journal of Symbolic Computation, 9:251–280.
5. Cormen, T., Leiserson, C., Rivest, R., and Stein, C. (2001). Introduction to Algorithms. 2nd Edition, McGraw Hill, New York.
6. Garey, M. R., Johnson, D. S. (1979). Computers and Intractability: A Guide to the Theory of NP-completeness. Freeman, New York.
7. He, J., Yao, X., and Li, J. (2005). A Comparative Study of Three Evolutionary Algorithms Incorporating Different Amount of Domain Knowledge for Node Covering Problems. IEEE Transactions on Systems, Man and Cybernetics, Part C. 35(2):266- 271.
8. Kehden, B., Neumann, F., Berghammer, R. (2005). Relational Implementation of Simple Parallel Evolutionary Algorithms. In: Proc. of the 8th International Conference on Relational Methods in Computer Science (RelMiCS 8), 137–146.
9. Leoniuk, B. (2001) ROBDD-based implementation of relational algebra with applications (in German). Diss., Univ. Kiel.
10. Michalewicz, Z. (2004). How to solve it: Modern heuristics. 2nd edition, Springer-Verlag, Berlin.
11. Milanese, U. (2003). On the implementation of a ROBDD-based tool for the manipulation and visualization of relations (in German). Diss., Univ. Kiel.
12. Raidl, G.R. and Julstrom, B.A. (2003). Edge sets: an effective evolutionary coding of spanning trees. IEEE Trans. on Evolutionary Computation 7, 225–239.
13. Schmidt, G., and Ströhlein, T. (1993). Relations and graphs. Discrete mathematics for computer scientists, EATCS Monographs on Theoret. Comp. Sci., Springer.
14. Schnorr, C.-P., and Subramanian, C. R. (1998). Almost Optimal (on the average) Combinatorial Algorithms for Boolean Matrix Product Witnesses, Computing the Diameter (Extended Abstract). RANDOM 1998: 218–231.
15. Soak, S.-M., Corne, D., and Ahn, B.-H. (2004). A Powerful New Encoding for Tree-Based Combinatorial Optimisation Problems. In Proc. of the 8th. Int. Conf. on Parallel Problem Solving from Nature (PPSN VIII). LNCS 3242, 430–439.
16. Wegener, I. (2000). Branching programs and binary decision diagrams: Theory and applications. SIAM Monographs on Discr. Math. and Appl., SIAM.

# New Computational Results for the Nurse Scheduling Problem: A Scatter Search Algorithm

Broos Maenhout[1] and Mario Vanhoucke[1,2]

[1] Faculty of Economics and Business Administration,
Ghent University, Gent, Belgium
[2] Operations & Technology Management Centre,
Vlerick Leuven Gent Management School, Gent, Belgium
{broos.maenhout, mario.vanhoucke}@ugent.be

**Abstract.** In this paper, we present a scatter search algorithm for the well-known nurse scheduling problem (NSP). This problem aims at the construction of roster schedules for nurses taking both hard and soft constraints into account. The objective is to minimize the total preference cost of the nurses and the total penalty cost from violations of the soft constraints. The problem is known to be NP-hard. The contribution of this paper is threefold. First, we are, to the best of our knowledge, the first to present a scatter search algorithm for the NSP. Second, we investigate two different types of solution combination methods in the scatter search framework, based on four different cost elements. Last, we present detailed computational experiments on a benchmark dataset presented recently, and solve these problem instances under different assumptions. We show that our procedure performs consistently well under many different circumstances, and hence, can be considered as robust against case-specific constraints.

**Keywords:** meta-heuristics; scatter search; nurse scheduling.

## 1 Introduction

Personnel scheduling problems are encountered in many application areas, such as public services, call centers, hospitals, and industry in general. For most of these organizations, the ability to have suitably qualified staff on duty at the right time is of critical importance when attempting to satisfy their customers' requirements and is frequently a large determinant of service organization efficiency [1,2]. This explains the broad attention given in literature to a great variety of personnel rostering applications [3,4]. In general, personnel scheduling is the process of constructing occupation timetables for staff to meet a time-dependent demand for different services while encountering specific workplace agreements and attempting to satisfy individual work preferences. The particular characteristics of different industries result in quite diverse rostering models which leads to the application of very different solution techniques to solve these models. Typically, personnel scheduling problems are highly constrained and complex optimization problems [4,5].

J. Gottlieb and G.R. Raidl (Eds.): EvoCOP 2006, LNCS 3906, pp. 159–170, 2006.

In this paper, a procedure is presented to solve the nurse scheduling problem (NSP) which involves the construction of duty rosters for nursing staff over a pre-defined period. Problem descriptions and models vary drastically and depend on the characteristics and policies of the particular hospital. Due to the huge variety of hard and soft constraints, and the several objective function possibilities, the nurse scheduling problem has a multitude of representations, and hence, many exact and heuristic procedures have been proposed to solve the NSP in various guises. Recent literature surveys [6,7] give an overview of all these procedures, and mention simulated annealing, tabu search and genetic algorithms as popular meta-heuristics for the NSP.

In constructing a nurse schedule, a set of nurses need to be assigned to days and shifts in order to meet the minimal coverage constraints and other case-specific constraints and to maximize the quality of the constructed timetable. Consequently, the NSP under study is acquainted with three main components, i.e.

- Each nurse needs to express his/her *preferences* as the aversion to work on a particular day and shift. According to [8], quantifying these nurses' preferences in the objective function maintains fairness and guarantees the quality of the nurse roster over the scheduling horizon.
- The *minimal coverage constraints* embody the minimal required nurses per shift and per day, and are inherent to any shift scheduling problem. These constraints are handled as soft constraints that can be violated at a certain penalty cost expressed in the objective function.
- The *case-specific constraints* are not inherent to any NSP instance but are rather *case-specific*, i.e. determined by personal time requirements, specific workplace conditions, national legislation, etc.... These constraints are handled as hard constraints which can not be violated at a certain penalty cost.

Hence, the objective of the NSP is to minimize the sum of the nurses' preferences and the total penalty cost of violating the minimal coverage requirements subject to different case-specific (hard) constraints. For a more formal description of the NSP, we refer to [9]. The problem is known to be NP-hard [10].

In section 2 of this paper, we briefly review the philosophy of the scatter search template provided by [11]. Moreover, we discuss and illustrate the underlying principles and the implementation of the scatter search framework for the nurse scheduling problem. In section 3, we present new computational results tested on the NSPLib dataset proposed by [12]. In section 4, conclusions are made and directions for future research are indicated.

## 2   Scatter Search for NSP

Scatter search [11] is a population-based meta-heuristic in which solutions are combined to yield better solutions using convex linear or non-linear combinations. This evolutionary meta-heuristic differs from other evolutionary approaches, such as genetic algorithms, by providing unifying principles for joining solutions based on generalized path constructions in Euclidian space and by utilizing strategic designs where other approaches resort to randomization. The scatter search methodology is very flexible, since each of its elements can be implemented in a variety of ways and degrees of sophistication. Hence, the scatter search template has been successfully applied in several application areas. However, to the best of our knowledge, the scatter search framework has been applied only once to personnel rostering, more precisely on a labour scheduling problem by [13]. In their paper, they describe the development and implementation of a decision support system for the optimization of the passenger flow by trading off service quality and labour costs at an airport. In their search for the minimal number of employees, their path relinking approach concentrates on the shifts which are staffed differently in the parent solutions.

For an overview of the basic and advanced features of the scatter search meta-heuristic, we refer to [14,15]. In the following, we describe our implementation of the scatter search approach to the nurse scheduling problem. The pseudo-code for our generic scatter search template to solve the NSP is written below.

```
Algorithm Scatter Search NSP
  Diversification Generation Method
  While Stop Criterion not met
    Subset Generation Method
    Subset Combination Method
    Improvement Method
    Reference Set Update Method
  Endwhile
```

**The Diversification Generation Method.** In this initialization step, a large pool of $P$ initial solution vectors is generated. A solution point is encoded by the nurse-day representation [6] which basically indicates the assignment of each nurse to a shift on each day. Since useful information about the structure of optimal solutions is typically contained in a suitably diverse collection of elite solutions, the initial solutions are generated in such a manner a critical level of diversity is guaranteed [14,15]. In order to generate a diverse set of initial solutions, we create $x$ solutions using a constructive heuristic and $P - x$ solutions in a random way. The constructive heuristic schedules the nurses in a random sequence taking both preference costs and penalty cost of violating the coverage constraints into account. This greedy heuristic is conceived as a minimum cost flow problem which represents all shifts on all days to which a particular nurse can be assigned to. Since not all case-specific constraints (e.g. the maximum number of assignments) can be modelled in the network, it has been implemented by a k-shortest path approach. Based on this initial population, a subset of the population elements are designated to be reference solutions. This reference set contains $b_1$ high quality solutions (*Refset₁*) and $b_2$ diverse solutions (*Refset₂*). The construction of *Refset₁* starts with the selection of the best $b_1$ solutions in terms of solution quality out of the $P$ initial solutions. In order to select the diverse solutions (*Refset₂*), the minimum distance between all remaining $P - b_1$ solutions and the $b_1$ solutions is calculated based on the adjacency degree of [16]. In pursuit of diversity, the $b_2$ solutions with maximal distance will be selected for membership of *Refset₂* while all other $P - b_1 - b_2$ solutions are disregarded.

**The Solution Generation Method.** After the initialization phase, scatter search operates on this reference set by combining pairs of reference solutions in a controlled, structured way. Two elements of the reference set are chosen in a systematic way to produce points both inside and outside the convex regions spanned by the reference solutions. [14] suggest to create new solutions out of all two-element subsets. Choosing the two reference solutions out of the same cluster stimulates intensification, while choosing them from different clusters stimulates diversification. Hence, in our scatter search, the solution method consists of the evaluation of all $b_1 \times b_1$, $b_1 \times b_2$ and $b_2 \times b_2$ combinations in a random sequence.

**The Solution Combination Method.** A new solution point is the result of a linear combination of two population elements in the reference set. The process of generating linear combinations of a set of reference solutions may be characterized as generating paths between solutions [17]. A path between solutions will generally yield new solutions that share a significant subset of attributes contained in both parent solutions, which can differ according to the path selected. The moves introduce attributes contributed by a guiding solution and/or reduce the distance between the initiating and the guiding solution. The goal is to capture the assignments that frequently or significantly occur in high quality solutions, and then to introduce some of these compatible assignments into other solutions that are generated by a heuristic combination mechanism.

Our specific combination method relies on problem-specific information of both the initiating and the guiding solution schedule to create a new schedule, and takes four criteria into account. Two of these criteria incorporate objective function related data into account, as follows:

- Preference costs: Since the overall objective is to minimize the nurses' aversion towards the constructed work timetable, the day/shift preference cost expressed by each nurse is an important determinant in constructing high-quality new solution points.
- Coverage information: In order to minimize the penalty cost of violating the minimal coverage constraints, the algorithm penalizes those shifts where the coverage constraints are violated. In doing so, the solution combination method biases the initiating solution towards a (more) feasible solution.

The other two criteria incorporate information to maintain the "good" characteristics of both the initiating and the guiding schedule, as follows:

- Critical shifts of the initiating solution: In directing the initiating solution towards the guiding solution, the algorithm prevents the removal of critical shifts from the initiating solution, which, in case of removal, would lead to an additional violation of the coverage constraints. In doing so, the algorithm aims at the construction of a new solution point that does not encounter any (additional) violations of the coverage constraints.
- Bias to the guiding solution: The algorithm guides the initiating solution to the assignments of the guiding solution, in order to decrease the distance between the two schedules by introducing attributes of the guiding solution.

These four elements will be carefully taken into account for each move from an initiating solution to a guiding solution. Since the re-linking process of two solutions out of the reference set can be based on more than one neighbourhood [15], our algorithm makes use of two types of neighbourhood moves: a nurse neighbourhood move or a day neighbourhood move.

In the *nurse neighbourhood move*, the schedule of a single nurse of the initiating solution is directed towards the schedule of the corresponding nurse in the guiding solution. Therefore, the algorithm relies on a k-shortest path approach to optimize the schedule for a particular nurse taking into account a weighted average of the four aforementioned elements. The scheduling of a single nurse at minimum cost over the complete scheduling horizon can be considered as a minimum cost flow problem and can be solved by any shortest path algorithm. Moreover, since not all case-specific constraints can be incorporated in a shortest path algorithm, a k-shortest path approach is implemented where the outcome of this algorithm (i.e. a nurse schedule) should be checked whether it is feasible or not with respect to all these constraints. If the outcome is not feasible, a $2^{nd}$ shortest path will be generated and checked for feasibility. This process continues until the shortest feasible pattern (i.e. the $k^{th}$ shortest path) for the nurse is found. The graph used for our algorithm consists of #days*#shifts nodes (plus two extra dummy nodes representing the start and end of the network) representing the daily shift assignments for the nurse under study. An arc $(a, b)$ is drawn to connect node $a$ representing a possible shift assignment on day $j$ to node $b$ representing a shift assignment on day $j + 1$. The distances between nodes are made up of a weighted average of the four abovementioned elements. In calculating the new schedule for a particular nurse, we rely on the algorithm of [18].

In the *day neighbourhood move*, the roster of a single day of the initiating solution is directed towards the roster of the corresponding day of the guiding solution, given the assignments of the nurses on all other days in the initiating solution. To that purpose, we transform a single day roster to a linear assignment problem (LAP) matrix, and solve it by means of the Hungarian method [19]. In constructing this matrix, we duplicate each shift column such that each shift has a number of columns that is equal to its coverage requirements. Moreover, we add dummy nurses and/or dummy shifts to allow under- or over-coverage of the coverage requirements. The number of extra dummy nurses equals the total daily nurse requirements, and penalizes under-coverage when a required shift column has to be assigned to a dummy nurse. The number of extra dummy shifts equals the total number of (non-dummy) nurses and allows over-coverage of the coverage requirements when nurses are assigned to dummy shifts. The cost of assigning non-dummy nurses to one of the dummy shifts is equal to the minimum cost of the (feasible) shifts a nurse can be assigned to. The LAP matrix contains costs associated with the four criteria the solution combination mechanism is based on. Furthermore, the LAP matrix excludes certain shift assignments to cope with the case-specific constraints, taking into account the fixed assignments of all other days of the current solution.

This solution combination method can be best illustrated on an example NSP instance with 5 nurses and a scheduling period of 4 days. We assume that each day consists of three working shifts (e.g. early ($s_1$), day ($s_2$), night ($s_3$)) and a free shift ($s_4$). The nurses' preferences as well as the minimal coverage requirements are displayed in the top table of figure 1. Since "$s_4$" is used to refer to a free shift, its daily coverage requirements equal zero. We assume some additional case specific constraints as follows: the number of assignments varies between a minimal value of 3 and a maximal value of 4. The consecutive working shifts vary between a

minimal value of 2 and a maximal value of 4. The assignment of nurses to maximal
one shift per day and the succession constraints are inherent to continuous personnel
scheduling. The latter constraint implies forbidden successive assignments between $s_3$
and $s_1$, $s_3$ and $s_2$ and $s_2$ and $s_1$.

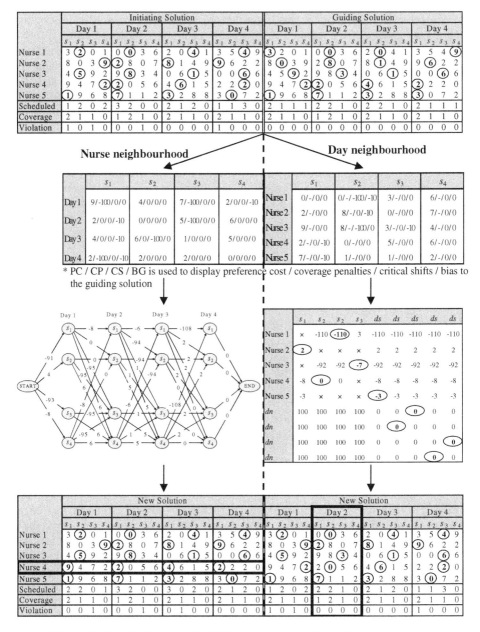

**Fig. 1.** The Solution Combination Method in nurse and day neighbourhood space

The left top table is assumed to be the initiating solution from $Refset_2$ with a total preference cost of 81 and four coverage violations (the specific assignments have been encircled). Since the algorithm penalizes each coverage violation with a penalty cost of 100, the total solution quality of the initiating solution equals 481. The right table is assumed to be the guiding solution from $Refset_1$ with a total preference cost of 70 and no coverage penalties. The adjacency degree measures the distance between schedules as the sum of zeros (identical day/shift assignment) and ones (different day/shift assignment), which leads to a total distance of 12.

In the remainder of this section, we illustrate the nurse neighbourhood move for nurse 4 (left part) and the day neighbourhood move for day 2 (right part) on our two parent solution schedules.

*Nurse neighbourhood move:* The left table below the parent solutions displays the calculations of the four elements of this section, i.e. the nurse's preference costs, the coverage penalties, the critical shifts and the assigned shifts of the guiding solution. The coverage penalty is set at 100, while the assigned shifts of the guiding solution have a negative cost of 10. The critical shifts are found for those assigned shifts of the initial schedule of nurse 4 when the difference between the number of scheduled nurses (row 'Scheduled') and the assignments of nurse 4 is lower than the minimum required number of nurses (row 'Coverage').

The sum of all costs in the table results in the cost values on the corresponding arcs of the network representing the scheduling of the nurse over the complete scheduling horizon. The graph counts 4 * 4 nodes and a start and an end dummy node. The shortest path in the network is $s_3 - s_3 - s_4 - s_1$ with a distance of -291. However, the path is infeasible since the constraint of minimal 2 consecutive working days is violated. Based on the same argument, the $2^{nd}$ shortest path, i.e. $s_1 - s_3 - s_4 - s_1$ with a distance of -289, is infeasible. The next shortest path is $s_1 - s_1 - s_1 - s_1$ with a distance of -213. This path is feasible for all case-specific constraints and leads to the newly constructed schedule at the left bottom part of the figure. The new solution point has a total solution quality of 386. The total preference cost has increased from 81 to 86, whereas the number of coverage violations has decreased from 4 to 3. The distance between the new initiating solution point and the guiding solution point has also decreased from 12 to 11.

*Day neighbourhood move:* The right table below the parent solutions displays the calculations of the four elements in a similar way as previously, but from the second day's point-of-view. The corresponding LAP matrix contains the sum of three of these four elements, i.e. the nurse's preference costs, the total critical shift cost and the cost of the assigned shifts of the guiding solutions. The coverage penalties have been incorporated implicitly in the structure of the LAP matrix since dummy nurses (*dn*) have been inserted which penalize the under-coverage of shifts. In contrast, the incorporation of dummy shifts (*ds*) allows the over-coverage of shifts. The case-specific constraints have been embedded in the LAP matrix by excluding some assignments (denoted by crossed cells). The optimal LAP solution has been encircled and leads to the newly constructed schedule at the right bottom part of the figure. The new solution point has a total solution quality of 374. The total preference cost has decreased from 81 to 74, and the number of coverage violations has also decreased from 4 to 3. The distance between the new initiating solution point and the guiding solution point has not changed and remains 12.

**The Improvement Method.** The improvement method applies heuristic processes to improve both total preference cost and coverage infeasibilities of the newly generated solution points. To that purpose, we implemented the three complementary local search algorithms of [9], each focusing on a different part of the scheduling matrix. The *pattern-based local search* aims at the optimization of the schedule for a particular nurse, given the schedules of all other nurses. The *day-based local search* optimizes the schedule for one day given the assignments of the nurses on all other days. The *schedule-based local search* aims to improve the quality of the schedule by swapping (parts of) schedules between nurses.

**The Reference Update Method.** After the application of the diversification and intensification process, the child solutions are added to the reference set if certain threshold values for the criteria which evaluate the merit of newly created solution points are met. A newly generated solution may become a member of the reference set either if the new solution has a better objective function value than the solution with the worst objective value in $Refset_1$ or if the diversity of the new solution with respect to the reference set is larger than the solution with the smallest distance value in $Refset_2$. In both cases the new solution replaces the worst and the ranking is updated to identify the new worst solution in terms of either quality or diversity. The reference set is dynamically updated. In contrast to a static update where the reference set is updated after combination of all generated sub-sets, a dynamic update evaluates each possible reference set entrance instantly. In this way, new "best" solutions can be combined faster and inferior solutions are eliminated faster. During the search, diversity in the reference set is maintained through the use of these artificial tiers in the reference set but also through a threshold distance depending on the problem size under study. The latter prevents the duplication of solution points in the reference set and/or the entrance of highly resembling solutions.

# 3   Computational Results

In this section, we present computational results for our scatter search procedure tested on the NSPLib problem instances of [12]. This testset contains 4 sub-sets with 25, 50, 75, and 100 nurses and a 7-days scheduling horizon (this so-called *diverse set* contains 4 * 7290 instances), and 2 sub-sets with 30 or 60 nurses and a 28-days scheduling horizon (this *realistic set* contains 2 * 960 instances). The nurses' preference structure and the coverage requirements of each sub-set are characterized by systematically varied levels of various NSP complexity indicators proposed in [12]. All sets have been extended by 8 mixes of case-specific constraints which appear frequently in literature [6], i.e. the minimum/maximum number of working assignments, the minimum/maximum number of assignments per shift type, the minimum/maximum consecutive working shifts, and the minimum/maximum consecutive working shifts per shift type. The testset, the sets of case-specific constraints and results can be downloaded from http://www. projectmanagement. ugent.be/nsp.php. The tests have been carried out under a stop criterion of 1,000 or

5,000 schedules (Toshiba SPA10 2.4Ghz processor). In the next sub-section, we compare different neighbourhood combination versions into detail. In section 3.2, we present best known solutions for our large dataset.

## 3.1 Day Neighbourhood or Nurse Neighbourhood

In order to test the performance of our two solution combination methods, the day neighbourhood (DNH) and nurse neighbourhood (NNH), we have randomly selected 576 instances from the 25- and 288 instances from the 30- nurse instances. Each neighbourhood move (day or nurse) contains a mix of the 4 elements, i.e. preference cost (PC), coverage penalty (CP), critical shift calculation (CS) and the bias to the guiding solution (BG). Both the nurse and the day neighbourhood moves will be compared with simple and straightforward moves, based on

- Complete replacement by the guiding solution (CGS): The day (DNH) or nurse (NNH) of the initiating solution will be completely replaced by the assignments of the guiding solution.
- Random cost (RAN): The cost matrix simply contains random numbers instead of the sum of PC, CP, CS and BG.
- Percentage of guiding solution (%GS): A randomly selected part of the day (DNH) or nurse (NNH) of the initiating solution will be replaced by the assignments of the guiding solution.

**Table 1.** Average solution quality (Avg_Sol) and Ranking for the various combination methods

| | Combination elements | | | | Nurse Neighbourhood | | | | Day Neighbourhood | | | |
|---|---|---|---|---|---|---|---|---|---|---|---|---|
| | | | | | N25 | | N30 | | N25 | | N30 | |
| | PC | CP | CS | BG | Avg_Sol | Ranking | Avg_Sol | Ranking | Avg_Sol | Ranking | Avg_Sol | Ranking |
| Solution Combination Method | × | - | - | - | 310.67 | 17 | 1,498.21 | 9 | 307.52 | 14 | 1,500.20 | 15 |
| | - | × | - | - | 312.26 | 18 | 1,528.22 | 18 | 307.37 | 11 | 1,501.21 | 16 |
| | - | - | × | - | 308.43 | 9 | 1,507.74 | 11 | 307.86 | 18 | 1,498.72 | 13 |
| | - | - | - | × | 306.67 | 5 | 1,496.02 | 6 | 307.00 | 5 | 1,495.93 | 9 |
| | × | × | - | - | 310.25 | 14 | 1,515.23 | 13 | 307.57 | 16 | 1,542.96 | 18 |
| | × | - | × | - | 309.18 | 11 | 1,496.80 | 7 | 307.67 | 17 | 1,498.66 | 12 |
| | × | - | - | × | **306.66** | **4** | **1,493.21** | **4** | 307.00 | 5 | 1,495.16 | 5 |
| | - | × | × | - | 310.40 | 15 | 1,519.58 | 15 | 307.50 | 13 | 1,498.87 | 14 |
| | - | × | - | × | 309.37 | 13 | 1,519.87 | 16 | 307.03 | 7 | 1,494.11 | 3 |
| | - | - | × | × | **306.57** | **3** | **1,492.17** | **3** | 307.27 | 10 | 1,495.43 | 7 |
| | × | × | × | - | 308.39 | 8 | 1,509.86 | 12 | 307.44 | 12 | 1,497.53 | 11 |
| | × | × | - | × | 309.26 | 12 | 1,518.92 | 14 | 306.97 | 4 | 1,496.35 | 10 |
| | × | - | × | × | **305.82** | **1** | **1,490.34** | **2** | 307.19 | 8 | 1,495.65 | 8 |
| | - | × | × | × | 308.99 | 10 | 1,494.23 | 5 | 307.21 | 9 | 1,495.24 | 6 |
| | × | × | × | × | **306.20** | **2** | **1,486.36** | **1** | 306.94 | 3 | 1,493.77 | 2 |
| CGS | - | - | - | - | 306.68 | 6 | 1,497.63 | 8 | 306.14 | 1 | 1,492.97 | 1 |
| RAN | - | - | - | - | 310.56 | 16 | 1,524.50 | 17 | 307.55 | 15 | 1,504.30 | 17 |
| %GS | - | - | - | - | 307.01 | 7 | 1,498.63 | 10 | 306.71 | 2 | 1,494.71 | 4 |

The results displayed in table 1 can be summarized as follows. First, the day-based neighbourhood (DNH) outperforms, on the average, the nurse-based neighbourhood (NNH). Indeed, 13 (10) of the 18 tests results in a DNH cost which is lower than its corresponding NNH cost. However, the best results can be obtained with the NNH method. The best result for the N25 (N30) instances amounts to 305.82 (1,486.36) which outperforms the best known results for the DNH method (306.94 and 1,492.97, respectively). Second, the top 4 results for the NNH have been displayed in bold, and show that the three elements, PC, CS and BG are relevant cost factors (3, 3 and 4 times used, respectively). The CP cost factor has been used only once in the top 4 results, and seems to be less important. Last, the simple CGS approach is the best approach for the DNH method, but is not able to outperform the best known results with the NNH method. The %GS has an average (good) performance for the NNH (DNH) method while the RAN method leads to rather poor results for both solution combination methods.

In section 3.2, we report best known solutions for all the data instances, based on the solution combination method NNH-PC/CS/BG for the diverse set with a 7-days scheduling horizon (N25, N50, N75 and N100) and the solution method NNH-PC/CP/CS/BG for the realistic set with a 28-days scheduling horizon (N30 and N60).

## 3.2  New Best Known Solutions

In order to benchmark our results and present best known state-of-the-art solutions, we have tested all 31,080 instances on all case-specific constraint files, resulting in 248,640 instances in total. We have truncated each test after a stop criterion of 1,000 and 5,000 schedules. Table 2 displays the results for the 5,000 schedule stop criterion. A similar table can be downloaded from www.projectmanagement.ugent.be/nsp.php with a 1,000 schedule stop criterion. The table displays the average solution quality, split up in the average total preference cost (Avg_Pref) and the average penalty cost (Avg_Pen) which is calculated as the average number of violations of the minimal coverage requirements times the penalty cost of 100, the required CPU time (Avg_CPU), the percentage of files for which a feasible solution has been found (%Feas), and the percentage deviation from the LP optimal solution which serves as a lower bound (%Dev_LP). The latter has been found by a simple and straightforward LP model, and has been used for similar tests by [9]. No LP lower bounds could be provided for the N30 en N60 sets, since the number of constraints exceeds the limits for the industrial LINDO optimization library, version 5.3 [20].

In order to fine-tune a number of parameters, we have run our procedure on a small subset of all instances under different parameter settings. In order to find an appropriate balance between the diversification and intensification process, we have combined the three proposed local search heuristics into a variable neighbourhood search. For the N25 instances, all nurses are subject to the pattern-based local search and all days are subject to the day-based local search. The schedule-based local search evaluates possible swaps between whole schedules and two-day sub-schedules. In our tests, 40% of all two-day sub-schedules are swapped between the nurses for all testsets. The population size ($b_1 + b_2$) has been set to 15 (10) for the N25, N50 and N30 (N75, N100 and N60) instances, respectively. Each time, 80% of these population elements ($b_1$) have been put into the $Refset_1$. Moreover, 40% of all nurses are subject to the NNH move for N25 and N50 instances, 60% for the N75 instances, 20% for the N100 instances and all nurses for the N30 and N60 instances.

The table reveals that the gap between the obtained solutions and the LP based lower bounds is in many cases very small. For some of the cases, the gap is somewhat larger which gives an indication of the constrainedness of the problem instances. Our scatter search algorithm is able to outperform (equalize) 40.5% and 62.06% (27.12% and 1.45%) of respectively the diverse and realistic instances solved by the electromagnetic procedure of [9] with a stop-criterion of 1,000 schedules, and 33.29% and 64.42% (45.02% and 1.74%) of respectively the diverse and realistic instances for the 5,000 schedules.

**Table 2.** Computational Results for the diverse and realistic dataset for 5,000 schedules

| | Diverse set | | | | | | | | | |
|---|---|---|---|---|---|---|---|---|---|---|
| | N25 | | | | | N50 | | | | |
| | Avg_Pref | Avg_Pen | %Feas | Avg_CPU | %Dev_LP | Avg_Pref | Avg_Pen | %Feas | Avg_CPU | %Dev_LP |
| Case 1 | 252.23 | 53.02 | 88.27% | 1.61 | 0.19% | 502.66 | 84.79 | 90.03% | 5.33 | 0.25% |
| Case 2 | 240.86 | 53.02 | 88.27% | 1.03 | 0.08% | 480.42 | 84.79 | 90.03% | 3.76 | 0.11% |
| Case 3 | 268.40 | 53.76 | 88.08% | 2.31 | 1.11% | 529.16 | 86.80 | 89.66% | 6.38 | 1.17% |
| Case 4 | 250.33 | 53.02 | 88.27% | 1.58 | 0.14% | 499.12 | 84.79 | 90.03% | 5.37 | 0.19% |
| Case 5 | 265.94 | 71.12 | 85.88% | 2.26 | 3.63% | 527.76 | 142.88 | 85.25% | 6.97 | 3.44% |
| Case 6 | 241.86 | 53.02 | 88.27% | 1.03 | 0.50% | 482.65 | 84.79 | 90.03% | 3.77 | 0.59% |
| Case 7 | 284.11 | 125.05 | 80.10% | 5.05 | 5.95% | 556.82 | 273.20 | 78.29% | 12.61 | 5.33% |
| Case 8 | 259.21 | 71.89 | 85.60% | 1.67 | 3.26% | 513.02 | 140.01 | 85.50% | 5.01 | 3.01% |
| | N75 | | | | | N100 | | | | |
| | Avg_Pref | Avg_Pen | %Feas | Avg_CPU | %Dev_LP | Avg_Pref | Avg_Pen | %Feas | Avg_CPU | %Dev_LP |
| Case 1 | 762.77 | 150.32 | 88.70% | 10.13 | 0.33% | 1,223.44 | 166.49 | 90.49% | 23.08 | 0.36% |
| Case 2 | 738.20 | 150.32 | 88.70% | 10.78 | 0.15% | 1,180.94 | 166.32 | 90.51% | 21.16 | 0.20% |
| Case 3 | 802.67 | 152.43 | 88.37% | 14.33 | 1.24% | 1,299.15 | 170.40 | 90.01% | 24.68 | 1.22% |
| Case 4 | 752.10 | 150.32 | 88.70% | 9.83 | 0.23% | 1,209.59 | 166.45 | 90.48% | 22.09 | 0.28% |
| Case 5 | 802.08 | 202.88 | 86.06% | 11.36 | 3.85% | 1,280.83 | 260.21 | 86.28% | 24.08 | 3.43% |
| Case 6 | 739.66 | 150.32 | 88.70% | 8.51 | 0.17% | 1,183.93 | 166.35 | 90.51% | 21.22 | 0.22% |
| Case 7 | 848.60 | 367.17 | 79.48% | 16.28 | 5.76% | 1,354.75 | 517.49 | 79.49% | 31.24 | 5.38% |
| Case 8 | 787.75 | 206.68 | 85.78% | 9.42 | 4.30% | 1,258.08 | 256.98 | 86.53% | 22.11 | 4.11% |
| | Realistic set | | | | | | | | | |
| | N30 | | | | | N60 | | | | |
| | Avg_Pref | Avg_Pen | %Feas | Avg_CPU | %Dev_LP | Avg_Pref | Avg_Pen | %Feas | Avg_CPU | %Dev_LP |
| Case 9 | 1,525.63 | 423.65 | 68.02% | 17.37 | - | 3125.66 | 743.65 | 68.54% | 37.99 | - |
| Case 10 | 1,429.45 | 392.50 | 69.58% | 7.52 | - | 2935.91 | 677.81 | 70.00% | 24.01 | - |
| Case 11 | 1,607.99 | 429.58 | 67.60% | 40.68 | - | 3277.91 | 758.54 | 68.44% | 73.00 | - |
| Case 12 | 1,465.36 | 392.40 | 69.38% | 8.95 | - | 3006.31 | 677.29 | 69.90% | 26.10 | - |
| Case 13 | 1,570.33 | 501.67 | 66.35% | 17.52 | - | 3211.65 | 906.46 | 67.92% | 37.66 | - |
| Case 14 | 1,444.48 | 394.17 | 69.48% | 7.64 | - | 2963.95 | 681.98 | 69.79% | 24.04 | - |
| Case 15 | 1,678.09 | 843.75 | 63.33% | 59.68 | - | 3424.43 | 1543.33 | 66.04% | 105.97 | - |
| Case 16 | 1,541.34 | 498.02 | 66.88% | 10.41 | - | 3154.49 | 882.92 | 68.23% | 27.44 | - |

# 4   Conclusions and Future Research

In this paper, we have presented a new scatter search procedure for the well-known nurse scheduling problem. To the best of our knowledge, the literature on scatter search for the nurse scheduling problem is completely void. This framework has only been applied once on a similar problem type of labour scheduling by [13].

We have investigated the use of two types of solution combination methods, based on the combinations of sub-schedules of nurses or days. Each method calculates the attractiveness of the move based on four criteria. We have shown that the scatter search algorithm leads to promising results and hence might have a bright future in the further development of meta-heuristic optimization algorithms. We have tested our procedure on a generated problem set NSPLib, under a strict test design with a strict stop criterion to facilitate comparison between procedures.

Our main future research intention is as follows. We will aim at the development of hybrid versions of different meta-heuristics, based on knowledge and concepts presented in this and many other research papers. A skilled combination of concepts of different meta-heuristics can provide a more efficient behaviour and a higher flexibility when dealing with real-world and large-scale problems.

# References

1. Thompson, G.M.: Improved implicit optimal modelling of the labour shift scheduling problem. Management Science 43 (1995) 595-607
2. Felici, G., and Gentile, C.: A polyhedral Approach for the Staff Rostering Problem. Management Science 50 (2004) 381-393
3. Ernst, A.T., Jiang, H., Krishamoorty, M., Owens, B., and Sier, D.: Staff scheduling and rostering: A review of applications, methods and models. European Journal of Operational Research 153 (2004) 3-27
4. Ernst, A.T., Jiang, H., Krishamoorty, M., Owens, B., and Sier, D.: An Annotated Bibliography of Personnel Scheduling and Rostering. Annals of Operations Research 127 (2004) 21-144
5. Glover, F., and McMillan, C.: The General Employee Scheduling Problem: An integration of MS and AI. Computers and Operations Research 13 (1986) 563-573
6. Cheang, B., Li, H., Lim, A., and Rodrigues, B.: Nurse rostering problems – a bibliographic survey. European Journal of Operational Research 151 (2003) 447-460
7. Burke, E.K., De Causmaecker, P., Vanden Berghe, G. and Van Landeghem, H.: The state of the art of nurse rostering. Journal of Scheduling 7 (2004) 441-499
8. Warner, H.W.: Scheduling Nursing Personnel According to Nursing Preference: A Mathematical Approach. Operations Research 24 (1976) 842-856
9. Maenhout, B., and Vanhoucke, M.: An Electromagnetism meta-heuristic for the nurse scheduling problem. Working paper 05/316 Ghent University (2005)
10. Osogami, T., and Imai, H.: Classification of Various Neighbourhood Operations for the Nurse Scheduling Problem. Lecture Notes in Computer Science, Vol. 1969 (2000) 72-83.
11. Glover, F: A Template for Scatter Search and Path Relinking. Lecture Notes in Computer Science, Vol. 1363 (1998) 13-54
12. Vanhoucke, M., and Maenhout, B.: Characterisation and Generation of Nurse Scheduling Problem Instances. Working paper 05/339 Ghent University (2005)
13. Casado, S., Laguna, M., and Pacheco, J.: Heuristical Labor Scheduling to Optimize Airport Passenger Flows. Journal of the Operational Research Society 56 (2005) 649-658
14. Glover, F., and Laguna, M.: Fundamentals of Scatter Search and Path Relinking. Control and Cybernetics 3 (2000) 653-684
15. Marti, R., Laguna, M., Glover, F.: Principles of scatter search. European Journal of Operational Research 169 (2006) 359-372
16. Aickelin, U.: Genetic Algorithms for Multiple-Choice Optimisation Principles. PhD, University of Wales Swansea (1999)
17. Glover, F., and Laguna, M.: Tabu Search. Kluwer Academic Publishers (1997)
18. Martins, E.Q.V., and Pascoal, M.M.B.: A new implementation of Yen's ranking loopless paths algorithm. 4OR – Quarterly Journal of the Belgian, French and Italian Operations Research Societies 1 (2003) 121-134
19. Kuhn, H.: The Hungarian method for the assignment problem. Naval Research Logistics 2 (1955) 83-97
20. Schrage, L.: LINDO: Optimization software for linear programming. LINDO Systems Inc.: Chicago, IL (1995)

# Fast EAX Algorithm Considering Population Diversity for Traveling Salesman Problems

Yuichi Nagata

Graduate School of Information Sciences,
Japan Advanced Institute of Science and Technology
nagatay@jaist.ac.jp

**Abstract.** This paper proposes an evolutionary algorithm (EA) that is applied to the traveling salesman problem (TSP). Existing approximation methods to address the TSP known to be state-of-the-art heuristics almost exclusively utilize Lin-Kernighan local search (LKLS) and its variants. We propose an EA that does not use LKLS, and demonstrate that it is comparable with these heuristics even though it does not use them. The proposed EA uses edge assembly crossover (EAX) that is known to be an efficient and effective crossover for solving TSPs. We first propose a modified EAX algorithm that can be executed more efficiently than the original, which is 2–7 times faster. We then propose a selection model that can efficiently maintain population diversity at negligible computational cost. The edge entropy measure is used as an indicator of population diversity.

The proposed method called EAX-1AB(ENT) is applied to TSP benchmarks up to instances of 13509 cities. Experimental results reveal that EAX-1AB(ENT) with a population of 200 can almost always find optimal solutions effectively in most TSP benchmarks up to instances of 5915 cities. In the experiments, a previously proposed EAs using EAX can find an optimal solution of usa13509 with reasonable computational cost due to the fast EAX algorithm proposed in this paper. We also demonstrate that EAX-1AB(ENT) is comparable to well-known LKLS methods when relatively small populations such as 30 are used.

## 1 Introduction

The traveling salesman problem (TSP) is a widely cited NP-hard optimization problem. Let $G = (V, E, w)$ be a weighted complete graph with n vertices, where $V$, $E$, and $w$ correspond to the set of vertices, the set of edges, and the weights of edges. The optimal solution is defined as the Hamilton cycle (tour) with the shortest tour length.

Many approximation methods of finding near optimal solutions have been proposed in the area of TSP research. In Johnson and McGeoch's surveys [1][2], the most efficient approximation methods for TSPs were based on Lin-Kernighan local searches (LKLS) [3]. The Iterated Lin-Kernighan (ILK) [1], for example, is a simple yet powerful improvement on LKLS. The Chained Lin-Kernighan (CLK)[4] are more sophisticated LKLS. Heslgaun [5] also proposed another type

J. Gottlieb and G.R. Raidl (Eds.): EvoCOP 2006, LNCS 3906, pp. 171–182, 2006.

of efficient LKLS (LKH) where candidates of edges searched by LKLS were effectively restricted. Iterated local searches using CLK and LKH are known to be one of the most efficient approximation methods for TSPs.

Many evolutionary algorithms (EAs) have also been applied to TSPs. Much effort has been devoted to designing effective crossovers suitable for TSPs because the performance of EAs is highly dependent on the design of crossovers. Consequently, several crossovers for TSPs have been proposed. Many researchers [6][7][8] have found that EAX crossover proposed by Nagata at el. [6] works particularly well, and several analyses have been conducted to explain why EAX works so well [9][10][12]. There have also been extensions of EAX [7][14][8] that have aimed at improving performance by making minor changes to EAX.

However, it has been found that EAs without LKLS are not comparable with state-of-the-art TSP heuristics based on LKLS. Methods that have incorporated LKLS into EAs have recently become known to be efficient heuristics for solving TSPs [11][12]. For example, Tsai et al. [12] proposed a hybrid algorithm composed of EAX and CLK, that was comparable with other LKLS-based approaches.

We attempted to design an EA to solve TSPs without using LKLS heuristics that would hopefully be comparable with state-of-the-art TSP heuristics based on LKLS. To achieve this, we used EAX as a crossover and employed an edge entropy measure to maintain population diversity. We propose a fast implementation of EAX in Section 2. A selection method that can maintain population diversity (edge entropy) at negligible computational cost is described in Section 3. Section 4 discusses our experiments and the results are compared with those of other LKLS-based approaches. Section 5 is the conclusion.

## 2    Fast Implementation of EAX

Here, we propose a fast implementation of EAX. First, we will briefly describe the algorithm of EAX [6] ( See an original paper for details ).

### 2.1    Outline of EAX

The following is an outline of EAX, where all steps correspond to the steps in Fig.1.

**Step 1.** A pair of parents is denoted as tour-A and tour-B, and $G_{AB}$ is defined as a graph constructed by merging tour-A and tour-B.

**Step 2.** Divide edges on $G_{AB}$ into *AB-cycles*, where an *AB-cycle* is defined as a closed loop on $G_{AB}$ that can be generated by alternately tracing edges of tour-$A$ and tour-$B$.

**Step 3.** Construct an *E-set* by selecting *AB-cycles* according to a given rule. *AB-cycle*s constructed of two edges are neglected.

**Step 4.** Generate an intermediate solution by applying *E-set* to tour-$A$, *i.e.*, by removing tour-$A$'s edges in the *E-set* from tour-$A$ and adding tour-B's edges in the *E-set* to it.

**Step 5.** Modify the intermediate solution to create a valid tour by connecting its sub-tours. Two sub-tours are connected by respectively deleting one edge from each sub-tour and adding two edges to connect them. Which sub-tours are connected and which edges are deleted are determined heuristically.

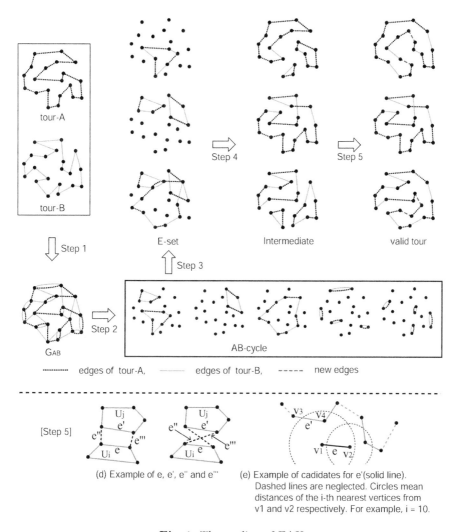

(d) Example of e, e', e" and e'"   (e) Example of cadidates for e'(solid line). Dashed lines are neglected. Circles mean distances of the i-th nearest vertices from v1 and v2 respectively. For example, i = 10.

**Fig. 1.** The outline of EAX

In Step 3, *E-set* can be constructed from any combination of *AB-cycle*s. The following two methods were proposed in previous reports [6][14].

**EAX-Rand:** *E-set* was constructed by randomly selecting *AB-cycle*s. The intermediate solution tends to include edges of tour-A and tour-B evenly.

**EAX-1AB:** *E-set* was constructed from a single *AB-cycle*. The intermediate solution tends to be similar to tour-A, i.e., children are generated by removing a small number of edges from tour-A and adding the same number of edges to it.

## 2.2   Fast Algorithm for EAX-1AB

Indeed, most of the computational cost of the EAX algorithm implemented by [6] is in Step 5 of the EAX algorithm. Following is the detail of Step 5.

**(5-1)** Let $U_i$ $(i = 1, \ldots, k)$ be the set of edges included in the i-th sub-tour, where $k$ is the number of sub-tours in an intermediate solution.
**(5-2)** Choose the smallest sub-tour from $|U_i|$ $(i = 1, \ldots, k)$, where $|U_i|$ is the number of edges in $U_i$. Let $U_r$ be a selected sub-tour.
**(5-3)** Find a pair of edges, $e \in U_r$ and $e' \in U_j$ $(j \neq r)$, so that it minimizes $\{-w(e) - w(e') + w(e'') + w(e''')\}$ where $e''$ and $e'''$ are determined to connect two sub-tours (see Fig.1(d)). Let $U_s$ be a sub-tour including edge $e'$. $U_r$ and $U_s$ are merged by $U_r := (U_r \cup U_s - \{e, e'\}) \cup \{e'', e'''\}$, and empty $U_s$. $U_s := U_k$ and subtract 1 from $k$.
**(5-4)** If $k$ is equal to 1, $U_1$ is a valid tour, then terminate, else go to (5-2).

For the purpose of reducing the computational cost in (5-3), the original EAX algorithm [6] restricted candidates for edge $e'$ around edge $e$ (see Fig.1(e)). In practice, for each edge $v_1 v_2 (= e) \in U_r$, edges $v_3 v_4$ such that either $v_3$ or $v_4$ is at most the i-th nearest from $v_1$ or $v_2$ are considered as candidates for edge $e'$. $i$ was set to 10 in the implementation.

The computational cost for (5-3) is not excessive compared to (5-1) if an intermediate individual is constructed from one large sub-tour and other small sub-tours, because only the smallest sub-tour is considered to have merged into other sub-tours. When EAX-1AB is used, changes in tour-A due to *E-set* are usually localized and satisfy these conditions. When EAX-1AB is used, the computational cost of (5-1) can also be reduced. Here, we propose a fast implementation of (5-1).

**[Fast algorithm for Step (5-1)]**
We used the data structure in Fig.2 to represent intermediate individuals. *'city'* is an array that represents the route for tour-A. *'pos'* is an array that is defined by $pos\,[city[\,i\,]\,] = i$ $(i = 1, \ldots, N)$, where $N$ is the number of cities.

**(a)** Arrays *city* and *pos* are prepared once for tour-A as shown in Fig.2(a).
**(b)** A new route for an intermediate individual is generated by deleting edges (solid lines) from tour-A and adding edges(dashed lines) to it according to *E-set*. An new route including sub-tours are represented as shown in Fig.2(b). The dashed lines cut the route of tour-A. The gray solid lines connect two corresponding cities. This procedure can efficiently be executed by using *pos*.

(c) Sub-tours are represented as shown in Fig.2(c). For example, sub-tour1 includes cities located from the 5-th to the 7-th position in the array *city*.
However, a naive algorithm requires a computational cost of $O(N)$ to obtain
these representations because all cities in the array *city* are traced according
to the new route. To reduce the computational cost, the positions of cities incident to added edges on the array *city* must be sorted as shown in Fig.2(c).
By using this, we can know both end positions for the blocks separated by
the dashed vertical lines in Fig.2(b). If the number of removed edges is much
smaller than $N$ (number of cities), this procedure s much faster than the
naive method.

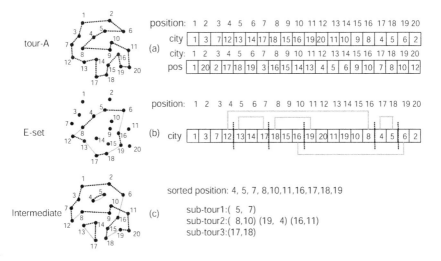

**Fig. 2.** Fast implementation of Step (5-i)

## 2.3   Effect of the Fast Algorithm for EAX-1AB

This subsection discusses our demonstration of the effects of the proposed algorithm for EAX-1AB. EAX-1AB was executed by using both naive and fast
algorithms. We also describe the results we obtained for EAX-Rand. These EAX
crossovers were incorporated with the greedy selection model described in 4.1 and
applied to several TSP benchmarks. Details on the experiments are described
in 4.1. Table 1 lists the results where we can see EAs using the fast algorithm
for EAX-1AB can be executed 2–7 times faster than the naive method. The
efficiency, especially, becomes increasingly prominent as the number of vertices
increases. Although running time of EAX-Rand became 1–2 times faster than
the naive method, the effect of the fast implementation is not so prominent as
in the case of EAX-1AB. EAs using EAX-1AB seem to obtain slightly better
quality solutions than EAX-Rand.

**Table 1.** Effects of the fast algorithm of EAX-1AB. 'time' means average execution time in seconds required for a run, where left and right values correspond to naive and fast algorithms. EAs were implemented in C++ and executed on 3.06 GHz Xeon single processor. Please refer Table 2 for 'opt', 'err.' and 'gen.'

| | | | | | EAX-Rand(Greedy) | | | | | | EAX-1AB(Greedy) | |
|---|---|---|---|---|---|---|---|---|---|---|---|---|
| instance | $N_p$ | $N_{ch}$ | opt. | err.(%) | gen. | time(s) | $N_p$ | $N_{ch}$ | opt. | err.(%) | gen. | time(s) |
| mv1084 | 300 | 50 | 11 | 0.0099 | 19.0 | 96 -> 65 | 300 | 30 | 24 | 0.0083 | 49.3 | 97 -> 28 |
| pcb1173 | 300 | 50 | 35 | 0.0033 | 21.3 | 117 -> 79 | 300 | 30 | 37 | 0.0020 | 63.3 | 71 -> 25 |
| u1432 | 300 | 50 | 12 | 0.0259 | 20.6 | 305 -> 257 | 300 | 30 | 12 | 0.0283 | 81.8 | 170 -> 59 |
| mv1748 | 300 | 50 | 23 | 0.0224 | 24.3 | 469 -> 330 | 300 | 30 | 18 | 0.0308 | 88.7 | 304 -> 72 |
| pr2392 | 300 | 50 | 29 | 0.0043 | 27.3 | 526 -> 335 | 300 | 30 | 40 | 0.0015 | 117.1 | 406 -> 93 |
| pcb3038 | 300 | 50 | 0 | 0.0165 | 38.6 | 1289 -> 504 | 300 | 30 | 8 | 0.0077 | 182.8 | 1273 -> 224 |
| fnl4461 | 300 | 50 | 0 | 0.0546 | 72.5 | 5355 -> 4067 | 300 | 30 | 2 | 0.0065 | 341.2 | 4372 -> 874 |
| rl5915 | 300 | 50 | 0 | 0.0124 | 31.2 | 2878 -> 1667 | 300 | 30 | 0 | 0.0091 | 138.3 | 3220 -> 475 |

# 3   Diversity Preserving GA with EAX

We proposed the fast algorithm for EAX-1AB in the previous section. Another important issue in this paper is a selection model. This section proposes a selection method that can efficiently and explicitly maintain population diversity.

## 3.1   Diversity of Population for TSP

Maintaining population diversity has been an important subject in the field of EAs. Several EAs have been applied to TSP that took population diversity into account [12][13][14]. Maekawa et al. [13] proposed a GA framework called TDGA where an edge entropy measure was used to evaluate population diversity. The edge entropy is defined as

$$H = - \sum_{e \in X} F(e)/N_p \log(F(e)/N_p), \tag{1}$$

where $X$ is a set of edges included in the current population, $N_p$ is the size of the population, and $F(e)$ is the number of edges $e$ in the current population. TDGA explicitly utilized $H$ in the selection for survival. Offspring with this method survive to minimize $L - TH$. $L$ is the average tour length for the population, and $T$ is a parameter having positive value. However, their method involves excessive computational cost to calculate $H$.

We employed $H$ as an indicator of population diversity in this paper, and propose a method of selection for survival that can efficiently calculate $H$ and utilize it.

## 3.2   Selection Model for Survival

We used the selection model described below which was used in [8][14].

**[Selection Model I ]**

**(0)** Set $N_p$ as the size of the population and $N_{ch}$ as the number of children generated from a pair of parents.

**(1)** Set $t = 0$. Generate $N_p$ solution candidates $x_1, x_2, \ldots, x_{N_p}$ with an appropriate method.
**(2)** Randomly shuffle the population; *i.e.*, the index is randomly assigned.
**(3)** For $i = 1, \ldots, N_p$, $x_i$ and $x_{i+1}$ are selected as a pair of parents. ( $x_i$ becomes tour-A and $x_{i+1}$ becomes tour-B in Fig.1. )
**(4)** For each pair of parents, generate $N_{ch}$ children from them with a crossover operator. Select the 'best' individual from $x_i$ and the children, that is denoted as $x_i'$. $x_i$ is replaced with $x_i'$.
**(5)** If termination conditions are satisfied, then stop, else increment $t$ and go to (2).

Figure 3(a) plots the population diversity $H$ against the average tour length $L$ in the population. These curves represent averaged data applied to pr2392 instances in the experiments described in 2.3. Although $H$ tends to decrease as $L$ decreases, $H$ should be maintained as high as possible for each value of $L$. EAX-1AB in the figure maintains $H$ slightly better than EAX-Rand.

In Step (4) of Selection Model I, each pair of parents, $\{x_i, x_{i+1}\}$, generates $N_{ch}$ children, and then the 'best' individual is selected as $x_i'$ to replace $x_i$. Let $y$ be a child generated from $\{x_i, x_{i+1}\}$. Also, let $\Delta L(y)$ and $\Delta H(y)$ be differences of $L$ and $H$ as the result of replacing $x_i$ with $y$. Figure 3(b) has an example of the distribution of $(\Delta L(y), \Delta H(y))$, which are generated from a pair of parents. An individual with the smallest $\Delta L(y)$ is selected as $x_i'$ in a greedy manner, which is marked with 'a' in Fig.3(b). From the point of view of accelerating population growth, $\Delta L(y)$ should be small. In the greedy selection, $y$ is evaluated as

$$eval_{Greedy}(y) = -\Delta L(y) \tag{2}$$

$\Delta H(y)$, on the other hand, should be large for the purpose of preserving population diversity. Consequently, there is a trade-off between $\Delta L(y)$ and $\Delta H(y)$. The five Pareto individuals in the figure are marked 'a'–'e'.

In Step (4), $x_i'$ should be selected from Pareto individuals. Although we could not determine the appropriate trade-off between $\Delta L(y)$ and $\Delta H(y)$, we employed

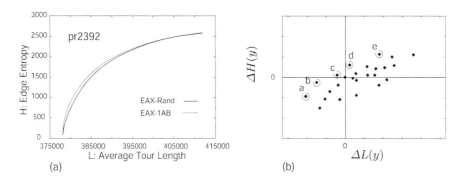

**Fig. 3.** Relations between $H$ and $L$

the following criteria to determine which individual should be selected as $x_i'$. An individual with the largest evaluation value is selected as $x_i'$.

$$eval_{ENT}(y) = \begin{cases} \Delta L(y)/\Delta H(y) \text{ if } \Delta H(y) < 0 \\ -\Delta L(y)/\epsilon \text{ if } \Delta H(y) \geq 0 \end{cases} \tag{3}$$

$\epsilon$ in this evaluation function has a sufficiently small positive value. Basically, child $y$ is evaluated by the amount of decrease in tour length per loss in population diversity. If there are children with $\Delta H(y)$ larger than zero and $\Delta L(y)$ less than zero, an individual with the smallest $\Delta L(y)$ is selected from these. Because $eval_{ENT}(x_i)$ is equal to 0, $x_i$ itself survives if there are no children with a positive evaluation value. For example in Fig.3(b), the individual marked 'c' has been selected as $x_i'$. If there are no individuals marked 'c', child 'b' is selected.

The proposed evaluation function is similar to that of EAX-Dis [14]. EAX-Dis employed the following evaluation function.

$$eval_{LocalDis}(y) = \begin{cases} \Delta L(y)/\Delta D(y) \text{ if } \Delta D(y) < 0 \\ -\Delta L(y)/\epsilon \text{ if } \Delta D(y) \geq 0 \end{cases} \tag{4}$$

$\Delta D(y)$ is defined by $d(x_i, y) - d(x_i, x_{i+1})$, where $d(a, b)$ is the number of different edges between individuals $a$ and $b$. $\Delta D(y)$ locally estimates the loss in population diversity caused by replacement of $x_i$ with $y$. Calculation of $\Delta D(y)$ is very easy.

### 3.3   Calculation of $\Delta H(y)$

$\Delta L(y)$ and $\Delta H(y)$ must be calculated to evaluate each child $y$ in Step (4) of Selection Model I. If EAX-1AB is used, they can be calculated at negligible computational cost because $x_i$ and $y$ are generally similar, i.e., $y$(child) is produced by removing a small number of edges from $x_i$(tour-A) and adding the same number of edges to it. Let $E_{ad}$ be a set of edges added to $x_i$ and $E_{re}$ be a set of edges removed from $x_i$ to generate $y$. $E_{ad}$ and $E_{re}$ are recorded in the EAX-1AB procedure. Then, $\Delta L(y)$ and $\Delta D(y)$ are calculated as

$$\Delta L(y) = \frac{1}{N_p} \{ \sum_{e \in E_{ad}} w(e) - \sum_{e \in E_{re}} w(e) \} \tag{5}$$

$$\Delta H(y) =$$
$$\sum_{e \in E_{ad}} \{-(F(e)+1)/N_p \log((F(e)+1)/N_p) + F(e)/N_p \log((F(e))/N_p)\}$$
$$+ \sum_{e \in E_{re}} \{-(F(e)-1)/N_p \log((F(e)-1)/N_p) + F(e)/N_p \log((F(e))/N_p)\} \tag{6}$$

where $w(e)$ is a weight (lenth) of an edge $e$.

After $x_i$ is replaced with $x_i'$, $F(e)$ is updated by adding 1 to $F(e)$ if $e \in E_{ad}$, and subtracting 1 from $F(e)$ if $e \in E_{re}$. When EAX-1AB is used, $|E_{ad}|$ and $|E_{re}|$ are usually much smaller than $N$. Therefore, the computational cost of calculating $eval_{ENT}(y)$ is negligible compared with the EAX-1AB procedure.

For example, the computational costs for calculating $\Delta H(y)$ is about 5 % in pr2392 instance. Because the selection for survival in TDGA are not restricted within a family , the computational cost of edge entropy becomes a bottleneck where the fast algorithm for EAX-1AB is used.

## 4   Experiments

### 4.1   Comparisons with Selection Methods

We compare three evaluation methods, i.e., $eval_{Greedy}(y)$, $eval_{LocalDis}(y)$, and $eval_{ENT}(y)$ on Selection Model I where the fast algorithm for EAX-1AB is used. Each method is denoted by EAX-1AB(Greedy), EAX-1AB(LocalDis), or EAX-1AB(ENT) and applied to several TSP benchmarks [15]. No mutation is used in order to focus on the evaluation methods.

Although EAX-1AB(Greedy) and EAX-1AB(LocalDis) used a population $N_p$ of 300, EAX-1AB(ENT) used 200 because EAs using EAX-1AB(ENT) converge slower than the others. $N_{ch}$ were set to 30. Fifty runs were executed for each instances. When the individuals of a population become similar at the end of evolution, the number of $AB$-$cycles$ formed by the EAX algorithm are less than $N_{ch}$. Therefore, EAX-Rand with greedy selection (EAX-Rand (Greedy)) is used after no improvements in the shortest tour length in the population are observed over 30 generations in Selection Model I. If the shortest tour length in the population stagnates over 50 generations, then terminate a run. In Step (1) of Selection Model I, the initial population was generated by 2-opt local search [1]. These algorithms were implemented in C++, and executed using a Xeon 3.06 GHz single processor with 2 GB RAM.

**Table 2.** Comparisons of performance of EAs using three selection methods. Column headed 'opt.' means number of trials that can reach optimal solutions. 'err.' means average excess from optimal tour length. 'gen.' means average generation (loop from Step (2) to (4) in Model I) required to reach best individual in each trial. 'time' means average execution time in seconds required for one run, where EAs were implemented in C++ and executed on 3.06 GHz Xeon single processor.

| instance | EAX-1AB(Greedy) Np = 300, Nch = 30 | | | | EAX-1AB(LocalDis) Np = 300, Nch = 30 | | | | EAX-1AB(ENT) Np = 200, Nch = 30 | | | |
|---|---|---|---|---|---|---|---|---|---|---|---|---|
| | opt. | err.(%) | gen. | time(s) | opt. | err.(%) | gen. | time(s) | opt. | err.(%) | gen. | time(s) |
| mv1084 | 24 | 0.0083 | 49.3 | 28 | 47 | 0.0011 | 62.2 | 36 | 50 | 0.0000 | 89.2 | 36 |
| pcb1173 | 37 | 0.0020 | 63.3 | 25 | 49 | 0.0000 | 85.1 | 35 | 50 | 0.0000 | 135.2 | 43 |
| u1432 | 12 | 0.0283 | 81.8 | 59 | 26 | 0.0105 | 110.8 | 62 | 35 | 0.0035 | 120.9 | 49 |
| vm1748 | 18 | 0.0308 | 88.7 | 72 | 50 | 0.0000 | 96.7 | 91 | 50 | 0.0000 | 153.4 | 90 |
| pr2392 | 40 | 0.0015 | 117.1 | 93 | 50 | 0.0000 | 164.2 | 123 | 50 | 0.0000 | 260.1 | 121 |
| pcb3038 | 8 | 0.0077 | 182.8 | 224 | 34 | 0.0013 | 262.1 | 288 | 50 | 0.0000 | 398.3 | 287 |
| fnl4461 | 2 | 0.0065 | 341.2 | 874 | 29 | 0.0013 | 535.3 | 1156 | 45 | 0.0001 | 786.5 | 1128 |
| rl5915 | 0 | 0.0091 | 138.3 | 475 | 12 | 0.0037 | 184.9 | 692 | 42 | 0.0014 | 326.6 | 608 |
| usa13509 | 0 | 0.0112 | 1164.6 | 7766 | 4 | 0.0015 | 1828.6 | 10534 | 0 | 0.0021 | 2561.7 | 12999 |

Table 2 lists the performance of the three methods. EAX-1AB(ENT) can reach optimal solutions more frequently than EAX-1AB(LocalDis) while having almost the same computational cost. However, the solution quality of usa13509 obtained by EAX-1AB(ENT) is worse than EAX-1AB(LocalDis). Especially, EAX-1AB(LocalDis) found an optimal solution. We guess that the edge entropy measure must be modified to address large size instances such as usa13509. EAX-1AB(Greedy) is fastest but the quality of solutions is significantly worse than the others, especially for large instances.

## 4.2   Comparisons with Other Optimization Methods

We compared EAX-1AB(ENT) with the heterogeneous selection evolutionary algorithm (HeSEA) [12]. In our survey on the application of EAs to TSPs, one of the best approaches was HeSEA. HeSEA is a hybrid that is constructed by integrating EAX and Chained Lin-Kernighan (CLK)[4]. We also considered other state-of-the-art TSP heuristic algorithms for comparison. From the results obtained by the "8-th DIMACS Implementation Challenge: The Traveling Salesman Problem" [2], we chose three representative methods that performed well. Iterated Lin-Kernighan (ILK) [1] is a simple yet powerful improvement to LKLS. Iterated local search of Chained Lin-Kernighan (CLK) [4] and Heslgaun's (LKH) [5] are known to be one of the most efficient approximation methods for TSPs.

EAX-1AB(ENT) was applied to 14 benchmarks [15] larger than 1000 cities under the same conditions as in the previous subsection, where the population was set 200 or 30. Fifty trials were executed for each instance. Table 3 lists the results for EAX-1AB(ENT) and the compared methods. Data of HeSEA is copied from [12], where twenty runs were executed for each instance. Data of ILK, CLK, and LKH are copied form "8-th DIMACS Implementation Challenge" [2], where only single run was executed for each instance. The err.(%) and time(s) are the average excess from the optima and the average running time in seconds for a run. The running times of EAX-1AB(ENT) and HeSEA are respectively based on Xeon 3.06 GHz single processor and Pentium IV 1.2 GHz processor. Running time for ILK, CLK and LKH are normalized running time on 500MHz Alpha processor. In our guess, Xeon 3.06 GHz processor is 4–5 times faster than 500MHz Alpha processor.

We can see that running time of EAX-1AB(ENT) with a population of 200 is roughly the same as that of HeSEA when differences in the machines are taken into consideration. Qualities of solutions are comparable. However, EAX-1AB(ENT) sometimes fails to find optimal solutions for u1432 and u2152 instances where cities are partially located within a reticular pattern. In this case, EAs may not work well because there exist several optimal solutions and EAs can not narrow the range of the promising search space.

EAX-1AB(ENT) with a population of 30 was compared to ILK, CLK, and LKH, because the proposed method is much slower if population of 200 is used. As we can see from the table, EAX-1AB(ENT) dominates ILK in some instances. The solution qualities of EAX-1AB(ENT) are often better than those of CLK, while running time of EAX-1AB(ENT) is larger than that of CLK, especially

**Table 3.** Comparisons of performance of EAX-1AB(ENT), HeSEA, ILK, CKL and LKH. The running times of EAX-1AB(ENT) and HeSEA are respectively based on Xeon 3.06 GHz single processor and Pentium IV 1.2 GHz processor. Runing time of ILK, CLK and LKH are normalized on 500MHz Alpha processor.

|  | EAX-1AB(ENT) [Np=200,Nch=30] | | EAX-1AB(ENT) [Np=30,Nch=30] | | HeSEA | | ILK [JM-N-b10] | | CLK [ABCC-10N] | | LKH [Helsgaun-N/10] | |
|---|---|---|---|---|---|---|---|---|---|---|---|---|
|  | err.(%) | time(s) | err.(%) | time(s) | err.(%) | time(s) | err.(%) | time(s) | err.(%) | time(s) | err.(%) | time(s) |
| mv1084 | 0.0000 | 36 | 0.02 | 4.2 | 0.0000 | 81 | 0.02 | 156.6 | 0.00 | 26.9 | 0.07 | 9.8 |
| pcb1173 | 0.0000 | 43 | 0.02 | 5.9 | 0.0000 | 85 | 0.01 | 66.3 | 0.16 | 19.7 | 0.18 | 8.5 |
| u1432 | 0.0035 | 49 | 0.12 | 7.0 | 0.0000 | 107 | 0.10 | 94.4 | 0.23 | 30.6 | 0.00 | 11.7 |
| d1655 | 0.0008 | 61 | 0.08 | 8.0 | N/A | | 0.01 | 604.4 | 0.38 | 27.6 | 0.00 | 6.7 |
| vm1748 | 0.0000 | 90 | 0.04 | 13.5 | 0.0000 | 141 | 0.00 | 352.0 | 0.05 | 51.5 | 0.02 | 21.3 |
| u2152 | 0.0003 | 93 | 0.14 | 13.8 | 0.0000 | 211 | 0.17 | 244.5 | 0.29 | 28.1 | 0.11 | 40.2 |
| pr2392 | 0.0000 | 121 | 0.02 | 19.5 | 0.0000 | 208 | 0.11 | 137.5 | 0.42 | 47.7 | 0.00 | 48.2 |
| pcb3038 | 0.0000 | 287 | 0.04 | 44.7 | 0.0000 | 612 | 0.12 | 257.0 | 0.26 | 66.6 | 0.03 | 84.2 |
| fnl4461 | 0.0001 | 1128 | 0.05 | 116.3 | 0.0005 | 2349 | 0.14 | 398.2 | 0.15 | 129.0 | 0.01 | 182.2 |
| rl5915 | 0.0014 | 608 | 0.09 | 85.7 | 0.0001 | 2773 | 0.02 | 1160.7 | 0.54 | 192.5 | 0.04 | 332.8 |
| usa13509 | 0.0021 | 12999 | 0.06 | 1645.2 | 0.0074 | 34984 | 0.16 | 4436.2 | 0.20 | 967.0 | 0.01 | 2631.1 |

in large instances. Although EAX-1AB(ENT) is dominated by LKH in most instances, the differences are not so large. As described in this comparisons, EAX-1AB(ENT) is quite comparable to state-of-the-art TSP heuristics even though it dose not use LKLS heuristics.

## 5  Conclusion

We proposed two ideas to improve the performances of EAs using EAX crossover. First, we proposed the fast implementation of EAX-1AB crossover. The key idea was that the change in edges caused by EAX would be localized, and the EAX algorithm would then be locally executed. EAs using EAX-1AB could be executed 2–7 times faster than the original algorithm (Table 1). We then proposed a selection model that could maintain population diversity at negligible computational cost. Here, edge entropy was used as an indicator of population diversity. The proposed selection model utilized this explicitly to maintain population diversity. Using this model, EAs with EAX-1AB could significantly improve the quality of solutions without increasing the execution time (Table 2). The proposed EA was called the EAX-1AB(ENT).

EA using EAX-1AB(ENT) was applied to several TSP benchmarks up to instances of 13509 cities. It was compared to other state-of-the-art TSP heuristics including HeSEA, ILK, CKL, and HLK. Those methods are all based on Lin-Kernighan local search (LKLS). The experiments demonstrated that the EAX-1AB(ENT) was quite comparable to these heuristics even thought it does not utilize LKLS (Table 3).

However, the solution quality of usa13509 obtained by EAX-1AB(ENT) is worth than EAX-1AB(LocalDisT). Especially, EAX-1AB(LocalDis) found an optimal solution. In the future work, the edge entropy measure must be modified to address large size instances.

# References

1. D. S. Johnson: Local Optimization and the Traveling Salesman Problem, Automata, Languages and Programming, Lecture note in Computer Science 442, Springer, Heidelberg, pp. 446-461.
2. 8-th DIMACS Implementation Challenge: The Traveling Salesman Problem, http://www.research.att.com/ dsj/chtsp.
3. S. Lin and B. Kernighan, Effective heuristic algorithms for the traveling salesman problem, Oper. Res., vol. 21, pp. 498- 516, 1973.
4. D. Applegate, R. Bixby, V. Chvatal, and W. Cook, "Finding tours in the TSP. Technical Report 99885, Forschungsinstitut fur Diskrete Mathematik, Universitat Bonn, 1999.
5. K. Helsgaun, "An effective implementation of the Lin-Kernighan traveling salesman heuristic, "Eur. J. Oper. Res., vol. 126, no.1, pp. 106-130, 2000.
6. Y. Nagata and S. Kobayashi, Edge Assembly Crossover: A High-power Genetic Algorithm for the Traveling Salesman Problem, Proc. of the 7th Int. Conference on Genetic Algorithms, pp. 450-457, 1997.
7. H.K. Tsai, J.M. Yang, and C.Y. Kao, Solving Traveling Salesman Problems by Combining Global and Local Search Mechanisms, Proc. of the the 2002 Congress on Evolutionary Computation, pp. 12920-1295, 2002.
8. K. Ideda and S. Kobayashi, Deterministic Multi-step Crossover Fusion: A Handy Crossover Composition for GAs, Proc. of the Seventh Int. Conference on Parallel Problem Solving from Nature, pp. 162-171, 2002.
9. J. Watson, C. Poss, D. Whitley et al., The Traveling Salesrep Problem, Edge Assembly Crossover, and 2-opt, Proc. of the Fifth Int. Conference on Parallel Problem Solving from Nature, pp. 823-833, 2000.
10. Y. Nagata, Criteria for Designing Crossovers for TSP, Proc. of the 2004 Congress on Evolutionary Computation, , pp. 1465-1472, 2004.
11. P. Merz and B. Freisleben: Genetic Local Search for the TSP: New Results, Proc. of the 1997 IEEE Int. Conf. on Evolutionary Computation, pp. 159-163 (1997).
12. H. K. Tsai, J. M. Yang, Y. F. Tsai, and C. Y. Kao, An Evolutionary Algorithm for Large Traveling Salesman Problem, IEEE Transaction on SMC-part B, vol. 34, no. 4, pp. 1718- 1729, 2004.
13. K. Maekawa, N. Mori, H. Kita, and H.Nishikawa, A Genetic Solution for the Traveling Salesman Problem by Means of a Thermodynamical Selection Rule, Proc. 1996 IEEE Int. Conference on Evolutionary Computation, pp. 529-534, 1996.
14. Y. Nagata, The EAX algorithm considering diversity loss, Proc. of the 8th Int. Conference on Parallel Problem Solving from Nature, pp. 332-341, 2004.
15. TSPLIB95, http://www.iwr.uni-heidelberg.de/iwr/compt/soft/TSPLIB95

# A Memetic Algorithm with Population Management (MA|PM) for the Capacitated Location-Routing Problem

Christian Prins[1], Caroline Prodhon[1], and Roberto Wolfler Calvo[1]

ISTIT, Université de Technologie de Troyes,
BP 2060, 10010 Troyes Cedex, France
{christian.prins, caroline.prodhon, roberto.wolfler}@utt.fr

**Abstract.** As shown in recent researches, in a distribution system, ignoring routes when locating depots may overestimate the overall system cost. The Location Routing Problem (LRP) overcomes this drawback dealing simultaneously with location and routing decisions. This paper presents a memetic algorithm with population management (MA|PM) to solve the LRP with capacitated routes and depots. MA|PM is a very recent form of memetic algorithm in which the diversity of a small population of solutions is controlled by accepting a new solution if its distance to the population exceeds a given threshold. The method is evaluated on three sets of instances, and compared to other heuristics and a lower bound. The preliminary results are quite promising since the MA|PM already finds the best results on several instances.

## 1 Introduction

Depot location and vehicle routing are crucial choices to reduce the logistic costs of companies, and often interdependent. Tackling separately these two levels of decision may lead to suboptimization [19]. The Location Routing Problem, LRP, appeared relatively recently in literature, is a combination of both levels. Given customers with known demands and possible depot locations, it consists of determining the depots to be opened and the vehicle routes connected to these depots, in order to cover the demands at minimum cost (see Section 2 for a formal definition).

As shown in [13], most early published papers consider either capacitated routes or capacitated depots, but not both [6, 10, 21]. In general, the LRP is formulated as a deterministic node routing problem (i.e., customers are located on nodes of the network). However, a few authors have studied stochastic cases [9, 5] and, more recently, arc routing versions [7, 8]. In the sequel, the case with capacities on both depots and routes is called *general LRP*. Albareda-Sambola *et al.* [1] proposed a two-phase Tabu Search (TS) heuristic for the LRP with one single route per capacitated open depot, tested on small instances only (at most 30 customers). Wu *et al.* [23] studied the general LRP with homogeneous or heterogeneous limited fleets. They divided the original problem into

J. Gottlieb and G.R. Raidl (Eds.): EvoCOP 2006, LNCS 3906, pp. 183–194, 2006.

two subproblems: a Location-Allocation Problem (LAP), and a Vehicle Routing Problem (VRP). Each subproblem is solved in a sequential and iterative manner by a Simulated Annealing (SA) algorithm with a tabu list to avoid cycling. An iterative "location first - route second" heuristic for a three-level LRP (factories-depots-customers) with capacity constraints on both depots and routes, and a maximum duration per route, was developed by Bruns and Klose [4]. Tuzun and Burke [22] developed another two-phase TS but for the LRP with capacitated routes and uncapacitated depots. The two phases are also dedicated to routing and location. The principle is to progressively increase the number of open depots until this deteriorates the total cost. These authors report results for up to 200 customers. Prins et al. have already developed two algorithms on the general LRP. The first one, [16], is a GRASP (Greedy Randomized Adaptive Search Procedure) with a memory on the depots used during a diversification phase. This information guides the search on the most promising depots during an intensification phase. The method is followed by a post-optimization based on a path relinking algorithm. The second one, [17], is a cooperative metaheuristic which alternates between a depot location phase and a routing phase, sharing some information. In the location phase, the routes are aggregated into super-customers to give a facility location problem solved by a Lagrangean relaxation of the assignment constraints. In the routing phase, the routes from the resulting multi-depot VRP are improved using a Granular Tabu Search (GTS). Barreto [3] developed a family of three-phase heuristics based on clustering techniques. Clusters of customers fitting vehicle capacity are formed in the first phase. A TSP is solved for each cluster in the second phase. Finally, in the third phase, the depots to open are determined by solving a facility location problem, in which the TSP cycles are aggregated into supernodes. Barreto also proposed a lower bound which is reached by the best heuristics on some small scale instances.

This paper deals with the general LRP with fixed costs to open a depot or a route. The objective is to determine the set of depots to open and the routes originating from each open depot, in order to minimize a total cost comprising the setup costs of depots and routes and the total cost of the routes. The proposed solution method, MA|PM, is a memetic algorithm (genetic algorithm hybridized with a local search procedure) with a population management technique based on a distance measure in the solution space. This last tool permits to control the diversity of a small population of good quality solutions. The paper is organized as follows. Section 2 defines the problem. The principle of MA|PM and how it is applied to the LRP is given in Section 3. Section 4 provides the global algorithm and the parameter used. To evaluate the first performances of the method, computational experiments are presented in Section 5 with some concluding remarks to close the paper.

## 2    Problem Definition

This paper deals with the general LRP with fixed costs to open a route or a depot. The data are based on a complete, weighted and undirected network

$G = (V, E, C)$. $V$ is a set of nodes comprising a subset $I$ of $m$ possible depot locations and a subset $J = V \backslash I$ of $n$ customers. The travelling cost between any two nodes $i$ and $j$ is given by $C_{ij}$. A capacity $W_i$ and an opening cost $O_i$ are associated to each depot site $i \in I$. Each customer $j \in J$ has a demand $d_j$. Identical vehicles of capacity $Q$ are available. When used, each vehicle incurs a fixed cost $F$ and performs one single route. The total number of vehicles used (or routes performed) is a decision variable.

The following constraints must hold:

- each demand $d_j$ must be served by one single vehicle;
- each route must begin and end at the same depot and its total load must not exceed vehicle capacity;
- the total load of the routes assigned to a depot must fit the capacity of that depot.

The total cost of a route includes the fixed cost $F$ and the costs of traversed edges. The objective is to find which depots should be opened and which routes should be constructed, in order to minimize the total cost (fixed costs of depots, plus total cost of the routes).

The LRP is obviously *NP*-hard since when $m = 1$, it reduces to the Vehicle Routing Problem (VRP), known to be *NP*-hard. It is even much more combinatorial than the VRP: in addition to the partition of customers into routes and the sequencing of each route, it involves the selection of open depots and the assignment of routes to these depots. Therefore only very small instances can be solved to optimality. A first non-trivial bound for the LRP with capacitated routes and depots has been proposed only recently [3] but it is too time-consuming beyond 50 customers.

## 3   MA|PM for the LRP

Evolutionary algorithms have been successfully applied to vehicle routing problems, especially the genetic algorithms hybridized with local search, also called memetic algorithms (MA), [11, 14]. Very recently, Sörensen and Sevaux [20] have proposed a new form called MA|PM or memetic algorithm with population management. MA|PM is characterized by a small population $P$, the improvement of new solutions by local search, and the replacement of the traditional mutation operator by a dynamic distance-based population management technique. Given a threshold $\Delta$, a new solution is accepted only if its distance to $P$ is at least $\Delta$. Otherwise, Sörensen and Sevaux proposed two options: either the offspring is mutated until its distance to $P$ reaches the threshold, or it is simply discarded. These authors described also several dynamic control policies for $\Delta$.

MA|PM has already been applied to the Capacitated Arc Routing Problem [18]. The results indicate that it converges faster than conventional memetic algorithms. Moreover, its general structure is simpler than other distance-based population metaheuristics such as scatter search or path relinking, and it is quite easy to upgrade an existing MA into an MA|PM.

The choice of MA|PM for the LRP has been inspired by these promising characteristics. The version studied here corresponds to the second option: children that do not match the threshold are simply discarded. Section 3.1 presents the structure of the chromosomes used, Section 3.2 explains the crossover phase, while Sections 3.3 and 3.4 respectively develop the local search and the population management technique.

## 3.1  Chromosomes and Evaluation

Defining a suitable chromosome encoding for the LRP is not trivial. Information about the depots, the assignment of customers and the order of deliveries to these customers has to be stored inside the chromosome. Moreover, it is important to design fixed length chromosomes because they are required by most crossovers and distance measures.

The adopted chromosome encoding comprises a depot status part $DS$ and a customer sequence part $CS$ (see Figure 1).

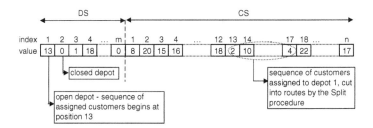

**Fig. 1.** Representation of an LRP solution as a chromosome

$DS$ (depot status) is a vector of $m$ numbers. $DS(i)$ represents the status of depot $i$ indicating whether it is closed (zero) or opened (non-zero value). If it is opened, $DS(i)$ is the index in $CS$ of the first customer assigned to depot $i$. $CS$ (customers sequence) contains the concatenation of the lists of customers assigned to the opened depots, without trip delimiters. So, $CS$ is a permutation of customers and has a fixed length.

The fitness $F(S)$ of a chromosome $S$ is the total cost of the associated LRP solution. $DS$ is used to deduce the cost of open depots and the list of customers assigned to each depot. Each such list can be optimally partitioned into trips using a procedure called Split. This procedure was originally developed by Prins for the VRP [14]. For a list of $p$ customers, it builds an auxiliary graph $H = (X, A, Z)$, where $X$ contains $p + 1$ nodes indexed from 0 to $p$, $A$ contains one arc $(i, j)$, $i < j$, if a trip servicing customer $S_{i+1}$ to $S_j$ (included) is feasible in terms of capacity. The weight $z_{ij}$ of $(i, j)$ is equal to the trip cost. The optimal splitting of the list corresponds to a min-cost path from node 0 to node $p$ in $H$. For the LRP, we apply Split to the list of customers of each depot.

## 3.2    Selection of Parents and Crossover

To generate an offspring, the first step is to select its parents. The first one comes from a binary tournament among the $\beta$ best solutions of the population, and the second one from a binary tournament on the whole population except the first selected parent. Then, basically, for two parents A and B, a one-point crossover is applied to the $DS$ vectors of the two parents and another one to their $CS$ vectors. The crossover for $DS$ works like for binary chromosomes. The one for $CS$ is adapted for permutations. The offspring C receives the sub-sequence of A located before the cutting point. B is then scanned from left to right, starting from the cutting point. The customers not yet in C are copied in $CS$ at the same position as in B to complete the offspring. Once arrived at the end of B, if there are still some vacant positions in $CS$, they are filled by non-inserted customers.

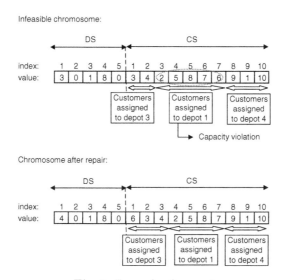

**Fig. 2.** Example of a repair

This operator may provide a chromosome corresponding to an infeasible solution, especially because of the capacity constraints on the depots. Thus, each child is tested and repaired in case of infeasibility (see figure 2). First of all, the repairing procedure checks if all the customers are assigned to a depot by verifying that the index 1 of $CS$ is in $DS$. It means that in the worst case, all the customers are assigned to a single depot. Otherwise, the first closed depot $i$ found is opened and $DS(i) = 1$. Then, the procedure looks for depots having a capacity violation. If one depot is found, that means that too much customers are assigned to it. Therefore, the algorithm scans backward the sequence of customers assigned to such a depot and removes them one by one until the capacity constraint of the depot holds. Removed customers are assigned to the first opened depot found having enough remaining capacity. If none exists, the closest depot is opened.

The offspring, if accepted by the population management system described in Section 3.4, replaces the worst chromosome in the population.

## 3.3   Local Search Procedures

MA|PM works on a small population of high-quality solutions [20]. This quality results from the application of a local search procedure. For the LRP, the offspring undergoes a local search LS1 based on the three following neighborhoods:

- *MOVE.* One customer is shifted from its current position to another position, in the same route or in a different route which may be assigned to the same depot or not, provided capacities are respected.
- *SWAP.* Two customers are exchanged. They may belong to the same route or, if residual capacities allow it, to two distinct routes sharing one common depot or not.
- *OPT.* This is a 2-opt procedure we proposed in [16] and in which two non-consecutive edges are removed, either in the same route or in two distinct routes assigned to a common depot or not. When they belong to different routes, there are various ways of reconnecting the trips. If they are from different depots, edges connecting the last customers of the two considered routes to their depot have to be replaced to satisfy the constraint imposing that a route must begin and finish at the same depot. This neighborhood is equivalent in the first case to the well-known 2-OPT move for the TSP [12].

LS1 must not be called systematically, to avoid a premature convergence and also because it is time-consuming with its $O(n^2)$ neighborhoods. In practice, it is applied to the offspring with a fixed probability $p_1$. Otherwise, a "light" version LS2 is executed with a given probability $p_2$. LS2 is identical to LS1, except that the moves involving two trips from two different depots are not evaluated.

Both LS1 and LS2 execute the first improving move found in the three neighbourhoods (not the best move) and stop when no such move can be found. Note that a depot can only be opened by the crossover and not by these procedures.

## 3.4   Population Management

The population management controls population diversity by filtering the offspring thanks to a distance measure in the solution space. Let $D(T, U)$ be the distance between two solutions $T$ and $U$ (chromosomes after the evaluation by Split). It can be defined as follows. For each pair $(i, j)$ of consecutive customers in $T$, four cases are considered:

- the pair $(i, j)$ or $(j, i)$ is found in $U$, so the pair is not broken, and does not contribute to the distance measure;
- $i$ and $j$ are no longer adjacent in $U$, but they are in a common trip: count 1 in the distance;
- $i$ and $j$ are in different trips in $U$, but assigned to one common depot: count 3;
- $i$ and $j$ are assigned to two different depots in $U$ : count 10.

Moreover, for each customer $i$ serviced at the beginning of a trip in $T$, count 10 if $i$ is assigned to another depot in $U$.

This measure is not exactly a distance in the mathematical sense, because the triangle inequality does not hold. However, note that $D(T, U) \geq 0$ and $T = U \iff D(T, U) = 0$. Moreover, this distance measure correctly detects equivalent solutions, e.g. when the trips are renumbered or when some of them are inverted (performed backward by the vehicle).

The distance of a solution $T$ to the current population $P$ is defined by:

$$d_P(T) = \min_{U \in P} D(T, U) \tag{1}$$

In MA|PM, a new solution $T$ may enter the population only if $d_P(T) \geq \Delta$, where $\Delta$ is a given threshold. If $\Delta = 0$, the algorithm behaves like a traditional MA. $\Delta \geq 1$ ensures distinct solutions in $P$ (note that the distance is integer). If $\Delta$ takes a large value, most children-solutions are rejected and the MA spends too much time in unproductive iterations. $\Delta$ can be dynamically adjusted between such extremes to control population diversity. Different control policies are suggested in [20]. In our MA|PM, $\Delta$ is initialized to a rather high value $\Delta_{max}$. If a series of $MaxNbRej$ successive rejections of the offsprings is reached, $\Delta$ decreases until accepting a new offspring in the population, but keeping the minimum value of $\Delta$ equal to 1. Then, $\Delta$ remains constant until a new series of successive rejections occurs. If a new best solution is found, $\Delta$ is reset to $\Delta_{max}$.

# 4   Algorithm and Parameters

## 4.1   General Structure

The algorithm begins by creating a population of $NbIndiv$ good solutions using two different randomized heuristics. The first half of the population comes from a randomized constructive heuristic based on the Nearest Neighbor Algorithm, followed by the local search LS1. The second half of the population comes from a greedy randomized heuristic based on an extended Clarke and Wright saving algorithm (ECWA) for the LRP [16], followed by LS1. All these initial solutions are distinct ($\Delta = 1$).

Periodically, the population undergoes a partial replacement of its solutions. This refreshment of the population occurs when a given number $MaxNbNoAdd$ of rejected offsprings has been reached. Every individual except the best one (having the best fitness) is replaced in the population. The algorithm stops when a given number $MaxNbAcc$ of accepted offsprings has been generated. An overview of the pseudocode of MA|PM for the LRP is given Algorithm 1..

The variable $BestCost$, $Cost(C)$, $NbAcc$, $NbRej$, $NbNoAdd$ respectively refers to the cost of the best solution found, the cost of the current offspring, the counter of accepted offsprings during the algorithm, the counter of successive iterations without improvement of the best solution, the counter of successive iterations without acceptation of offspring in the population and the counter of rejected offsprings in a population.

**Algorithm 1.** Overview of the algorithm

1: BestCost := $+\infty$
2: NbAcc := 0
3: NbRej := 0
4: $\Delta := \Delta_{max}$
5: GenPop(P)
6: //Main loop
7: **repeat**
8:   //Second loop
9:   **repeat**
10:     Selection(A,B)
11:     Crossover(A,B,C)
12:     **if** (random < P1) **then**
13:       $LS1(C)$
14:     **else**
15:       **if** (random < P2) **then**
16:         $LS2(C)$
17:       **end if**
18:     **end if**
19:     **if** (cost(C) < BestCost) **then**
20:       $\Delta := \Delta_{max}$
21:       BestCost := cost(C)
22:       BestSoln := C
23:     **end if**
24:     **if** ($d_P(C) < \Delta$) **then**
25:       NbRej := NbRej + 1
26:       NbNoAdd := NbNoAdd + 1
27:     **else**
28:       NbRej:= 0
29:       NbAcc := NbAcc +1
30:       AddInPop(C)
31:     **end if**
32:     **if** (NbRej > MaxNbRej) **then**
33:       $\Delta := \text{Max}(1, \Delta - 1)$
34:     **end if**
35:   **until** (NbNoAdd > MaxNbNoAdd and NbAcc > MaxNbAcc)
36:   $\Delta := \Delta_{max}$
37:   NbNoAdd := 0
38:   GenPop2(P,BestSoln)
39: **until** (NbAcc > MaxNbAcc)
40: Return (BestSoln)

The function *GenPop, Selection, Crossover, AddInPop, GenPop* and *Return* respectively generates a population, selects the parents, creates a feasible offspring, replaces the worst individual in the population by an accepted offspring, refreshes the population and returns the best solution found. LS1, LS2 refer to the local searches.

## 4.2   Parameters

After a preliminary testing phase on the instances, the parameters used in this metaheuristic are set to the values provided in Table 1.

Table 1. Parameters of MA|PM for the LRP

| Parameter | Value | Parameter | Value |
|---|---|---|---|
| NbIndiv | $\lceil((n+m)/5)+1\rceil$ | $\beta$ | $NbIndiv \cdot 1/3$ |
| MaxNbAcc | $(n+m) \cdot 10$ | $\Delta_{max}$ | $\lceil(n+m)/10)+10\rceil$ |
| MaxNbNoAdd | $\lceil(n+m) \cdot 2.5\rceil$ | $p_1$ | 0.5 |
| MaxNbRej | NbIndiv | $p_2$ | 0.7 |

# 5   Results

MA|PM for the LRP is evaluated on three sets of randomly generated Euclidean instances briefly described in subsection 5.1. Tables of results are provided and commented in subsection 5.2. Some remarks close the paper in subsection 5.3.

## 5.1   Instances

*The first set* contains 30 LRP instances with capacitated routes and depots, elaborated for a previous work on a multi-start local search method (MSLS) [15], and used for a GRASP [16] and LRGTS [17]. It contains the largest instances with capacitated depots (200 customers). *The second set* comprises 36 instances with uncapacitated depots used by Tuzun and Burke [22] to evaluate a tabu search and in which $n \in \{100, 150, 200\}$, $m \in \{10, 20\}$. *The third set* of 13 instances was gathered by Barreto in his recent thesis on clustering heuristics for the LRP [3]. It may be downloaded at [2]. These files either come from the LRP literature or are obtained by adding several depots to classical VRP instances. The algorithm is coded in Visual C++ and tested on a Dell PC Optiplex GX260, with a 2.4 GHz Pentium 4, 512 MB of RAM and Windows XP. Here the results are not given for each instance but on average by instances size.

## 5.2   Commented Results

Contrary to classical GAs, the results of a memetic algorithm are very robust, i.e. they are only slightly affected by the random number generator used. This can be explained by the local search, which converts the offspring into local optima. This is why the results reported come from one single run. On the Tables, the *Gap* columns indicate the deviations in % to MA|PM taken as reference. Hence, the negative gaps correspond to improvements compared to MA|PM.

Table 2 provides a comparison between MA|PM and three other methods MSLS [15], GRASP [16] and LRGTS [17] on the 30 instances initially designed for MSLS. MSLS [15] is basically a multi-start local search algorithm that uses some

**Table 2.** Results for the first set with capacitated depots

| n | m | $MA \mid PM$ | | $LRGTS$ | | | $GRASP$ | | | $MSLS$ | | |
|---|---|---|---|---|---|---|---|---|---|---|---|---|
| | | Cost | CPU | Cost | CPU | Gap | Cost | CPU | Gap | Cost | CPU | Gap |
| 20 | 5 | 45087 | 0 | 45104 | 0 | 0.0 | 45144 | 0 | 0.1 | 45530 | 0 | 0.9 |
| 50 | 5 | 74188 | 3 | 74515 | 1 | 0.4 | 74701 | 2 | 0.7 | 79077 | 2 | 6.2 |
| 100 | 5 | 201094 | 32 | 200544 | 5 | -0.1 | 202210 | 22 | 0.7 | 203548 | 27 | 1.3 |
| 100 | 10 | 249248 | 29 | 241178 | 13 | -2.7 | 257048 | 35 | 3.2 | 267780 | 51 | 8.0 |
| 200 | 10 | 422228 | 308 | 423046 | 73 | 0.2 | 447657 | 422 | 6.1 | 487236 | 744 | 15.0 |
| Average | | | | | | -0.4 | | | 2.2 | | | 6.6 |

**Table 3.** Results for the third set with uncapacitated depots

| n | m | $MA \mid PM$ | | $LRGTS$ | | | $GRASP$ | | | $TS$ | | |
|---|---|---|---|---|---|---|---|---|---|---|---|---|
| | | Cost | CPU | Cost | CPU | Gap | Cost | CPU | Gap | Cost | CPU | Gap |
| 100 | 10 | 1174.3 | 34 | 1179.1 | 7 | 0.6 | 1191.4 | 23 | 1.5 | 1215.7 | 4 | 3.5 |
| 100 | 20 | 1180.7 | 42 | 1185.0 | 7 | 0.5 | 1196.0 | 35 | 1.3 | 1205.4 | 3 | 2.0 |
| 150 | 10 | 1587.2 | 138 | 1600.4 | 24 | 1.0 | 1625.0 | 92 | 2.4 | 1639.6 | 11 | 3.5 |
| 150 | 20 | 1453.2 | 167 | 1449.6 | 25 | -0.1 | 1466.6 | 127 | 1.1 | 1474.3 | 11 | 1.6 |
| 200 | 10 | 1983.0 | 375 | 1987.7 | 65 | 0.3 | 2020.7 | 232 | 1.9 | 2051.6 | 21 | 3.7 |
| 200 | 20 | 1814.7 | 422 | 1783.1 | 73 | -1.3 | 1839.3 | 337 | 1.4 | 1814.1 | 20 | 0.3 |
| Average | | | | | | 0.2 | | | 1.6 | | | 2.4 |

randomness in its search. It is the slowest method (164.9 seconds on average) and, in spite of 2 best solutions (also found by other algorithms), it displays an average deviation of 6.6% to MA|PM. The relative slowness can be explained by the time-consuming look-ahead process used to try closing some depots. The other heuristics are faster and also more efficient than MSLS. Thanks to the granularity system, LRGTS is by far the fastest method, nearly 5 and 4 times respectively faster than GRASP and MA|PM, and 9 times faster than MSLS, while finding the best solution for 11 out of the 30 instances. Its better average solution quality can be explained by the cooperation between the location and routing phases. Nevertheless, MA|PM outperforms GRASP and is able to compete with LRGTS on many instances. It even outperforms LRGTS on 16 instances. In fact, LRGTS does slightly better on average (-0.4%) mainly because it saves 9% on two very hard instances, for which MA|PM is not so good. Without these two instances, the ranking would be inverted and LRGTS would be at 0.2% to MA|PM. Table 3 provides a comparison between MA|PM and GRASP [16], LRGTS [17] and TS [22] on the second set of instances. MA|PM becomes the method providing the best results on average. It also seems more effective than LRGTS on small and medium size instances, especially with uncapacitated depots, while LRGTS is still superior on most of the largest instances.

Finally, due to lack of room, results for the third set of instances are not reported in a Table. We have compared Barreto's clustering heuristic CH [3], GRASP [16], LRGTS [17] and MA|PM with the lower bound provided by Barreto [3], and they are respectively at 5.3%, 3.3%, 3.6% and 3.7% to this bound.

GRASP, LRGTS and MA|PM are here almost as good as each other, with a little advantage to GRASP.

## 5.3   Conclusion

In this paper, a new metaheuristic for the Location Routing Problem (LRP) with both capacitated depots and vehicles is presented. The method is an evolutionary algorithm called memetic algorithm with population management (MA|PM). It is a genetic method hybridized with local search techniques and a distance measure permitting to control the diversification of the solutions. The method is not yet definitively tuned but it has been already tested on three sets of small, medium and large scale instances with up to 200 customers, and compared to other heuristics and a lower bound. The solutions obtained show that this algorithm is promising as it already outperforms other metaheuristics except one and is able to find good quality solutions on various kinds of instances.

### Acknowledgments

This research is partly supported by the Champagne-Ardenne Regional Council and by the European Social Fund.

# References

1. M. Albareda-Sambola, J.A Díaz, and E. Fernández. A compact model and tight bounds for a combined location-routing problem. *Computers and Operations Research*, 32(3):407–428, 2005.
2. S. S. Barreto. http://sweet.ua.pt/ iscf143, 2004.
3. S. S. Barreto. *Análise e Modelização de Problemas de localização-distribuição [Analysis and modelling of location-routing problems]*. PhD thesis, University of Aveiro, campus universitário de Santiago, 3810-193 Aveiro, Portugal, October 2004. [In Portuguese].
4. A. Bruns and A. Klose. An iterative heuristic for location-routing problems based on clustering. In *Proceedings of the second International Workshop on Distribution Logistics*, Oegstgeest, The Netherlands, 1995.
5. Y. Chan, W.B. Carter, and M.D. Burnes. A multiple-depot, multiple-vehicle, location-routing problem with stochastically processed demands. *Computers and Operations Research*, 28:803–826, 2001.
6. T. W. Chien. Heuristic procedures for practical-sized uncapacitated location-capacitated routing problems. *Decision Sciences*, 24(5):995–1021, 1993.
7. G. Ghiani and G. Laporte. Location-arc routing problems. *OPSEARCH*, 38:151–159, 2001.
8. N. Labadi. *Problèmes tactiques et stratégiques en tournées sur arcs*. PhD thesis, Université de Technologie de Troyes, 2003.
9. G. Laporte, F. Louveaux, and H. Mercure. Models and exact solutions for a class of stochastic location-routing problems. *European Journal of Operational Research*, 39:71–78, 1989.

10. G. Laporte, Y. Norbert, and S. Taillefer. Solving a family of multi-depot vehicle routing and location-routing problems. *Transportation Science*, 22(3):161–167, 1988.
11. C.M.R.R. Lima, M.C. Goldbarg, and E.F.G Goldbarg. A memetic algorithm for heterogeneous fleet vehicle routing problem. *Electronic Notes in Discrete Mathematics*, (18):171–176, 2004.
12. S. Lin and B.W. Kernighan. An effective heuristic algorithm for the traveling salesman problem. *Operations Research*, 21:498–516, 1973.
13. H. Min, V. Jayaraman, and R. Srivastava. Combined location-routing problems: A synthesis and future research directions. *European Journal of Operational Research*, 108:1–15, 1998.
14. C. Prins. A simple and effective evolutionary algorithm for the vehicle routing problem. *Computers and Operations Research*, 2004.
15. C. Prins, C. Prodhon, and R. Wolfler Calvo. Nouveaux algorithmes pour le problème de localisation et routage sous contraintes de capacité. In A. Dolgui and S. Dauzère-Pérès, editors, *MOSIM'04*, volume 2, pages 1115–1122, Ecole des Mines de Nantes, September 2004. Lavoisier.
16. C. Prins, C. Prodhon, and R. Wolfler Calvo. A reactive grasp and path relinking algorithm for the capacitated location routing problem. In *IESM'05*, Marrakech, Marroco, May 2005.
17. C. Prins, C. Prodhon, A. Ruiz, P. Soriano, and R. Wolfler Calvo. A cooperative lagrangean relaxation-granular tabu search heuristic for the capacitated location routing problem. In *MIC'05*, Vienna, Austria, August 2005.
18. C. Prins, M. Sevaux, and K. Sörensen. A genetic algorithm with population management (*GA | PM*) for the carp. In *Tristan V (5th Triennal Symposium on Transportation Analysis*, Le Gosier, Guadeloupe, 13-18 June 2004.
19. S. Salhi and G. K. Rand. The effect of ignoring routes when locating depots. *European Journal of Operational Research*, 39:150–156, 1989.
20. K. Sörensen and M. Sevaux. *MA | PM*: memetic algorithms with population management. *Computers and Operations Research*, (33):1214–1225, 2006.
21. R. Srivastava. Alternate solution procedures for the locating-routing problem. *OMEGA International Journal of Management Science*, 21(4):497–506, 1993.
22. D. Tuzun and L.I. Burke. A two-phase tabu search approach to the location routing problem. *European Journal of Operational Research*, 116:87–99, 1999.
23. T.H. Wu, C. Low, and J.W Bai. Heuristic solutions to multi-depot location-routing problems. *Computers and Operations Research*, 29:1393–1415, 2002.

# The Core Concept for the Multidimensional Knapsack Problem*

Jakob Puchinger[1], Günther R. Raidl[1], and Ulrich Pferschy[2]

[1] Institute of Computer Graphics and Algorithms,
Vienna University of Technology, Vienna, Austria
{puchinger, raidl}@ads.tuwien.ac.at
[2] Institute of Statistics and Operations Research,
University of Graz, Austria
pferschy@uni-graz.at

**Abstract.** We present the newly developed core concept for the Multidimensional Knapsack Problem (MKP) which is an extension of the classical concept for the one-dimensional case. The core for the multidimensional problem is defined in dependence of a chosen efficiency function of the items, since no single obvious efficiency measure is available for MKP. An empirical study on the cores of widely-used benchmark instances is presented, as well as experiments with different approximate core sizes. Furthermore we describe a memetic algorithm and a relaxation guided variable neighborhood search for the MKP, which are applied to the original and to the core problems. The experimental results show that given a fixed run-time, the different metaheuristics as well as a general purpose integer linear programming solver yield better solution when applied to approximate core problems of fixed size.

## 1   Introduction

The Multidimensional Knapsack Problem (MKP) is a well-studied, strongly NP-hard combinatorial optimization problem occurring in many different applications. It can be defined by the following ILP:

$$(MKP) \qquad \text{maximize} \quad z = \sum_{j=1}^{n} p_j x_j \qquad (1)$$

$$\text{subject to} \quad \sum_{j=1}^{n} w_{ij} x_j \leq c_i, \quad i = 1, \ldots, m \qquad (2)$$

$$x_j \in \{0, 1\}, \quad j = 1, \ldots, n. \qquad (3)$$

Given are $n$ items with profits $p_j > 0$ and $m$ resources with capacities $c_i > 0$. Each item $j$ consumes an amount $w_{ij} \geq 0$ from each resource $i$. The goal is to

---

* This work is supported by RTN ADONET under grant 504438 and the Austrian Science Fund (FWF) under grant P16263-N04.

J. Gottlieb and G.R. Raidl (Eds.): EvoCOP 2006, LNCS 3906, pp. 195–208, 2006.

select a subset of items with maximum total profit, see (1); chosen items must, however, not exceed resource capacities, see (2). The 0–1 decision variables $x_j$ indicate which items are selected.

A comprehensive overview on practical and theoretical results for the MKP can be found in the monograph on knapsack problems by Kellerer et al. [8]. Besides exact techniques for solving small to moderately sized instances, see [8], many kinds of metaheuristics have already been applied to the MKP. To our knowledge, the method currently yielding the best results, at least for commonly used benchmark instances, was described by Vasquez and Hao [18] and has recently been refined by Vasquez and Vimont [19]. Various other metaheuristics have been described for the MKP [5, 3], including several variants of hybrid evolutionary algorithms (EAs); see [16] for a survey and comparison of EAs for the MKP.

We first introduce the core concept for KP, and then expand it to MKP with respect to different efficiency measures. We then give some results of an empirical study of the cores of widely-used benchmark instances and present the application of the general ILP-solver CPLEX to MKP cores of fixed sizes. Furthermore we present a Memetic Algorithm (MA) and a Relaxation Guided Variable Neighborhood Search (RGVNS) applied to cores of hard to solve benchmark instances. We finally conclude by summarizing our work.

## 2   The Core Concept

The core concept was first presented for the classical 0/1-knapsack problem [1], which led to very successful KP algorithms [9, 11, 12]. The main idea is to reduce the original problem by only considering a core of items for which it is hard to decide if they will occur in an optimal solution or not, whereas the variables for all items outside the core are fixed to certain values.

### 2.1   The Core Concept for KP

The one-dimensional 0/1-knapsack problem (KP) considers items $j = 1, \ldots, n$, associated profits $p_j$, and weights $w_j$. A subset of these items has to be selected and packed into a knapsack having a capacity $c$. The total profit of the items in the knapsack has to be maximized, while the total weight is not allowed to exceed $c$. Obviously, KP is the special case of MKP with $m = 1$.

If the items are sorted according to decreasing efficiency values

$$e_j = \frac{p_j}{w_j}, \tag{4}$$

it is well known that the solution of the LP-relaxation consists in general of three consecutive parts: The first part contains variables set to 1, the second part consists of at most one split item $s$, whose corresponding LP-values is fractional, and finally the remaining variables, which are always set to zero, form the third part. For most instances of KP (except those with a very special structure of profits and weights) the integer optimal solution closely corresponds to this partitioning in the sense that it contains most of the highly efficient items of the first part, some items with medium efficiencies near the split item, and almost no items with low efficiencies from the third part. Items of medium efficiency constitute the so called core.

Balas and Zemel [1] gave the following precise definition of the core of a one-dimensional 0/1-knapsack problem, based on the knowledge of an optimal integer solution $x^*$. Assume that the items are sorted according to decreasing efficiencies and let

$$a := \min\{j \mid x_j^* = 0\}, \qquad b := \max\{j \mid x_j^* = 1\}. \tag{5}$$

The core is given by the items in the interval $C = \{a, \ldots, b\}$. It is obvious that the split item is always part of the core.

The KP Core (KPC) problem is defined as

$$(\text{KPC}) \qquad \text{maximize} \quad z = \sum_{j \in C} p_j x_j + \tilde{p} \tag{6}$$

$$\text{subject to} \quad \sum_{j \in C} w_j x_j \leq c - \tilde{w}, \tag{7}$$

$$x_j \in \{0, 1\}, \quad j \in C, \tag{8}$$

with $\tilde{p} = \sum_{j=1}^{a-1} p_j$ and $\tilde{w} = \sum_{j=1}^{a-1} w_j$. The solution of KPC would suffice to compute the optimal solution of KP, which, however, has to be already partially known to determine $C$. Pisinger [12] reported experimental investigations of the exact core size. He moreover studied the hardness of core problems, giving also a model for their expected hardness in [13].

The first class of core algorithms is based on solving a core problem with an approximate core of fixed size $c = \{s - \delta, \ldots, s + \delta\}$ with various choices of $\delta$, e.g. with $\delta$ being a constant or $\delta = \sqrt{n}$. An example is the MT2 algorithm by Martello and Toth [9].

Since it is impossible to estimate the core size in advance, Pisinger proposed two expanding core algorithms. Expknap [11] uses branch and bound for enumeration, whereas Minknap [12] (which enumerates at most the smallest symmetrical core) uses dynamic programming. For more details on core algorithms for KP we refer to Kellerer et al. [8].

## 2.2  The Core Concept for MKP

The previous definition of the core for KP can be expanded to MKP without major difficulties. The main problem, however, lies in the fact that there is no obvious efficiency measure.

**Efficiency Measures for MKP.** Consider the most obvious form of efficiency for the MKP which is a direct generalization of the one-dimensional case:

$$e_j(\text{simple}) = \frac{p_j}{\sum_{i=1}^{m} w_{ij}}. \tag{9}$$

Different orders of magnitude of the constraints are not considered and a single constraint may dominate the others. This drawback can easily be avoided by scaling:

$$e_j(\text{scaled}) = \frac{p_j}{\sum_{i=1}^{m} \frac{w_{ij}}{c_i}}. \tag{10}$$

Taking into account the relative contribution of the constraints Senju and Toyoda [17] get:

$$e_j(\text{st}) = \frac{p_j}{\sum_{i=1}^{m} w_{ij} \left(\sum_{j=1}^{n} w_{ij} - c_i\right)}. \tag{11}$$

For more details on efficiency values we refer to Kellerer et al. [8] where a general form of efficiency is defined by introducing relevance values $r_i$ for every constraint:

$$e_j(\text{general}) = \frac{p_j}{\sum_{i=1}^{m} r_i w_{ij}}. \tag{12}$$

The relevance values $r_i$ can also be seen as kind of surrogate multipliers. Pirkul calculates good multipliers heuristically [10]. Fréville and Plateau [4] suggested setting

$$r_i = \frac{\sum_{j=1}^{n} w_{ij} - c_i}{\sum_{j=1}^{n} w_{ij}}, \tag{13}$$

giving the efficiency value $e_j(\text{fp})$. Setting the relevance values $r_i$ to the values of an optimal solution to the dual problem of the MKP's LP-relaxation was a successful choice in [3], yielding the efficiency value $e_j(\text{duals})$.

**The MKP Core.** Since there are several possibilities of defining efficiency measures for MKP, the core and the core problem have to be defined depending on a specific efficiency measure $e$. Let $x^*$ be an optimal solution and assume that the items are sorted according to decreasing efficiency $e$, then let

$$a_e := \min\{j \mid x_j^* = 0\}, \qquad b_e := \max\{j \mid x_j^* = 1\}. \tag{14}$$

The core is given by the items in the interval $C_e := \{a_e, \ldots, b_e\}$, and the core problem is defined as

$$(\text{MKPC}_e) \qquad \text{maximize} \quad z = \sum_{j \in C} p_j x_j + \tilde{p} \tag{15}$$

$$\text{subject to} \quad \sum_{j \in C} w_{ij} x_j \leq c_i - \tilde{w}_i, \quad i = 1, \ldots, m \tag{16}$$

$$x_j \in \{0, 1\}, \quad j \in C, \tag{17}$$

with $\tilde{p} = \sum_{j=1}^{a-1} p_j$ and $\tilde{w}_i = \sum_{j=1}^{a-1} w_{ij}$, $i = 1, \ldots, m$.

In contrast to KP, the solution of the LP-relaxation of MKP in general does not consist of a single fractional split item. But up to $m$ fractional values give rise to a whole *split interval* $S_e := \{s_e, \ldots, t_e\}$, where $s_e$ and $t_e$ are the first and the last index of variables with fractional values after sorting by efficiency $e$.

Note that depending on the choice of the efficiency measure, the split interval can also contain variables with integer values. Moreover, the sets $S_e$ and $C_e$ can have almost any relation to each other, from inclusion to disjointness. For a "reasonable" choice of $e$, we expected them, however, to overlap to a large extent.

If the dual solution values of the LP-relaxation are taken as relevance values, the split interval $S_e$ resulting from the corresponding efficiency values $e_j(\text{duals})$ can be precisely characterized. Let $x^{LP}$ be the optimal solution of the LP-relaxation of MKP.

**Theorem 1.**

$$x_j^{LP} = \begin{cases} 1 & \text{if } e_j > 1, \\ \in [0,1] & \text{if } e_j = 1, \\ 0 & \text{if } e_j < 1. \end{cases} \tag{18}$$

*Proof.* The dual LP associated with the LP-relaxation of MKP is given by

$$(\text{D(MKP)}) \qquad \text{minimize} \quad \sum_{i=1}^{m} c_i u_i + \sum_{j=1}^{n} v_j \tag{19}$$

$$\text{subject to} \quad \sum_{i=1}^{m} w_{ij} u_i + v_j \geq p_j, \quad j = 1, \ldots, n \tag{20}$$

$$u_i, v_j \geq 0, \quad i = 1, \ldots, m, \ j = 1, \ldots, n, \tag{21}$$

where $u_i$ are the dual variables corresponding to the capacity constraints (2) and $v_j$ correspond to the inequalities $x_j \leq 1$. For the optimal primal and dual solutions the following complementary slackness conditions hold (see any textbook on linear programming, e.g. [2]):

$$x_j \left( \sum_{i=1}^{m} w_{ij} u_i + v_j - p_j \right) = 0 \tag{22}$$

$$v_j (x_j - 1) = 0 \tag{23}$$

Recall that $e_j(\text{duals}) = \frac{p_j}{\sum_{i=1}^{m} u_i w_{ij}}$. Hence, $e_j > 1$ implies $p_j > \sum_{i=1}^{m} w_{ij} u_i$, which means that (20) can only be fulfilled by $v_j > 0$. Now, (23) immediately yields $x_j = 1$, which proves the first part of the theorem.

If $e_j < 1$, there is $p_j < \sum_{i=1}^{m} w_{ij} u_i$ which together with $v_j \geq 0$ makes the second factor of (22) strictly positive and requires $x_j = 0$. This proves the theorem since nothing has to be shown for $e_j = 1$. $\square$

It follows from Theorem 1 that $S_e \subseteq \{j \mid e_j = 1, \ j = 1, \ldots, n\}$. It should be noted that the theorem gives only a structural result which does not yield any direct algorithmic advantage to compute the primal solution $x^{LP}$ since it requires knowing the dual optimal solution.

# 3 Experimental Study of MKP Cores and Core Sizes

## 3.1 MKP Cores and Efficiency Measures

In order to analyze the core sizes in dependence on different efficiency values, we performed an empirical in-depth examination on smaller instances of Chu and Beasley's benchmark library[1]. Chu and Beasley [3] generated the instances as suggested by Fréville and Plateau [4]. The instance classes consist of ten instances each with $n \in \{100, 250, 500\}$ items, $m \in \{5, 10, 30\}$ constraints, and tightness ratios $\alpha = c_i / \sum_{j=1}^{n} w_{ij}$, $\alpha \in \{0.25, 0.5, 0.75\}$.

For the empirical results presented in this section, we used the smaller instances, which could be solved to proven optimality in reasonable time using the ILP-solver CPLEX, with $n = 100$ and $m \in \{5, 10\}$, and $n = 250$ and $m = 5$.

In Table 1 we examine cores devised using the scaled efficiency $e(\text{scaled})$, the efficiency $e(\text{st})$, the efficiency $e(\text{fp})$ as defined in equations (12) and (13), and finally the efficiency $e(\text{duals})$ setting the relevance values $r_i$ of equation (12) to

**Table 1.** Split intervals, core sizes and their mutual coverages and distances for different efficiency values (average percent values taken from 10 instances and average over all problem classes)

| $n$ | $m$ | $\alpha$ | $e(\text{scaled})$ | | | | | $e(\text{st})$ | | | | |
|---|---|---|---|---|---|---|---|---|---|---|---|---|
| | | | $|S_e|$ | $|C_e|$ | ScC | CcS | $C_{\text{dist}}$ | $|S_e|$ | $|C_e|$ | ScC | CcS | $C_{\text{dist}}$ |
| 100 | 5 | 0.25 | 23.40 | 30.50 | 72.69 | 94.71 | 4.05 | 27.20 | 30.20 | 78.85 | 88.11 | 4.80 |
| | | 0.5 | 29.50 | 37.60 | 71.93 | 88.45 | 5.95 | 27.00 | 35.60 | 69.88 | 89.01 | 5.90 |
| | | 0.75 | 24.30 | 27.00 | 72.61 | 83.13 | 5.05 | 22.80 | 25.20 | 77.72 | 84.08 | 4.30 |
| 250 | 5 | 0.25 | 17.44 | 22.40 | 77.20 | 97.38 | 1.88 | 17.12 | 22.20 | 76.91 | 94.62 | 2.46 |
| | | 0.5 | 22.88 | 29.44 | 71.71 | 94.25 | 3.44 | 23.76 | 30.88 | 74.95 | 94.69 | 4.04 |
| | | 0.75 | 11.44 | 17.84 | 56.14 | 88.45 | 4.60 | 11.96 | 16.64 | 63.82 | 85.86 | 3.62 |
| 100 | 10 | 0.25 | 42.60 | 38.30 | 92.62 | 84.39 | 4.35 | 43.30 | 38.20 | 88.78 | 79.36 | 5.55 |
| | | 0.5 | 39.40 | 45.20 | 80.80 | 91.20 | 5.30 | 44.40 | 46.50 | 85.43 | 88.49 | 5.65 |
| | | 0.75 | 37.50 | 34.80 | 94.29 | 86.42 | 2.55 | 38.60 | 36.20 | 93.04 | 87.16 | 2.10 |
| Average | | | 27.61 | 31.45 | 76.67 | 89.82 | 4.13 | 28.46 | 31.29 | 78.82 | 87.93 | 4.27 |

| $n$ | $m$ | $\alpha$ | $e(\text{fp})$ | | | | | $e(\text{duals})$ | | | | |
|---|---|---|---|---|---|---|---|---|---|---|---|---|
| | | | $|S_e|$ | $|C_e|$ | ScC | CcS | $C_{\text{dist}}$ | $|S_e|$ | $|C_e|$ | ScC | CcS | $C_{\text{dist}}$ |
| 100 | 5 | 0.25 | 24.70 | 30.10 | 75.50 | 91.94 | 4.20 | 5.00 | 20.20 | 28.12 | 100.00 | 3.30 |
| | | 0.5 | 27.10 | 35.80 | 70.36 | 89.74 | 6.35 | 5.00 | 22.10 | 27.49 | 100.00 | 3.45 |
| | | 0.75 | 23.20 | 26.10 | 74.47 | 84.22 | 4.55 | 5.00 | 19.60 | 26.95 | 100.00 | 3.20 |
| 250 | 5 | 0.25 | 16.92 | 21.72 | 76.87 | 95.63 | 2.24 | 2.00 | 12.68 | 18.16 | 100.00 | 2.46 |
| | | 0.5 | 22.96 | 29.68 | 74.79 | 95.02 | 3.56 | 2.00 | 12.20 | 18.45 | 100.00 | 1.38 |
| | | 0.75 | 11.40 | 17.12 | 59.00 | 87.27 | 4.06 | 2.00 | 10.40 | 20.18 | 100.00 | 1.56 |
| 100 | 10 | 0.25 | 42.10 | 38.20 | 90.41 | 83.74 | 4.75 | 10.00 | 23.20 | 46.57 | 100.00 | 2.90 |
| | | 0.5 | 41.90 | 45.60 | 84.52 | 90.85 | 5.15 | 9.80 | 25.70 | 48.17 | 95.00 | 3.15 |
| | | 0.75 | 37.90 | 35.30 | 94.55 | 86.96 | 2.40 | 9.70 | 18.80 | 55.74 | 99.00 | 2.75 |
| Average | | | 27.58 | 31.07 | 77.83 | 89.49 | 4.14 | 5.61 | 18.32 | 32.20 | 99.33 | 2.68 |

---

[1] http://people.brunel.ac.uk/~mastjjb/jeb/info.html

the optimal dual variable values of the MKP's LP-relaxation. Listed are average values of the sizes (in percent of the number of items) of the split interval ($|S_e|$) and of the exact core ($|C_e|$), the percentage of how much the split interval covers the core (ScC) and how much the core covers the split interval (CcS), and the distance (in percent of the number of items) between the center of the split interval and the center of the core ($C_{dist}$).

As expected from Theorem 1, the smallest split intervals, consisting of the fractional variables only are derived with $e$(duals). They further yield the smallest cores. Using one of the other efficiency measures results in significantly larger split intervals and cores. Furthermore, the smallest distances between the centers of the split intervals and the cores are produced by $e$(duals) for almost all the subclasses. The most promising information for devising approximate cores are therefore available from the split intervals generated with $e$(duals), on which we will concentrate our further investigations.

## 3.2   A Fixed Core Approach

In order to evaluate the influence of core sizes on solution quality and run-times, we propose a fixed core size algorithm, where we solve approximate cores using

**Table 2.** Solving cores of different sizes exactly (average over 10 instances and average over all problem classes)

| $n$ $m$ $\alpha$ | no core | | $\delta = 0.1n$ | | | $\delta = 0.15n$ | | |
|---|---|---|---|---|---|---|---|---|
| | $\bar{z}$ | $\bar{t}[s]$ | $\%_{opt}$ | # | $\%t$ | $\%_{opt}$ | # | $\%t$ |
| 100 5 0.25 | 24197 | 21 | 0.097 | 5 | 1 | 0.034 | 7 | 9 |
| 0.5 | 43253 | 27 | 0.053 | 4 | 1 | 0.018 | 6 | 6 |
| 0.75 | 60471 | 6 | 0.038 | 5 | 4 | 0.021 | 7 | 17 |
| 250 5 0.25 | 60414 | 1474 | 0.008 | 7 | 36 | 0.003 | 9 | 81 |
| 0.5 | 109293 | 1767 | 0.002 | 8 | 21 | 0.000 | 10 | 63 |
| 0.75 | 151560 | 817 | 0.000 | 10 | 17 | 0.000 | 10 | 47 |
| 100 10 0.25 | 22602 | 189 | 0.473 | 1 | 0 | 0.152 | 4 | 1 |
| 0.5 | 42661 | 97 | 0.234 | 3 | 0 | 0.084 | 5 | 1 |
| 0.75 | 59556 | 29 | 0.036 | 6 | 0 | 0.015 | 8 | 3 |
| Average | 63778 | 492 | 0.105 | 5.4 | 9 | 0.036 | 7.3 | 25 |
| $n$ $m$ $\alpha$ | $\delta = 0.2n$ | | | $\delta = 2m + 0.1n$ | | | $\delta = 2m + 0.2n$ | | |
| | $\%_{opt}$ | # | $\%t$ | $\%_{opt}$ | # | $\%t$ | $\%_{opt}$ | # | $\%t$ |
| 100 5 0.25 | 0.015 | 9 | 32 | 0.015 | 9 | 32 | 0.000 | 10 | 62 |
| 0.5 | 0.002 | 9 | 24 | 0.002 | 9 | 24 | 0.002 | 9 | 64 |
| 0.75 | 0.001 | 9 | 39 | 0.001 | 9 | 39 | 0.000 | 10 | 61 |
| 250 5 0.25 | 0.000 | 10 | 82 | 0.003 | 9 | 69 | 0.000 | 10 | 91 |
| 0.5 | 0.000 | 10 | 67 | 0.000 | 10 | 59 | 0.000 | 10 | 73 |
| 0.75 | 0.000 | 10 | 72 | 0.000 | 10 | 40 | 0.000 | 10 | 61 |
| 100 10 0.25 | 0.002 | 9 | 10 | 0.000 | 10 | 46 | 0.000 | 10 | 66 |
| 0.5 | 0.030 | 8 | 13 | 0.022 | 8 | 60 | 0.000 | 10 | 75 |
| 0.75 | 0.011 | 9 | 22 | 0.000 | 10 | 54 | 0.000 | 10 | 70 |
| Average | 0.007 | 9.2 | 40 | 0.005 | 9.3 | 47 | 0.000 | 9.9 | 69 |

the general purpose ILP-solver CPLEX 9.0. We performed the experiments on a 2.4 GHz Pentium 4 computer.

In analogy to KP, the approximate core is generated by adding $\delta$ items on each side of the center of the split interval. We created the cores with the $e$(duals) efficiency. The values of $\delta$ were chosen in accordance with the results of the previous section, where an average core size of about $0.2n$ was observed. Table 2 lists average objective values and run-times for the original problem, and percentage gaps to the optimal solution ($\overline{\%_{\text{opt}}} = 100 \cdot (z^* - z)/z^*$), the number of times the optimum was reached ($\#$), as well as the average run-times $\overline{\%t}$ (in percent of the run-time required for solving the original problem) for cores of different sizes.

Observing the results of CPLEX applied to cores of different sizes, we see that smaller cores can be solved substantially faster and the obtained solution values are only slightly worse than the optimal ones given by the *no core* column. The best results with respect to average run-times were achieved with $\delta = 0.1n$, the run-time could be reduced by factors going from 3 to 1000, whereas, most importantly, the obtained objective values are very close to the respective optima (0.1% on average). Solving the bigger cores needs more run-time, but almost all of the optimal results could be reached, with still significant time savings.

# 4   Applying Metaheuristics to the Core

The question of how the reduction to MKP cores influences the performance of metaheuristics arises due to the observed differences in run-times and solution qualities of the previous section. Furthermore the core concept might enable us to find better solutions for larger instances which cannot be solved to optimality. We therefore study a memetic algorithm and a relaxation guided variable neighborhood search for solving the MKP, applied to MKP cores. The results of Section 3.2 indicate that the obtained solutions should be good approximations of the overall MKP optimum.

## 4.1   A Memetic Algorithm

The MA which we consider here is based on Chu and Beasley's principles and includes some improvements suggested in [6, 15, 16]. The framework is steady-state and the creation of initial solutions is guided by the LP-relaxation of the MKP, as described in [6]. Each new candidate solution is derived by selecting two parents via binary tournaments, performing uniform crossover on their characteristic vectors $x$, flipping each bit with probability $1/n$, performing repair if a capacity constraint is violated, and always performing local improvement. If such a new candidate solution is different from all solutions in the current population, it replaces the worst of them.

Both, repair and local improvement, are based on greedy first-fit strategies and guarantee that any resulting candidate solution lies at the boundary of the feasible region, where optimal solutions are always located. The repair procedure considers all items in a specific order $\Pi$ and removes selected items

($x_j = 1 \rightarrow x_j = 0$) as long as any capacity constraint is violated. Local improvement works vice-versa: It considers all items in the reverse order $\overline{\Pi}$ and selects items not yet appearing in the solution as long as no capacity limit is exceeded.

Crucial for these strategies to work well is the choice of the ordering $\Pi$. Items that are likely to be selected in an optimal solution must appear near the end of $\Pi$. Following the results of Section 3.1 we determine $\Pi$ by ordering the items according to $e$(duals), as it has also been previously done in [3].

## 4.2   A Relaxation Guided VNS for the MKP

Relaxation Guided Variable Neighborhood Search (RGVNS) [14] is a recently developed Variable Neighborhood Search (VNS) [7] variant where the neighborhood-order of Variable Neighborhood Descent (VND) is dynamically determined by solving relaxations of the neighborhoods. The RGVNS used here for the MKP, is a slightly improved version of the approach described in [14].

**Representation and Initialization.** Solutions are directly represented by binary strings, and all our neighborhoods are defined on the space of feasible solutions only. We denote by $I_1(x^f) = \{j \mid x_j^f = 1\}$ the index-set of the items contained in the knapsack of a current solution $x^f$ and by $I_0(x^f) = \{j \mid x_j^f = 0\}$ its complement. The initial solution for the RGVNS is generated using a greedy first-fit heuristic, considering the items in a certain order, determined by sorting the items according to decreasing values of the solutions to the MKP's LP-relaxation; see [16].

**ILP Based Neighborhoods.** We want to force a certain number of items of the current feasible solution $x^f$ to be removed from or added to the knapsack. This is realized by adding neighborhood-defining constraints depending on $x^f$ to the ILP formulation of the MKP.

In the first neighborhood, *ILP-Remove-and-Fill IRF*$(x^f, k)$, we force precisely $k$ items from $I_1$ to be removed from the knapsack, and any combination of items from $I_0$ is allowed to be added to the knapsack as long as the solution remains feasible. This is accomplished by adding the equation $\sum_{j \in I_1(x^f)} x_j = \sum_{j \in I_1(x^f)} x_j^f - k$ to (1)–(3).

In the second neighborhood, *ILP-Add-and-Remove IAR*$(x^f, k)$, we force precisely $k$ items not yet packed, i.e. from $I_0$, to be included in the knapsack. To achieve feasibility any combination of items from $I_1$ may be removed. This is achieved by adding $\sum_{j \in I_0(x^f)} x_j = k$ to (1)–(3).

As relaxations $IRF^R(x^f, k)$ and $IAR^R(x^f, k)$ we use the corresponding LP-relaxations in which the integrality constraints (3) are replaced by $0 \le x_j \le 1$, $j = 1, \ldots, n$. For searching the (integer) neighborhoods we use a general purpose ILP-solver (CPLEX) with a certain time limit.

**Classical Neighborhoods.** As first neighborhood we use a simple swap $SWP(x^f)$, where a pair of items $(x_i^f, x_j^f)$, with $i \in I_1$ and $j \in I_0$, is exchanged, i.e. $x_i^f := 0$ and $x_j^f := 1$. Infeasible solutions are discarded. Note that this neighborhood is contained in both, $IRF(x^f, 1)$ and $IAR(x^f, 1)$. Its main advantage is that it can be explored very fast.

Based on the ideas of Chu and Beasley [3] and as another simplification of IRF and IAR but an extension of SWP, we define two additional neighborhoods based on greedy concepts. In the first case, the *Remove-and-Greedy-Fill* neighborhood $RGF(x^f, k)$, $k$ items are removed from $x^f$; i.e. a $k$-tuple of variables from $I_1(x^f)$ is flipped. The resulting solution is then locally optimized as described in Section 4.1. In the second case, the *Add-and-Greedy-Repair* neighborhood $AGR(x^f, k)$, $k$ items are added to $x^f$; i.e. $k$ variables from $I_0(x^f)$ are flipped. The resulting solution, which is usually infeasible, is then repaired and locally improved as previously described.

**Relaxation Guided VND.** The Relaxation Guided VND (RGVND) is based on the previously defined neighborhoods and follows a best neighbor strategy. The faster to solve neighborhoods are ordered as follows: $\mathcal{N}_1 := SWP(x^f)$, $\mathcal{N}_2 := RGF(x^f, 1)$, $\mathcal{N}_3 := AGR(x^f, 1)$. If none of these neighborhoods leads to an improved solution, we solve the LP-relaxations of $IRF(x^f, k)$ and $IAR(x^f, k)$ for $k = 1, \ldots, k_{\max}$, where $k_{\max}$ is a prespecified upper limit on the number of items we want to remove or add. The neighborhoods are then sorted according to decreasing LP-relaxation solution values. Ties are broken by considering smaller $k$s earlier. Only the first $\beta_{\max}$ ILP-based neighborhoods are explored before shaking.

**Shaking.** After RGVND has terminated, shaking is performed for diversification. It flips $\kappa$ different randomly selected variables of the currently best solution and applies greedy repair and local improvement as previously described for the MA. As usual in general VNS, $\kappa$ runs from 1 to some $\kappa_{\max}$ and is reset to 1 if an improved solution was found. Furthermore, the whole process is iterated until a termination criterion, in our case the CPU-time, is met.

## 5   Computational Experiments

We present several computational experiments where we evaluated the influence of differently sized cores on the performance of CPLEX, the presented MA, and RGVNS. The algorithms were given 500 seconds per run. Since the MA converges much earlier, it was restarted every 1 000 000 generations, always keeping the so-far best solution in the population. In RGVNS, CPLEX was given a maximum of 5 seconds for exploring the ILP-based neighborhoods, $k_{\max}$ and $\beta_{\max}$ were set to 10, and $\kappa_{\max} = n$. We used the hardest instances of the Chu and Beasley benchmark set, i.e. those with $n = 500$ items and $m \in \{5, 10, 30\}$ constraints. As before, CPLEX 9.0 was used and we performed the experiments on a 2.4 GHz Pentium 4 computer.

In Table 3 we display the results of CPLEX applied to cores of different sizes. For comparison CPLEX was also applied to the original problem with the same time limit. We list averages over ten instances of the percentage gaps to the optimal objective value of the LP-relaxation ($\overline{\%_{\mathrm{LP}}} = 100 \cdot (z^{\mathrm{LP}} - z)/z^{\mathrm{LP}}$), the number of times this core size yielded the best solution of this algorithm ($\#$), and the number of explored nodes of the branch and bound tree.

First, it can be noticed that CPLEX applied to approximate cores of different sizes yields, on average, better results than CPLEX applied to the original problem. Second, the number of explored nodes increases with decreasing problem/core size. The best average results are obtained with higher core sizes.

**Table 3.** Solving cores of different sizes with CPLEX (avgerage over 10 instances and average over all problem classes, $n = 500$)

| $m$ | $\alpha$ | no core | | | $\delta = 0.1n$ | | | $\delta = 0.15n$ | | | $\delta = 0.2n$ | | |
|---|---|---|---|---|---|---|---|---|---|---|---|---|---|
| | | $\overline{\%_{\mathrm{LP}}}$ | $\#$ | $\overline{\mathrm{Nnodes}}$ | $\overline{\%_{\mathrm{LP}}}$ | $\#$ | $\overline{\mathrm{Nnodes}}$ | $\overline{\%_{\mathrm{LP}}}$ | $\#$ | $\overline{\mathrm{Nnodes}}$ | $\overline{\%_{\mathrm{LP}}}$ | $\#$ | $\overline{\mathrm{Nnodes}}$ |
| 5 | 0.25 | 0.080 | 5 | 5.50E5 | 0.075 | 9 | 1.00E6 | 0.076 | 9 | 9.85E5 | 0.076 | 8 | 8.34E5 |
| | 0.5 | 0.040 | 6 | 5.06E5 | 0.039 | 7 | 1.05E6 | 0.039 | 9 | 1.00E6 | 0.039 | 9 | 8.38E5 |
| | 0.75 | 0.025 | 6 | 5.36E5 | 0.024 | 10 | 1.05E6 | 0.025 | 8 | 1.02E6 | 0.025 | 8 | 9.04E5 |
| 10 | 0.25 | 0.206 | 1 | 3.15E5 | 0.198 | 5 | 1.10E6 | 0.195 | 6 | 6.99E5 | 0.198 | 4 | 5.68E5 |
| | 0.5 | 0.094 | 4 | 3.01E5 | 0.088 | 8 | 1.11E6 | 0.090 | 6 | 6.95E5 | 0.092 | 5 | 5.73E5 |
| | 0.75 | 0.066 | 4 | 3.05E5 | 0.065 | 5 | 1.07E6 | 0.064 | 7 | 6.83E5 | 0.065 | 7 | 5.59E5 |
| 30 | 0.25 | 0.598 | 2 | 1.11E5 | 0.621 | 0 | 4.22E5 | 0.566 | 4 | 3.06E5 | 0.537 | 6 | 2.28E5 |
| | 0.5 | 0.258 | 2 | 1.15E5 | 0.246 | 3 | 4.50E5 | 0.243 | 4 | 3.28E5 | 0.250 | 2 | 2.38E5 |
| | 0.75 | 0.158 | 2 | 1.12E5 | 0.151 | 6 | 4.48E5 | 0.160 | 1 | 3.14E5 | 0.151 | 5 | 2.36E5 |
| Average | | 0.169 | 3.6 | 3.17E5 | 0.167 | 5.9 | 8.55E5 | 0.162 | 6.0 | 6.70E5 | 0.159 | 6.0 | 5.53E5 |

In Table 4 the results of the MA applied to approximate cores of different sizes are shown. In order to evaluate the benefits of using a core-based approach, we also applied the MA to the original problem. The table lists ($\overline{\%_{\mathrm{LP}}}$), the number of times this core size yielded the best solution of this algorithm ($\#$), and the average numbers of MA iterations.

As observed with CPLEX, the use of approximate cores consistently increases the achieved solution quality. The core size has a significant influence on the number of iterations performed by the MA, which can be explained by the smaller size of the problem to be solved. This also seems to be a reason for the better results, since more candidate solutions can be examined in the given run-time. Furthemore, the search space of the MA is restricted to a highly promising part of the original search space. The best average results were obtained with $\delta = 0.15n$. The smaller approximate cores yield better results on average.

In Table 5, the results of RGVNS when applied to approximate cores of different sizes are shown together with the results of RGVNS on the original problem. The table also displays the average total number of iterations performed by RGVND inside RGVNS.

The results obtained by RGVNS applied to the smaller approximate cores clearly dominate the results obtained without core and with $\delta = 0.2n$. This can be explained by the fact that CPLEX is used in RGVNS, and that it is able

**Table 4.** Solving cores of different sizes with the MA (avgerage over 10 instances and average over all problem classes, $n = 500$)

| $m$ | $\alpha$ | no core | | | $\delta = 0.1n$ | | | $\delta = 0.15n$ | | | $\delta = 0.2n$ | | |
|---|---|---|---|---|---|---|---|---|---|---|---|---|---|
| | | $\%_{LP}$ | # | Niter | $\%_{LP}$ | # | Niter | $\%_{LP}$ | # | Niter | $\%_{LP}$ | # | Niter |
| 5 | 0.25 | 0.078 | 6 | 1.40E7 | 0.073 | 10 | 5.08E7 | 0.074 | 9 | 4.07E7 | 0.074 | 9 | 3.33E7 |
| | 0.5 | 0.040 | 6 | 1.35E7 | 0.039 | 9 | 5.07E7 | 0.039 | 9 | 4.07E7 | 0.040 | 7 | 3.33E7 |
| | 0.75 | 0.025 | 7 | 1.46E7 | 0.024 | 9 | 5.07E7 | 0.024 | 10 | 4.08E7 | 0.024 | 9 | 3.34E7 |
| 10 | 0.25 | 0.208 | 5 | 1.26E7 | 0.202 | 5 | 4.54E7 | 0.202 | 6 | 3.62E7 | 0.208 | 4 | 2.90E7 |
| | 0.5 | 0.099 | 2 | 1.21E7 | 0.093 | 6 | 4.51E7 | 0.091 | 8 | 3.59E7 | 0.093 | 5 | 2.89E7 |
| | 0.75 | 0.066 | 6 | 1.31E7 | 0.065 | 8 | 4.53E7 | 0.067 | 4 | 3.59E7 | 0.068 | 4 | 2.87E7 |
| 30 | 0.25 | 0.604 | 1 | 9.10E6 | 0.573 | 5 | 3.08E7 | 0.575 | 5 | 2.39E7 | 0.569 | 6 | 1.92E7 |
| | 0.5 | 0.254 | 3 | 8.10E6 | 0.257 | 1 | 3.08E7 | 0.246 | 7 | 2.37E7 | 0.253 | 3 | 1.90E7 |
| | 0.75 | 0.159 | 4 | 8.12E6 | 0.156 | 5 | 3.14E7 | 0.157 | 3 | 2.35E7 | 0.157 | 5 | 1.96E7 |
| Average | | 0.170 | 4.4 | 1.17E7 | 0.165 | 6.4 | 4.23E7 | 0.164 | 6.8 | 3.35E7 | 0.165 | 5.8 | 2.72E7 |

**Table 5.** Solving cores of different sizes with RGVNS (avgerage over 10 instances and average over all problem classes, $n = 500$)

| $m$ | $\alpha$ | no core | | | $\delta = 0.1n$ | | | $\delta = 0.15n$ | | | $\delta = 0.2n$ | | |
|---|---|---|---|---|---|---|---|---|---|---|---|---|---|
| | | $\%_{LP}$ | # | Niter | $\%_{LP}$ | # | Niter | $\%_{LP}$ | # | Niter | $\%_{LP}$ | # | Niter |
| 5 | 0.25 | 0.088 | 4 | 230 | 0.080 | 5 | 208 | 0.080 | 6 | 223 | 0.082 | 4 | 230 |
| | 0.5 | 0.043 | 5 | 236 | 0.040 | 7 | 215 | 0.040 | 8 | 226 | 0.040 | 7 | 239 |
| | 0.75 | 0.027 | 5 | 246 | 0.026 | 8 | 230 | 0.026 | 8 | 252 | 0.026 | 7 | 240 |
| 10 | 0.25 | 0.230 | 0 | 225 | 0.198 | 7 | 200 | 0.211 | 2 | 193 | 0.210 | 3 | 205 |
| | 0.5 | 0.108 | 1 | 209 | 0.096 | 5 | 201 | 0.096 | 3 | 199 | 0.100 | 1 | 205 |
| | 0.75 | 0.069 | 2 | 208 | 0.066 | 7 | 207 | 0.066 | 7 | 211 | 0.066 | 4 | 214 |
| 30 | 0.25 | 0.595 | 5 | 202 | 0.599 | 3 | 196 | 0.593 | 4 | 191 | 0.609 | 5 | 195 |
| | 0.5 | 0.263 | 3 | 197 | 0.260 | 0 | 198 | 0.254 | 6 | 189 | 0.261 | 3 | 197 |
| | 0.75 | 0.168 | 2 | 191 | 0.158 | 5 | 191 | 0.164 | 3 | 187 | 0.164 | 2 | 191 |
| Average | | 0.177 | 3.0 | 216 | 0.169 | 5.2 | 205 | 0.170 | 5.2 | 208 | 0.173 | 4.0 | 213 |

to find better solutions when dealing with smaller problem sizes. Interestingly, the number of iterations stays about the same for the different settings. The reason is that CPLEX is given the same constant time limit for searching for the neighborhoods within RGVND.

Comparing our results to the best known solutions [19], we are able to reach the best solutions for $m = 5$, and stay only 0.5% below these solutions for $m \in \{10, 30\}$, requiring 500 seconds, whereas in [19] up to 33 hours were needed.

# 6   Conclusions

We presented the new core concept for the multidimensional knapsack problem, extending the core concept for the classical one-dimensional 0/1-knapsack problem. An empirical study of the exact core sizes of widely used benchmark instances with different efficiency measures was performed. The efficiency value using dual-variable values as relevance factors yielded the smallest possible split-intervals and the smallest cores.

We further studied the influence of restricting problem solving to approximate cores of different sizes, and observed significant differences in terms of run-time when applying the general-purpose ILP-solver CPLEX to approximate cores or to the original problem, whereas the objective values remained very close to the respective optima.

We finally applied CPLEX and two metaheuristics to approximate cores of hard to solve benchmark instances and observed that using approximate cores of fixed size instead of the original problem clearly and consistently improves the solution quality when using a fixed run-time.

In the future, we want to further examine the MKP core concept and possibly extend it to other combinatorial optimization problems.

# References

1. E. Balas and E. Zemel. An algorithm for large zero-one knapsack problems. *Operations Research*, 28:1130–1154, 1980.
2. D. Bertsimas and J. N. Tsitsiklis. *Introduction to Linear Optimization*. Athena Scientific, 1997.
3. P. C. Chu and J. Beasley. A genetic algorithm for the multiconstrained knapsack problem. *Journal of Heuristics*, 4:63–86, 1998.
4. A. Fréville and G. Plateau. An efficient preprocessing procedure for the multi-dimensional 0–1 knapsack problem. *Discrete Applied Mathematics*, 49:189–212, 1994.
5. F. Glover and G. Kochenberger. Critical event tabu search for multidimensional knapsack problems. In I. Osman and J. Kelly, editors, *Metaheuristics: Theory and Applications*, pages 407–427. Kluwer Academic Publishers, 1996.
6. J. Gottlieb. On the effectivity of evolutionary algorithms for multidimensional knapsack problems. In C. Fonlupt et al., editors, *Proceedings of Artificial Evolution: Fourth European Conference*, volume 1829 of *LNCS*, pages 22–37. Springer, 1999.
7. P. Hansen and N. Mladenović. An introduction to variable neighborhood search. In S. Voss, S. Martello, I. Osman, and C. Roucairol, editors, *Metaheuristics, Advances and Trends in Local Search Paradigms for Optimization*, pages 433–458. Kluwer, 1999.
8. H. Kellerer, U. Pferschy, and D. Pisinger. *Knapsack Problems*. Springer, 2004.
9. S. Martello and P. Toth. A new algorithm for the 0-1 knapsack problem. *Management Science*, 34:633–644, 1988.
10. H. Pirkul. A heuristic solution procedure for the multiconstraint zero-one knapsack problem. *Naval Research Logistics*, 34:161–172, 1987.
11. D. Pisinger. An expanding-core algorithm for the exact 0–1 knapsack problem. *European Journal of Operational Research*, 87:175–187, 1995.
12. D. Pisinger. A minimal algorithm for the 0–1 knapsack problem. *Operations Research*, 45:758–767, 1997.
13. D. Pisinger. Core problems in knapsack algorithms. *Operations Research*, 47:570–575, 1999.

14. J. Puchinger and G. R. Raidl. Relaxation guided variable neighborhood search. In P. Hansen, N. Mladenović, J. A. M. Pérez, B. M. Batista, and J. M. Moreno-Vega, editors, *Proceedings of the 18th Mini Euro Conference on Variable Neighborhood Search*, Tenerife, Spain, 2005.

15. G. R. Raidl. An improved genetic algorithm for the multiconstrained 0–1 knapsack problem. In D. Fogel et al., editors, *Proceedings of the 5th IEEE International Conference on Evolutionary Computation*, pages 207–211. IEEE Press, 1998.

16. G. R. Raidl and J. Gottlieb. Empirical analysis of locality, heritability and heuristic bias in evolutionary algorithms: A case study for the multidimensional knapsack problem. *Evolutionary Computation Journal*, 13(4), to appear 2005.

17. S. Senju and Y. Toyoda. An approach to linear programming with 0–1 variables. *Management Science*, 15:196–207, 1968.

18. M. Vasquez and J.-K. Hao. A hybrid approach for the 0–1 multidimensional knapsack problem. In *Proceedings of the Int. Joint Conference on Artificial Intelligence 2001*, pages 328–333, 2001.

19. M. Vasquez and Y. Vimont. Improved results on the 0-1 multidimensional knapsack problem. *European Journal of Operational Research*, 165:70–81, 2005.

# Multiobjective Scheduling of Jobs
# with Incompatible Families on
# Parallel Batch Machines

Dirk Reichelt and Lars Mönch

Technical University of Ilmenau,
Institute of Information Systems, Helmholtzplatz 3,
98684 Ilmenau, Germany
{Dirk.Reichelt, Lars.Moench}@tu-ilmenau.de

**Abstract.** We consider scheduling heuristics for batching machines from semiconductor manufacturing. A batch is a collection of jobs that are processed at the same time on the same machine. The processing time of a batch is given by the identical processing time of the jobs within one incompatible family. We are interested in minimizing total weighted tardiness and makespan at the same time. In order to solve this problem, i.e. generate a Pareto-front, we suggest a multiobjective genetic algorithm. We present results from computational experiments on stochastically generated test instances that show the good solution quality of the suggested approach.

## 1 Introduction

A great majority of academic scheduling research deals with single-objective problems. This research focuses on multiobjective scheduling of batch processing machines found in the diffusion and oxidation areas of a semiconductor wafer fabrication facility (wafer fab). The processing times of these batching operations are extremely long (10 hours) when compared to other operations (1-2 hours) in a wafer fab. Mathirajan and Sivakumar [9] state that the effective scheduling of these operations is important to achieve good system performance. Though several jobs can be processed simultaneously in these batch processing machines, process restrictions require that only jobs belonging to the same family be processed together at one time. The jobs to be processed have different priorities/weights, due-dates and ready times. In the presence of unequal ready times, it is sometimes advantageous to form a non full batch while in other situations it is a better strategy to wait for future job arrivals in order to increase the fullness of the batch.

We model diffusion and oxidation operations as parallel batch processing machines with incompatible job families. The performance measures to be simultaneously minimized are total weighted tardiness and makespan. Total weighted tardiness is related to on-time delivery performance whereas a small makespan indicates a high system utilization.

J. Gottlieb and G.R. Raidl (Eds.): EvoCOP 2006, LNCS 3906, pp. 209–221, 2006.

There are many papers that deal with batching (cf. the two recent survey papers [9, 14]). Most of the papers consider single-objective batching problems. One known exception is [6] where the simultaneous minimization of earliness and tardiness of jobs on a batch processing machine is discussed as a bicriteria scheduling problem. Several papers discuss multicriteria scheduling problems for parallel machines. We refer, for example, to [1]. In this paper an integrated convex preference measure is introduced. Hybrid genetic algorithms of MPGA (**M**ulti-**P**opulation **G**enetic **A**lgorithm) [2] and MOGA (**M**ulti-**O**bjective **G**enetic **A**lgorithm) [12] type are used. A comparison of the new measure with existing measures is performed. In this paper we extend the previous work [11] of the second present author for scheduling of batch machines from a single-objective situation to a bicriteria situation.

This paper is organized as follows. In Section 2 we describe the problem. Then, we describe the solution technique to solve the multiobjective scheduling problem. We present results of computational experiments in Section 4.

## 2   Problem Setting

The assumptions involved in the scheduling of parallel batch processing machines with incompatible job families and unequal ready times of the jobs to minimize total weighted tardiness and makespan are:

1. Jobs of the same family have the same processing times.
2. All the batch processing machines are identical in nature.
3. Once a machine is started, it cannot be interrupted, i.e. no preemption is allowed.

We use the following notation throughout the rest of the paper.

1. Jobs fall into $fam$ different incompatible families that cannot be processed together.
2. There are $n$ jobs that have to be scheduled.
3. There are $m$ identical machines in parallel.
4. There are $n_j$ jobs of family $j$ to be scheduled. We have $n = \sum_{j=1}^{fam} n_j$.
5. Job $i$ of family $j$ is represented as $ij$.
6. The priority weight for job $i$ of family $j$ is represented as $w_{ij}$.
7. The due date of job $i$ of family $j$ is represented as $d_{ij}$.
8. The processing time of jobs in family $j$ is represented as $p_j$.
9. The ready time of job $i$ in family $j$ is represented as $r_{ij}$.
10. The batch processing machine capacity is $B$ jobs.
11. Batch $k$ of family $j$ is represented by $B_{kj}$.
12. The completion time of job $i$ of family $j$ is denoted by $C_{ij}$.
13. The makespan is defined as $C_{max} := \max_{ij}\{C_{ij}\}$. Minimizing the makespan usually leads to a high utilization of the machines.
14. The weighted tardiness of job $ij$ is represented as $w_{ij}T_{ij} = w_{ij}(C_{ij} - d_{ij})^+$, where we use the notation $z^+ := \max(z, 0)$ throughout the rest of the paper.

The total weighted tardiness of all jobs is defined as $TWT := \sum_{ij} w_{ij} T_{ij}$. Minimizing total weighted tardiness of the jobs implies a high customer satisfaction level because of a good on-time delivery of the jobs.

Using the $\alpha|\beta|\gamma$ notation from the scheduling literature [5] this problem can be represented as:

$$Pm|r_{ij}, batch, incompatible|TWT, C_{\max}. \tag{1}$$

Here, we use the notation $P_m$ for an identical parallel machine environment. The batching with incompatible job families is denoted by *batch incompatible*. Finally, the $\gamma$ field is used to indicate the two used objectives, TWT and makespan.

The single machine total weighted tardiness scheduling problem is NP-Hard [13]. Hence, the parallel batch machine scheduling problem with TWT objective is also NP-Hard. The parallel machine scheduling problem with makespan objective is NP-Hard [13]. Therefore, the parallel batch machine problem with makespan objective is also NP-Hard. Finally, problem (1) is NP-Hard because it would be NP-Hard if each of the two objectives would be optimized separately [13, 15]. Hence, we have to look for efficient heuristics.

# 3 Solution Approach for the Multiobjective Batching Problem

In this section we briefly describe the overall three-phase solution approach. It contains a batch formation, a batch assignment, and a batch sequencing phase respectively. Then we characterize the NSGA-II metaheuristic that is used for the second phase. Furthermore, we also describe the usage of a more advanced version of a hybrid multiobjective metaheuristic combining the NSGA-II algorithm and a local search technique in the assignment phase.

## 3.1 Three-Phase Scheduling Approach

We extend the batch scheduling approach from [11] to the multiobjective situation. In the first phase, we have to form batches. In the second phase, we assign each batch from the first phase to one of the parallel machine. The third phase is responsible for sequencing the batches on each machine.

We start from a fixed point of time $t$ and consider a time window $(t, t + \Delta t)$ in the first phase. Let the set

$$\bar{M}(j, t, \Delta t) := \{ij | r_{ij} \leq t + \Delta t\} \tag{2}$$

denote the set of jobs from family $j$ whose ready times $r_{ij}$ are smaller than $t + \Delta t$. Let $I_{ij}$ be an index-based criterion for ranking these jobs and let $pos(ij)$ be the position of job $ij$ based on $I_{ij}$. From the set $\bar{M}(j, t, \Delta t)$ we derive the subset:

$$\tilde{M}(j, t, \Delta t, thres) := \{ij | ij \in \bar{M}(j, t, \Delta t) \quad \text{and} \quad pos(ij) \leq thres\}, \tag{3}$$

where *thres* is the maximum number of jobs in this new set. The ranking of jobs in the set $\bar{M}(j, t, \Delta t)$ is obtained by the Apparent Tardiness Cost (ATC)-based criterion (cf. [13])

$$I_{ij,ATC}(t) := \left( \frac{w_{ij}}{p_j} \right) \exp \left( -\frac{(d_{ij} - p_j + (r_{ij} - t))^+}{\kappa \bar{p}} \right),  \tag{4}$$

where $\kappa$ is a look-ahead parameter and $\bar{p}$ is the average processing time of the remaining unscheduled jobs. By $t$ we denote the time for decision-making. It is well known that the application of ATC-based rules leads to schedules with small TWT [13]. The jobs in the set $\bar{M}(j, t, \Delta t)$ are ranked in the non-decreasing order of $I_{ij,ATC}$ and the first *thres* are chosen for the subset $\tilde{M}(j, t, \Delta t, thres)$.

All possible batching combinations are considered in this subset. To choose a single batch from these batches, we use the index $I_{bj}(t)$:

$$I_{bj}(t) := \sum_{i=1}^{n_{bj}} \left( \frac{w_{ij}}{p_j} \right) \exp \left( -\frac{(d_{ij} - p_j - t + (r_{bj} - t)^+)^+}{\kappa \bar{p}} \right).  \tag{5}$$

We denote the number of jobs in the batch by $n_{bj}$. Furthermore, we set $r_{bj} := \max_{ij \in B_{bj}}(r_{ij})$. The $I_{bj}(t)$ index is called the BATC (Batched Apparent Tardiness Cost) index. The batch with the highest value of $I_{bj}(t)$ is scheduled.

The second phase is based on genetic algorithms and is explained in more detail in Sections 3.2 and 3.3.

After the assignment step performed in the second phase the batches are sequenced on each single batch machine by using the BATC index (5).

## 3.2  NSGA-II for the Second Phase

The NSGA-II algorithm [4] is a well-established metaheuristic for multiobjective combinatorial optimization problems based on the principles of genetic algorithms (GAs). NSGA-II evaluates each solution by its domination to each other population member. Using a nondominated sorting procedure the algorithm splits the complete population into several fronts in each iteration. A front for a population member (a solution $y$) is determined by the number of solutions that dominate $y$ and the solutions dominated by $y$. Taking this property into account any member of a front is Pareto-optimal to each other solution of the front it belongs to. Clearly, the solutions of the first front are nondominated by any other solution of the population. In order to ensure elitism NSGA-II joins in each iteration the parent and offspring populations. The fitness of an individual is assigned according to its front. The new population is generated by iteratively adding solutions from the first front. The population is filled up with solutions from the remaining fronts when the size of the first front is smaller as the population size. To ensure diversification NSGA-II uses a crowding comparison procedure for the tournament section [10]. The same procedure is used to select the solutions from the last front that do not fit completely into the new generation.

## 3.3  Combining NSGA-II with Local Search for the Second Phase

A hybrid evolutionary approach based on the NSGA-II algorithm is proposed by Deb and Goel in [3]. Each solution in the Pareto-front is improved by a local search when the NSGA-II algorithm is finished. The local search moves the outcomes of NSGA-II closer to the true Pareto-front $\mathcal{Y}_{true}$.

In this paper, we adopt the concept proposed in [3] for our multiobjective batching approach. A local search procedure tries to improve the objectives for every solution from the NSGA-II Pareto-front. The local search procedure uses a single-objective fitness function instead of the multiobjective NSGA-II metaheuristic. We use a weighted objective approach in order to transfer the multiobjective problem into a problem with single-objective fitness function.

**Require:** $\mathcal{Y}_{known}, steps_{LS}$
1: **for all** $y_{work} \in \mathcal{Y}_{known}$ **do**
2:     $\mathcal{Y}_{result} = \mathcal{Y}_{result} \cup_{nondom} y_{work}$
3:     calculate $\lambda_{TWT}$ and $\lambda_{C_{max}}$
4:     calculate $f(y_{work})$
5:     $counter = 0$
6:     **while** $counter < steps_{LS}$ **do**
7:         $y_{neighbor} = move(y_{work})$
8:         calculate $f(y_{neighbor})$
9:         **if** $f(y_{neighbor}) < f(y_{work})$ **then**
10:             $y_{work} = y_{neighbor}$
11:             $counter = 0$
12:         **else**
13:             $counter++$
14:         **end if**
15:     **end while**
16:     $\mathcal{Y}_{result} = \mathcal{Y}_{result} \cup_{nondom} y_{work}$
17: **end for**
18: return $\mathcal{Y}_{result}$

**Fig. 1.** local search procedure

Therefore, we generate for each objective a weight $\lambda_k$ according to its location within the Pareto-front. The weights for each objective are determined by

$$\lambda_{TWT}(y) := \frac{f_{TWT}^{max} - f_{TWT}(y)}{f_{TWT}^{max} - f_{TWT}^{min}},$$
$$\lambda_{C_{max}}(y) := \frac{f_{C_{max}}^{max} - f_{C_{max}}(y)}{f_{C_{max}}^{max} - f_{C_{max}}^{min}}, \quad (6)$$

where we denote by $f_k^{max}$ and $f_k^{min}$ the maximum and the minimum value of the objective function $f_k$. Additionally, the weights are normalized. If $f_k^{max} = f_k^{min}$ is valid in expression (6) then we set $\lambda_k = 1$.

The weights for a solution $y$ take into account its location within the Pareto-front, i.e., for a solution $y$ with $f_{TWT}(y) = f_{TWT}^{min}(y)$ and $f_{C_{max}}(y) = f_{C_{max}}^{max}(y)$ holds $\lambda_{TWT}(y) = 1$ and $\lambda_{C_{max}}(y) = 0$. In this situation, a local search hillclimber accepts any solution with a smaller TWT without considering $C_{max}$.

The single-objective fitness function $f$ using the weights calculated in (6) is defined in equation (7):

$$f(y) := \lambda_{TWT}(y) \cdot f_{TWT}(y) + \lambda_{C_{max}} \cdot f_{C_{max}}(y). \quad (7)$$

The complete local search procedure used in this research is given by Figure 1. The local search procedure is called with a solution front $\mathcal{Y}_{known}$ and the maximum number of local search steps denoted by $steps_{LS}$. Finally, a set of

Pareto-optimal solutions is returned by the procedure. We introduce an external archive $\mathcal{Y}_{result}$ that stores all nondominated solutions found during the local search. The operator $\cup_{nondom}$ denotes the update procedure for $\mathcal{Y}_{result}$. $\cup_{nondom}$ adds a solution to $\mathcal{Y}_{result}$ if it dominates solutions in $\mathcal{Y}_{result}$ or if it is Pareto-optimal for solutions in $\mathcal{Y}_{result}$. Solutions in $\mathcal{Y}_{result}$ dominated by the added solutions are removed. The local search iterates over all solutions in $\mathcal{Y}_{known}$. We first determine the weights $\lambda_{TWT}$ and $\lambda_{C_{max}}$ and calculate the fitness value for the current solution $y_{work}$. The inner loop (line 6) performs the local search. We consider for each solution $y_{work}$ $steps_{LS}$ solutions in its neighborhood. We define the neighborhood of a solution as the set of solutions obtained by simply swapping two batches across different machines. The operator $move(y_{work})$ generates a solution $y_{neighbor}$ in the neighborhood of $y_{work}$ by randomly exchanging two batches between two machines. The fitness for $y_{neighbor}$ is calculated using expression (7). A new solution is accepted as current solution if $f(y_{neighbor}) < f(y_{work})$. Afterward, the local search restarts with the new current solution $y_{work}$.

Figure 2 illustrates the local search for a Pareto-front $\mathcal{Y}_{known}$ generated by NSGA-II. The arrows connect the solutions generated by the local search with the initial solution in $\mathcal{Y}_{known}$. Due to the application of the local search procedure described above we improve the solutions $y_A$, $y_B$, and $y_C$. Solution $y_A$ has the lowest TWT value and the highest $C_{max}$ value. The weights used within the fitness function are $\lambda_{TWT}(y_A) = 1$ and $\lambda_{C_{max}}(y_A) = 0$.

The highest TWT value and the lowest $C_{max}$ value are obtained for solution $y_C$. For solution $y_C$ the weights are $\lambda_{TWT}(y_C) = 0$ and $\lambda_{C_{max}}(y_C) = 1$. The corresponding weights for solution $y_B$ are $\lambda_{TWT}(y_B) = 0.86$ and $\lambda_{C_{max}}(y_B) = 0.14$. Figure 2 shows that the local search generates for $y_A$ two new solutions $y_1$ and $y_2$ and only one new (improved) solution for $y_B$ and $y_C$. Figure 2 clearly states that the objective for the local search of $y_A$ (determined by the weights) is to minimize TWT without considering $C_{max}$. Therefore, any solution with a smaller

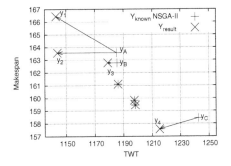

**Fig. 2.** Example Improvement of Solutions from $\mathcal{Y}_{known}$ by Local Search

$TWT$ is accepted by the local search procedure, even if $C_{max}$ is higher than $f_{C_{max}}(y_{work})$. For solution $y_C$ the local search is focused on minimizing $C_{max}$ without considering $TWT$. Analyzing the local search results for solutions in the 'middle' of the input Pareto-front we can conclude that the local search for this problem is unable to find solutions in the neighborhood with a smaller fitness value. Finally, the results show that a Pareto-front generated by a NSGA-II can be improved by applying a local search procedure for each solution in $\mathcal{Y}_{known}$.

# 4   Results of Computational Experiments

In this section, we describe the design of experiments. Then we discuss the results of computational experiments.

## 4.1   Design of Experiments

We use a similar design of experiments as in [11]. We expect that the quality of the results is influenced by the number of parallel machines, the number of jobs and the number of incompatible families. We perform ten indepen-

**Table 1.** Design of Experiments

| Factor | Level | Count |
|--------|-------|-------|
| Number of Machines | $m = 2, 3, 4$ | 3 |
| Number of Jobs | $n = 60, 80, 100$ | 3 |
| Batch Size | $B = 4, 8$ | 2 |
| Number of Families | $fam = 3$ | 1 |
| Family Processing Time | 2 with probability of 0.2 | |
| | 4 with a probability of 0.2 | |
| | 10 with a probability of 0.3 | 1 |
| | 16 with a probability of 0.2 | |
| | 20 with a probability of 0.1 | |
| Weight $w_{ij}$ per job | U(0,1) | |
| Ready Times | $r_{ij} \sim U(0, \frac{\alpha}{mB} \sum p_{ij})$ | 9 |
| Due Dates | $d_{ij} \sim U(0, \frac{\beta}{mB} \sum p_{ij})$ | |
| | $\alpha = 0.25, 0.50, 0.75$ | |
| | $\beta = 0.25, 0.50, 0.75$ | |
| | Total parameter combinations | 162 |
| | Number of problem instances | 10 |
| | Total problems | 1620 |

dent runs for each test instance with both NSGA-II and NSGA-II with local search. We denote the NSGA-II approach with local search for abbreviation with NSGA-II-LS. We summarize the design used in Table 1. The Pareto-optimal solutions out of all runs are aggregated to the solution Pareto-front $\mathcal{Y}_{known}$ for each approach. A near-to-optimal Pareto-front was generated by $\mathcal{Y}_{true} := \mathcal{Y}_{known}(\text{NSGA-II}) \cup_{nondom} \mathcal{Y}_{known}(\text{NSGA-II-LS})$. Throughout all experiments we use $\Delta t = 2h$, $thres = 10$ and $steps_{LS} = 30$. We encoded a solution for the GAs by an integer string with the size equal to the number of batches formed in the first phase. Each gene in the chromosome represents the maschine used for processing the batch. The population size is 100. The initial population is generated by randomly assigning a machine to each batch. The GAs stops after 300 generations. A uniform crossover [10] with crossover probability of 1 is used. The mutation operator randomly exchanges two batches between two machines.

We use a mutation probablity $p_{mut} = 0.02$. The parameters for the GAs are determined by trial and error based on intensive computational experiments. Both NSGA-II and NSGA-II-LS are implemented by using MOMHLib++ [8] under the linux operating system. The experiments are performed on a Pentium IV, 2 GHz PC.

### 4.2     Performance Metrics for the Assessment of the Algorithms

In the single-objective case an algorithm outperforms a second one if it generates a solution with an improved objective value (with eventually fewer computational effort). As shown in Section 3 a multiobjective algorithm, however, generates a set of Pareto-optimal solutions (a Pareto-front) instead of a single solution.

Performance metrics are required in order to evaluate and compare the outcome of several multiobjective algorithms. A common concept for evaluating the outcomes of a multiobjective algorithm is an approximation-based evaluation. Approximation-based performance metrics evaluate the approximation of a known (near-to-optimal) Pareto-front $\mathcal{Y}_{true}$ by a Pareto-front $\mathcal{Y}_{known}$ [7]. In this paper, we use cardinality-based and distance-based metrics.

Veldhuizen proposed some cardinality-based performance metrics in [16]. We use the following cardinality-based performance metrics for the assessment for the two algorithms described in Section 3. The metric Overall Non-dominated Vector Generation (ONVG) is the number of solutions in a Pareto-front $\mathcal{Y}_{known}$. It is defined by

$$ONVG := |\mathcal{Y}_{known}|. \tag{8}$$

The metric Overall True Non-dominated Vector Generation (OTNVG) is given by the number of solutions from $\mathcal{Y}_{known}$ that are in $\mathcal{Y}_{true}$. A high value for OTNVG indicates that many solutions of $\mathcal{Y}_{known}$ are used to generate $\mathcal{Y}_{true}$. This measure is defined through the expression

$$OTNVG := |\{y|y \in \mathcal{Y}_{known} \cap \mathcal{Y}_{true}\}|. \tag{9}$$

The third metric is called Overall Non-dominated Vector Generation Ratio (ONVGR). It is the ratio of ONVG and $|\mathcal{Y}_{true}|$. More formally, this measure is defined as follows:

$$ONVGR := ONVG/|\mathcal{Y}_{true}|. \tag{10}$$

We denote the error ratio for the solutions from $\mathcal{Y}_{known}$ that are not in $\mathcal{Y}_{true}$ by Error. The cardinality-based metric Error is defined as follows:

$$Error := \frac{\sum_{i=1}^{|\mathcal{Y}_{known}|} e_i}{ONVG}, \tag{11}$$

where

$$e_i := \begin{cases} 0 & : \quad \text{if} \quad y_i \in \mathcal{Y}_{known} \cap \mathcal{Y}_{true} \\ 1 & : \quad \text{otherwise} \end{cases}.$$

An Error value close to one means that only a small number of solution from $\mathcal{Y}_{known}$ is in $\mathcal{Y}_{true}$. $Error = 0$ is desirable. Jaszkiewicz points out in [7] that cardinality-based performance measures are not enought to assess the approximation of a Pareto-front $\mathcal{Y}_{true}$ by a Pareto-front $\mathcal{Y}_{known}$. Thus, we use additionally distance-based performance measures. Veldhuizen [16] proposes the distance measure $dist$ representing the average distance $dist$ between solutions $y \in \mathcal{Y}_{known}$ and their closest solution $y' \in \mathcal{Y}_{true}$. The distance $dist$ is calculated by

$$dist := \frac{\sqrt{\sum\limits_{y \in \mathcal{Y}_{known}} \left(\min\limits_{y' \in \mathcal{Y}_{known}} d(y, y')\right)^2}}{ONVG}. \tag{12}$$

Here the distance $d(y, y')$ between a solutions $y \in \mathcal{Y}_{known}$ and $y' \in \mathcal{Y}_{known}$ is given by

$$d^2(y, y') := \sum_{j=1}^{J} (\theta_k(f_k(y) - f_k(y')))^2, \tag{13}$$

where $\theta_k = (f_k^{max} - f_k^{min})^{-1}$ for $j = 1, ..., J$ are weights for normalizing the objective value $f_k$. If $f_k^{max} = f_k^{min}$ then $\theta_k(f_k(y) - f_k(y')) = 1$ holds. Furthermore, we use the distance metrics proposed by Jaszkiewicz in [7] that consider the distance between the solutions in $\mathcal{Y}_{true}$ and the closest neighbors in $\mathcal{Y}_{known}$. In this paper, we use the following distance performance metrics:

$$dist_1 := \frac{1}{|\mathcal{Y}_{true}|} \sum_{y \in \mathcal{Y}_{true}} \{\min_{y' \in \mathcal{Y}_{known}} \{d(y', y)\}\},$$
$$dist_2 := \max_{y \in \mathcal{Y}_{true}} \{\min_{y' \in \mathcal{Y}_{known}} \{d(y', y)\}\}. \tag{14}$$

Here the distance $d(y, y')$ is calculated using equation (13). The distance $dist_1$ shows the average distance between a solution $y \in \mathcal{Y}_{true}$ and its closest neighbor $y' \in \mathcal{Y}_{known}$. The worst case is provided by the maximum distance between a solution $y \in \mathcal{Y}_{true}$ and the closest neighbor $y' \in \mathcal{Y}_{known}$. It is covered by $dist_2$. A measure for the uniformity of the approximation is determined by

$$dist_3 = dist_2/dist_1. \tag{15}$$

If $dist_1 = 0$ then $dist_3 = 1$ holds.

## 4.3   Results

Table 2 presents the obtained computational results. Each row of Table 2 compares the solution front $\mathcal{Y}_{known}$ obtained by using NSGA-II and NSGA-II-LS with the near-to-optimal Pareto-front $\mathcal{Y}_{true}$. A solution front $\mathcal{Y}_{known}$ is generated by the Pareto-optimal solutions out of all runs. For each solution front $\mathcal{Y}_{known}$ we show the metrics introduced in Section 4.2. Furthermore, Table 2 shows the average computational time for a single run in seconds.

The first two rows in Table 2 provide an overall comparison of the two algorithms investigated in this paper. The values for each performance metric

represent a average value over all test problems with the given specific of the
first row. $B = 4$ means for example that we take the average over all test instances with a batch size of four.

The results clearly show that NSGA-II-LS outperforms the generic NSGA-II
for all performance metrics. The overall results show that the values obtained for
$ONVG$ in $\mathcal{Y}_{known}$ are similar for both approaches. Furthermore, the OTNVG
value for NSGA-II-LS points out that this approach contributes more solutions
for $\mathcal{Y}_{true}$ than NSGA-II. The value for Error supports this observation. The distance metrics clearly identify the NSGA-II-LS solution front as the front with
the closest distance to $\mathcal{Y}_{known}$. The smaller $dist$ value for the NSGA-II-LS approach shows that all solutions from $\mathcal{Y}_{known}$(NSGA-II-LS) have a closer distance
to the nearest neighbor in $\mathcal{Y}_{true}$ than solutions in $\mathcal{Y}_{known}$(NSGA-II). The approximation of $\mathcal{Y}_{true}$ by $\mathcal{Y}_{known}$ is expressed by the distance measures $dist_1$,
$dist_2$, and $dist_3$. These distance measures again show that $\mathcal{Y}_{known}$(NSGA-II-LS)
outperforms $\mathcal{Y}_{known}$(NSGA-II). The values $dist_1$ and $dist_2$ indicate that for the
NSGA-II-LS approach the average and maximum distance between a solution
in $\mathcal{Y}_{true}$ and the closest solution in $\mathcal{Y}_{known}$ is smaller. The results for different

**Table 2.** Computational Results

|  | Algorithm | $ONVG$ | $OTNVG$ | $ONVGR$ | $Error$ | $dist$ | $dist_1$ | $dist_2$ | $dist_3$ | $Time$ |
|---|---|---|---|---|---|---|---|---|---|---|
| overall | NSGA-II | 4.30 | 2.15 | 1.01 | 0.46 | 0.12 | 0.19 | 0.41 | 2.60 | 9.41 |
|  | NSGA-II LS | 4.41 | 2.55 | 1.03 | 0.39 | 0.10 | 0.16 | 0.34 | 2.56 | 9.45 |
| m=3 | NSGA-II | 3.71 | 1.89 | 1.02 | 0.45 | 0.14 | 0.21 | 0.41 | 2.30 | 10.66 |
|  | NSGA-II LS | 3.87 | 2.11 | 1.07 | 0.42 | 0.13 | 0.19 | 0.36 | 2.20 | 10.70 |
| m=4 | NSGA-II | 4.50 | 2.21 | 1.01 | 0.48 | 0.12 | 0.19 | 0.41 | 2.69 | 9.24 |
|  | NSGA-II LS | 4.51 | 2.69 | 1.01 | 0.37 | 0.09 | 0.14 | 0.33 | 2.69 | 9.27 |
| m=5 | NSGA-II | 4.70 | 2.34 | 1.00 | 0.46 | 0.11 | 0.18 | 0.42 | 2.81 | 8.34 |
|  | NSGA-II LS | 4.84 | 2.85 | 1.01 | 0.38 | 0.09 | 0.14 | 0.33 | 2.80 | 8.37 |
| n=60 | NSGA-II | 4.32 | 2.21 | 0.99 | 0.45 | 0.11 | 0.17 | 0.38 | 2.64 | 6.09 |
|  | NSGA-II LS | 4.55 | 2.65 | 1.04 | 0.39 | 0.09 | 0.14 | 0.29 | 2.64 | 6.10 |
| n=80 | NSGA-II | 4.19 | 2.07 | 1.00 | 0.47 | 0.13 | 0.20 | 0.42 | 2.49 | 9.28 |
|  | NSGA-II LS | 4.39 | 2.50 | 1.08 | 0.40 | 0.11 | 0.17 | 0.37 | 2.61 | 9.31 |
| n=120 | NSGA-II | 4.41 | 2.16 | 1.04 | 0.48 | 0.14 | 0.21 | 0.44 | 2.67 | 12.87 |
|  | NSGA-II LS | 4.28 | 2.51 | 0.97 | 0.38 | 0.10 | 0.16 | 0.36 | 2.44 | 12.93 |
| B=4 | NSGA-II | 4.18 | 2.04 | 1.03 | 0.47 | 0.13 | 0.21 | 0.44 | 2.45 | 11.93 |
|  | NSGA-II LS | 4.30 | 2.45 | 1.06 | 0.39 | 0.11 | 0.17 | 0.36 | 2.40 | 11.96 |
| B=8 | NSGA-II | 4.43 | 2.25 | 0.99 | 0.46 | 0.11 | 0.17 | 0.39 | 2.75 | 6.90 |
|  | NSGA-II LS | 4.52 | 2.65 | 1.00 | 0.39 | 0.10 | 0.14 | 0.32 | 2.73 | 6.93 |

values for $m$ point out that the number of solutions in $\mathcal{Y}_{known}$ increases for
both approaches when $m$ increases. Comparing the cardinality-based metrics
for different number of machines indicates that NSGA-II-LS performs better
than NSGA-II. The distance metrics also identify the NSGA-II-LS approach
as the superior approach. The distance-based performance measure results for

NSGA-II-LS with different number of machines show that the distance between $\mathcal{Y}_{known}$(NSGA-II-LS) and $\mathcal{Y}_{true}$ is similar for all $m$. Thus, the results obtained here show that the number of machines has no impact on the solution quality. Analyzing these facts we have to take into account that the near-to-optimal Pareto-front is only generated by the nondominated solutions from NSGA-II and NSGA-II-LS. A Pareto-front $\mathcal{Y}_{known}$ with solutions from more approaches might lead to other results. The results for different number of jobs show that when the problem becomes harder, i.e., in case of more jobs, NSGA-II-LS generates fewer solutions and has a lower contribution to $\mathcal{Y}_{true}$. We can conclude that NSGA-II-LS outperforms NSGA-II for all investigated values of $n$. With an increase of the batch size $B$ the first phase generates fewer batches. Thus, the decision space in the second phase becomes smaller. The results in Table 2 show that both approaches generate more solutions for $B = 8$. The values for $OTNVG$ show that both approaches contribute more solutions for $\mathcal{Y}_{known}$ with a higher $B$. The distance-based performance measures show that a larger $B$ has only a small impact on the distance between $\mathcal{Y}_{known}$ and $\mathcal{Y}_{true}$.

Clearly, the additional local search causes a higher computational effort. A comparison of the values for the column *Time* shows that NSGA-II-LS runs slower than NSGA-II. Taking into account the higher solution quality obtained by NSGA-II-LS this is a reasonable additional effort. An other remarkable fact for the solutions in $\mathcal{Y}_{true}$ is that the average TWT values of the solutions are similar to the average TWT values obtained by using the single-objectice approach from [11].

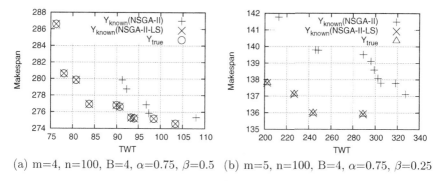

(a) m=4, n=100, B=4, $\alpha$=0.75, $\beta$=0.5   (b) m=5, n=100, B=4, $\alpha$=0.75, $\beta$=0.25

**Fig. 3.** Comparison of $\mathcal{Y}_{true}$ and Pareto-fronts Obtained by NSGA-II and NSGA-II-LS

Figure 3 shows the Pareto-fronts for two test problems. Each plot shows the Pareto-fronts $\mathcal{Y}_{known}$(NSGA-II), $\mathcal{Y}_{known}$(NSGA-II-LS), and $\mathcal{Y}_{true}$. Both plots emphasis the analytic results shown in Table 2. All solutions in $\mathcal{Y}_{known}$(NSGA-II) are dominated by solution in $\mathcal{Y}_{known}$(NSGA-II-LS). Therfore, for both problems the near-to-optimal Pareto-front $\mathcal{Y}_{true}$ is formed by solutions from $\mathcal{Y}_{known}$(NSGA-II-LS).

# 5 Conclusions and Future Research

In this paper, we discuss a multiobjective scheduling problem for parallel batch machines. We suggest the usage of a genetic algorithm in order to determine Pareto efficient solution. Then, in a second step, we incorporate a local search technique in the overall solution scheme in order to improve the solution quality. We present the results of computational experiments that show that the NSGA-II combined with local search techniques outperforms the pure NSGA-II algorithm.

There are several directions for future research. First of all it seems to be possible to replace the genetic algorithm based approaches by other local search techniques like simulated annealing or tabu search. It seems to be possible to consider more sophisticated local search techniques based on dominance properties. Here, a swapping of jobs across batches based on their potential contribution to TWT and/or makespan reduction should be done.

# References

1. Carlyle, W.M., Kim, B., Fowler, J.W., Gel, E.S.: Comparison of Multiple Objective Genetic Algorithms for Parallel Machine Scheduling Problems. In: Zitzler, E. (eds.): Proceedings EMO 2001. Lecture Notes in Computer Science Vol. 1993 Springer-Verlag Berlin Heidelberg New York (2001) 473–485
2. Cochran, J.K., Horng, S.-M., Fowler, J.W.: A Multi Population Genetic Algorithm to Solve Parallel Machine Scheduling Problems with Sequence Dependent Setups. Computers & Operations Research **30**(7) (2003) 1087–1102
3. Deb, K., Goel, T.:A Hybrid Multi-Objective Evolutionary Approach to Engineering Shape Design. In: Zitzler, E. (eds.): Proceedings of the First International Conference on Evolutionary Multi-Criterion Optimization Springer-Verlag Berlin (2001) 385–399
4. Deb, K., Pratap, A., Agarwal, S., Meyarivan, T.: A Fast and Elitist Multiobjective Genetic Algorithm: NSGA-II. IEEE Transactions on Evolutionary Computation **6**(2) (2002) 182–197
5. Graham, R. L., Lawler, E. L., Lenstra, J. K., Rinnooy Kann, A. H. G.: Optimization and Approximation in Deterministic Sequencing and Scheduling: a Survey. Annals of Discrete Mathematics **5** (1979) 287–326
6. Gupta, A. K., Sivakumar, A. I.: On-time Delivery Pareto Controllability for Batch Processing in Semiconductor Manufacturing. In: Proceedings 3rd International Conference on Modeling and Analysis of Semiconductor Manufacturing Singapore (2005) 13–20
7. Jaszkiewicz, A.: Evaluation of Multiple Objective Metaheuristics. In: Metaheuristics for Multiobjective Optimisation, Lecture Notes in Economics and Mathematical Systems Vol. 535 Springer-Verlag Berlin (2004) 65–89
8. Jaszkiewicz, A.: MOMHLib++: Multiple Objective MetaHeuristics Library in C++. http://www-idss.cs.put.poznan.pl/~jaszkiewicz/MOMHLib (2005)
9. Mathirajan, M., Sivakumar, A.I.: Scheduling of Batch Processors in Semiconductor Manufacturing - a Review. Singapure MIT Alliance (SMA) 2003 Symposium National University of Singapure https://dspace.-mit.edu/retrieve/3521/IMST021 (2003)

10. Michalewicz, Z.: Genetic Algorithms + Data Structures = Evolution Programs. 3rd edn. Springer-Verlag, Berlin Heidelberg New York (1996)
11. Mönch, L., Balasubramanian, H., Fowler, J. W., Pfund, M.: Heuristic Scheduling of Jobs on Parallel Batch Machines with Incompatible Job Families and Unequal Ready Times. Computers & Operations Research **32** (2005) 2731–2750
12. Murata, T., Ishibuchi, H., Tanaka, H.: Multi-objective Genetic Algorithm and its Application to Flow Shop Scheduling. Computers and Industrial Engineering **30**(4) (1996) 957–968
13. Pinedo, M.: Scheduling: Theory, Algorithms, and Systems. 2nd edition Prentice Hall New Jersey (2002)
14. Potts, C. N., Kovalyov, M. Y.: Scheduling with Batching: a Review. European Journal of Operational Research **120** (2000) 228–249
15. T'kindt, V., Billaut, J.-C.: Multicriteria Scheduling: Theory, Models and Algorithms. Springer-Verlag, Berlin Heidelberg New York (2002)
16. Van Veldhuizen, D. A.: Multiobjective Evolutionary Algorithms: Classifications, Analyses, and New Innovations. Air Force Institute of Technology Department of Electrical and Computer Engineering Ohio (1999)

# A Memetic Algorithm for the Biobjective Minimum Spanning Tree Problem

Daniel A. M. Rocha, Elizabeth F. Gouvêa Goldbarg, and Marco César Goldbarg

Depto de Informática e Matemática Aplicada,
Universidade Federal do Rio Grande do Norte,
Campus Universitário, 59072-970, Natal, Brazil
danielrocha@digizap.com.br
{beth, gold}@dimap.ufrn.br

**Abstract.** Combinatorial optimization problems with multiple objectives are, in general, more realistic representations of practical situations than their counterparts with a single-objective. The bi-objective minimum spanning tree problem is a NP-hard problem with applications in network design. In this paper a memetic algorithm is presented to solve this problem. A computational experiment compares the proposed approach with AESSEA, a known algorithm of the literature. The comparison of the algorithms is done with basis on the binary additive $\varepsilon$-indicator. The results show that the proposed algorithm consistently produces better solutions than the other method.

## 1 Introduction

Combinatorial optimization problems with multiple objectives are natural extensions of single-objective problems. Although single-objective combinatorial problems model a great number of applications, in many real situations they can be very simplistic representations, since they can not deal with conflicting points equivalent to the multiple objectives. A problem that arises when considering more suitable models to represent actual applications, as the case of multiple objectives, is that, in general, those extensions have an increasing in complexity with relation to their counterparts with a single-objective. This is the case, for instance, of the Minimum Spanning Tree and the Linear Assignment Problem [5].

The general multi-objective minimization problem (with no restrictions) can be stated as:

$$\text{"minimize" } f(x) = (f_1(x), ..., f_k(x)), \text{ subjected to } x \in X$$

where, x is a discrete value solution vector and X is a finite set of feasible solutions. Function f(x) maps the set of feasible solutions X in $\Re^k$, $k > 1$ being the number of objectives.

Once there is not only a single solution for the problem, the word minimize has to be understood in another context. Let x, y $\in$ X, then x dominates y, written x $\succ$ y, if and only if $\forall i$, $i=1,...k$, $f_i(x) \leq f_i(y)$ and $\exists i$, such that $f_i(x) < f_i(y)$. The set of optimal solutions $X^* \subseteq X$ is called Pareto optimal. A solution $x^* \in X^*$ if there is no x $\in$ X such that x $\succ$ x*. The non-dominated solutions are said also to be efficient solutions. Thus,

J. Gottlieb and G.R. Raidl (Eds.): EvoCOP 2006, LNCS 3906, pp. 222–233, 2006.

to solve a multi-criteria problem, one is required to find the set of efficient solutions. Solutions of this set can be divided in two classes: the supported and nonsupported efficient solutions. The supported efficient solutions can be obtained by solving the minimization problem with a weighted sum of those objectives. More formally [6],

$$\text{Minimize } \sum_{i=1,\dots,k} \lambda_i f(x_i) \text{, where } \sum_{i=1,\dots,k} \lambda_i = 1 \text{, } \lambda_i > 0 \text{, I} = 1,\dots,k$$

The nonsupported efficient solutions are those which are not optimal for any weighted sum of objectives. This set of solutions is a major challenge for researchers.

On the last three decades, a great effort has been dedicated to the research of multi-criteria problems. Some classes of exact algorithms are listed in the paper of Ehrgott and Gandibleux [6], where a number of applications are also reported.

Since exact approaches are able to solve only small instances within a reasonable computing time, approximation algorithms, mainly based upon metaheuristic techniques, have been proposed to solve multi-criteria problems [7]. Among those approaches, the Evolutionary Algorithms are one of the most popular. A survey of evolutionary algorithms for multi-objective problems is presented by Coello [4].

The Memetic Algorithms, introduced by Moscato [20], are evolutionary algorithms which unify the diversification potential of Genetic Algorithms with methods specialized in intensify the search in certain regions of the solution space. It has been observed that for a wide range of applications those algorithms perform better than classical Genetic Algorithms [19].

Section 3 presents a Memetic Algorithm to solve the Biobjective Minimum Spanning Tree Problem, presented in section 2. A direct encoding of the trees and a crossover operator based upon the work of Raidl [24] are utilized. A Tabu Search procedure is proposed, as part of the main algorithm, as an intensification tool.

A computational experiment compares the proposed algorithm with AESSEA, another evolutionary approach proposed previously for the same problem [16]. The algorithms are applied to 39 instances with number of nodes ranging from 10 to 500, generated in accordance with the method introduced by Knowles [14].

A challenging issue in multi-criteria optimization is to define quantitative measures for the performance of different algorithms. Zitzler et al. [28] proposed a general binary indicator which is utilized in this work for the task of algorithms comparison. Their binary additive $\varepsilon$-indicator is considered to compare the quality of the sets of solutions generated by each algorithm.

Finally some concluding remarks are drawn in section 5.

## 2   The Multi-objective MST

A spanning tree of a connected undirected graph $G = (N, E)$ is an acyclic subgraph of $G$ with $n$ - 1 edges, where $n = |N|$. If $G$ is a weighted graph, a minimum spanning tree, MST, of $G$ is spanning tree for which the summation of the weights of its edges is minimum over all spanning trees of $G$. The MST is a well known combinatorial optimization problem with applications in distinct areas such as, networks design and clustering. Its theoretical importance comes from the fact that it may be utilized in approximation algorithms for other combinatorial optimization problems such as the

Traveling Salesman Problem [12] and the Steiner Tree [13]. The MST is solvable in polynomial time and the classical algorithms presented for it are due to Prim [21], Kruskal [18] and Borüvka [3]. The history of this problem is presented in the work of Graham and Hell [11]. A survey of the MST problem and algorithms is presented in the paper of Bazlamaçci and Hindi [2].

The MST problem is polynomial, but constraints often render it NP-hard, as described by Garey and Johnson [9]. Examples include the degree-constrained minimum spanning tree, the maximum-leaf spanning tree, and the shortest-total-path-length spanning tree problems.

Another difficult variant of this problem is the Multi-criteria Minimum Spanning Tree, mc-MST. Given a graph $G = (N,E)$, a vector of non negative weights $w_{ij} = (w_{ij}^1,..., w_{ij}^k)$, $k > 1$, is assigned to each edge $(i,j) \in E$. Let $S$ be the set of all possible spanning trees, $T = (N_T,E_T)$, of $G$ and $W = (W^1,..., W^k)$, where

$$W^q = \sum_{(i,j) \in E_T} w_{ij}^q , q=1,...,k.$$

The problem seeks $S^* \subseteq S$, such that $T^* \in S^*$ if and only if $\nexists\, T \in S$, such that $T \succ T^*$. In this work the biobjective problem is considered, although the proposed algorithm can be adapted to consider $k > 2$ objectives.

An important application of the mc-MST with two objectives is in the area of network design, where the edge weights can be associated, for instance, to reliability restrictions and installation costs.

Aggarwal et al. [1] showed that the 0-1 knapsack problem can be polynomially reduced to the bicriterion spanning tree. Therefore, the biobjective spanning tree problem belongs to the NP-hard class.

Recently, a number of works have been dedicated to this problem [16], [23], [26], [27]. Among the heuristic approaches presented for the problem, two evolutionary algorithms were introduced by Zhou and Gen [27] and Knowles and Corne [16].

Zhou and Gen[27] present a genetic algorithm which solution representation is based upon Prüfer's code [22]. The algorithm is applied to instances with $n$ between 10 and 50 and a comparison is done with an enumeration method proposed in the same work. Later, Knowles and Corne [17] showed that the enumeration algorithm was not correct.

Knowles and Corne [16] present the algorithm AESSEA (Archived Elitist Steady State Evolutionary Algorithm) in two versions: AESSEA+Prüfer and AESSEA+ Direct/RPM. The first utilizes the Prüfer method to encode solutions and the operators suggested in the work of Zhou and Gen[27]. The latter utilizes versions of the encoding and operators presented by Raidl [24] for the degree constrained minimum spanning tree problem. The algorithms are tested in a set of instances with $n$ ranging from 10 and 50. The experiments show that the second version of AESSEA is more efficient than the first version.

A discussion of representations for minimum spanning tree problems is presented by Raidl [24] and Raidl and Julstrom [25].

## 3 The Algorithm

The method proposed in this paper to solve the biobjective MST is a genetic algorithm hybridized with a tabu search procedure. The population, with size #*size_pop*, is generated in two steps. At first, at most #*max_rmckrus* individuals are generated with a random greedy method based upon Kruskal's algorithm for the single-objective MST [18]. The other part of the population is generated with the RandomWalk method [24].

Kruskal's algorithm is modified to deal with edges with two weights by means of a scalarizing vector $\lambda$. Thus the two weights of each edge $(i,j)$ are replaced by a single value resultant from the inner product $\lambda w_{ij}$. The version of Kruskal's algorithm implemented in this work is based upon the constructive phase of GRASP [8] where, for each decision step, a restricted candidate list, RCL, is formed and an element of this list is randomly chosen to be added to the solution. A high level pseudo-code of the algorithm r-mc-Kruskal is given in the following. It receives two parameters: the scalarizing vector $\lambda$ and a random number, *num*, generated in the interval [0,0.1]. The edges are sorted in non-decreasing order of their values (scalarized weights) in a list $L$. Let $e_1$ the edge with the minimum associated value $a_1$, then the restricted candidate list is built with edges $e_j$ such that their associated values $a_j \leq (1+num)a_1$. The algorithm, iteratively, constructs this list, chooses randomly an edge of the RCL and removes this edge from $L$. If a cycle is not created with the inclusion of the chosen edge, then it is add to the solution. The algorithm stops when a spanning tree is obtained.

```
procedure r-mc-Kruskal(λ,num)
     L ← sort(E,λ)
     repeat
          w ← weight of the first edge of L
          RCL ← restricted_list(L,w,num)
          e ← random_choice(RCL)
          if (Sol ∪ {e}) is acyclic
          then
               Sol ← Sol ∪ {e}
          L ← L\{e}
     until (|Sol| = n-1)
```

A fixed number of different scalarizing vectors, #*max_rmckrus*, generates solutions for the initial population. If the same solution is generated more than once (by different scalarizing vectors) only one of them remains in the population.

The memetic algorithm keeps an archive with non-dominated solutions, Global_Arc. The number of non-dominated solutions in this archive is not limited. Since a solution is generated the algorithm checks Global_Arc and if it is necessary, the archive is updated with the new solution. If the new generated solution, s, is non-dominated in relation to the solutions in Global_Arc, it is added to the archive. If s dominates any solutions in Global_Arc, then those are removed.

Different scalarizing vectors are generated and passed to r-mc-Kruskal. The values of $\lambda$'s components varies between (0,1) and (1,0) in intervals with fixed size given by 1/#*max_rmckrus*.

Unlike mc-Prim [16], r-mc-Kruskal is not guaranteed to generate only Pareto optimal solutions, since the edge with the lowest assigned value is not necessarily chosen during the iterations. However, this method is able to create non supported efficient solutions, due to the same reason.

The pseudo-code of the memetic algorithm, Mem-MC-MST is given in the following. Individuals are encoded with the direct representation presented by Raidl [24]. The algorithm runs for a fixed number of iterations, #*max_gen*. At the beginning of each iteration, a tabu search procedure is called for the whole population. This search procedure is described below. An auxiliary population, $P'$, initially empty, is created to keep the offspring of the current generation. Each generation, #*size_pop* children are created.

The selection scheme is a variation of the binary tournament [10]. Two pairs of individuals are randomly chosen from the current population, each of them to compete in one criterion. That is, the first and second parents of a given child are the winners of the first and second tournaments in respect to one and other criterion. This is implemented in function *binary_tournament()*.

Parents are mated by means of the edge-crossover [24], generating one offspring.

The inclusion of the new generated individual in $P'$ is decided regarding two conditions. First, the solution encoded on the individual must be not dominated by the solutions of Global_Arc. Function *belong(x,arc)* verifies if a solution $x$ is non-dominated with respect to the solutions of the file *arc*, returning a "true" value if it occurs. Second, the individual must be in a less crowd grid location than both parents. Function *m_grid(arc,x,y)* returns a "true" value if solution $x$ is in a less crowd region with respect to the solutions of *arc* than solution $y$. This strategy was introduced by Knowles and Corne [15], where it is described in details. If those conditions are satisfied then Global_Arc is updated with the new offspring in function *update()*. Otherwise, the offspring is discarded and a new tentative is done. At most, #*max_offsp* tentatives are tried. If no offspring satisfies those conditions, a solution of Global_Arc is chosen at random and copied to offspring. The offspring is included in the auxiliary population.

When $P'$ is complete, it replaces the current population. Finally, the algorithm verifies convergence by checking if Global_Arc has not been updated for #*max_it* iterations. If it is the case, then half of the current population is replaced by new solutions generated by the RandomWalk method [24]. The substituted parcel is chosen at random. The following summarizes the memetic algorithm.

```
Algorithm Mem-MC-MST

    P ← generate_pop(#max_rmckrus,#size_pop)

    for i ← 1 to #max_gen do

      for each p ∈ P do

        Local_Arc ← {s | s ∈ Global_Arc and s⊁p}

        p ← tabu_search(p,Local_Arc)

      end_for

      P' ← {}

      for j ←1 to #size_pop do

        q ← 0

        do

          Choose at random p₁, p₂, p₃, p₄ from P

          parent₁ ← binary_tournament(p₁,p₂,objective_1)

          parent₂ ← binary_tournament(p₃,p₄,objective_2)

          offspringⱼ ← crossover(parent₁,parent₂)

          if belong(offspringⱼ,Global_Arc) or
(m_grid(Global_Arc,  offspringⱼ,parent₁) and
m_grid(Global_Arc,offspringⱼ,parent₂)), then

              q ← #max_offsp + 1

              update(Global_Arc,offspringⱼ)

          else q ← q + 1

        while(q < #max_offsp)

        if q = #max_offsp, then

            offspringⱼ ← random_choice(Global_Arc)

        P' ← P'∪ {offspringⱼ}

      end_for

    P ← P'

    if |Global_Arc| does not change in #max_it itera-
tions, then Diversify(P)

    end_for

return(Global_Arc)
```

```
procedure tabu_search(s,Local_Arc)

  tabu_list ← {}

  repeat

    r ← 0

    do

      r ← r+1

      Choose  at  random  a  pair  of  edges  (vᵢ,vᵢ₊₁),(vⱼ,vⱼ₊₁)
      from s

      remove((vᵢ,vᵢ₊₁),(vⱼ,vⱼ₊₁)) from s

      relink(vᵢ,vⱼ,vᵢ₊₁,vⱼ₊₁), producing s'

      if s'≻s then s ← s'

      else

      if ((((vᵢ,vⱼ) ∉ tabu_list) and ((vₖ,vᵣ)∉ tabu_list))

              and m_grid(Local_Arc,s',s)) then

          s ← s'

    while ( r < #max_neighbors ) and ( s not modified )

    if s was modified then

      include(tabu_list,(vᵢ,vⱼ),(vₖ,vᵣ),#tabutenure)

      update(Local_Arc,s)

      update(Global_Arc,s)

  until ((#max_tabu iterations) or (s not modified))
```

The tabu search procedure receives two parameters: a solution s and an archive with the solutions of Global_Arc that do not dominate s.

The neighborhood structure considers s' a neighbor of solution s, if s' is generated by the removal of two edges from s, $(v_i,v_{i+1})$ and $(v_j,v_{j+1})$, $i+1<j$, and the re-linking of those terminal vertices. There is only one way to re-link the terminal vertices and it is implemented in function *relink()*. The function checks whether the inclusion of edge $(v_i,v_j)$ creates a cycle or not. If a cycle is not created then edges $(v_i,v_j)$ and $(v_{i+1},v_{j+1})$ are added, otherwise the edges included in the solution are $(v_i,v_{j+1})$ and $(v_{i+1},v_j)$.

If *s'* dominates *s*, then *s* is updated with the new current solution. Otherwise, the algorithm verifies if none of the two edges that entered in *s'* have a "tabu" status and if *s'* is in a less crowd region of the location grid regarding the solutions in Local_Arc. If it is the case, then *s* is also updated with *s'*.

To speed the process, on each step, at most #*max_neighbors* solutions are generated and two edges that leave a certain solution s are chosen at random.

If $s$ is modified then the tabu list, *tabu_list*, is updated with the edges that were removed from $s$. These edges stay with a "tabu" status for *#tabutenure* iterations. If necessary, the archives of solutions are also updated.

# 4  Computational Experiments

The proposed algorithm was compared with the algorithm called AESSEA+Direct/RPM presented by Knowles and Corne [16]. The algorithm was implemented in C and both algorithms of the computational test ran on a Pentium IV (2.8 GHz and 1 Gb of RAM) with Linux. The algorithms were applied to thirty-nine instances generated in accordance with the method described in the work of Knowles [14]. Three groups of thirteen instances belonging to the classes concave, correlated and anti-correlated were generated as complete graphs with two objectives. Each class has an instance with $n = 10$, 25 and 50, and two instances with $n = 100$, 200, 300, 400 and 500. The correlated and anti-correlated instances require a correlation factor $\alpha$ and the concave instances require two parameters, $\zeta$ and $\eta$, to be generated. Table 1 summarizes the parameters utilized to generate the set of test instances.

**Table 1.** Parameters to generate the instances for the computational experiment

| Instances | Concave | | Correlated | Anti-correlated |
| --- | --- | --- | --- | --- |
| | $\zeta$ | $\eta$ | $\alpha$ | $\alpha$ |
| 10 | 0.1 | 0.25 | 0.7 | -0.7 |
| 25 | 0.05 | 0.2 | 0.7 | -0.7 |
| 50 | 0.03 | 0.125 | 0.7 | -0.7 |
| 100_1 | 0.01 | 0.02 | 0.3 | -0.3 |
| 100_2 | 0.02 | 0.1 | 0.7 | -0.7 |
| 200_1 | 0.05 | 0.2 | 0.3 | -0.3 |
| 200_2 | 0.08 | 0.1 | 0.7 | -0.7 |
| 300_1 | 0.03 | 0.1 | 0.3 | -0.3 |
| 300_2 | 0.05 | 0.125 | 0.7 | -0.7 |
| 400_1 | 0.025 | 0.125 | 0.3 | -0.3 |
| 400_2 | 0.04 | 0.2 | 0.7 | -0.7 |
| 500_1 | 0.02 | 0.1 | 0.3 | -0.3 |
| 500_2 | 0.03 | 0.15 | 0.7 | -0.7 |

The parameters for AESSEA reported by its authors [16] were utilized in the experiment. The following parameters were utilized in Mem-MC-MST: #max_rmckrus = 110, #size_pop = 150, #max_gen = 40, #max_offsp = 10, #max_it = 6, #max_neighbors = 5, #max_tabu= 30, #tabutenure = 5. The archive with non-dominated solution has unlimited size and the adaptive grid has 1024 positions. Ten independent runs of each algorithm were done for each instance.

The comparison of the sets of solutions generated by each algorithm was done with basis on the binary additive $\varepsilon$-indicator, $I_{\varepsilon+}$[28]. Given two sets of solutions $A$ and $B$, a value $I_{\varepsilon+}(A,B) < 0$, indicates that every solution of $B$ is strictly dominated by at least one solution of $A$. Values $I_{\varepsilon+}(A,B) \leq 0$ and $I_{\varepsilon+}(B,A) > 0$, indicates that every solution of $B$ is weakly dominated by at least one solution of $A$. Values $I_{\varepsilon+}(A,B) > 0$ and $I_{\varepsilon+}(B,A) > 0$, indicates that neither $A$ weakly dominates $B$ nor $B$ weakly dominates $A$. The results of the comparisons are showed in table 2 for correlated instances, table 3 for anti-correlated instances and table 4 for concave instances. Columns of the three tables show the instance, the indicators $I_{\varepsilon+}(A,B)$ and $I_{\varepsilon+}(B,A)$, where $A$ denotes Mem-MC-MST and $B$ denotes AESSEA, the runtime and the number non-dominated solutions found by both algorithms.

From the thirteen instances of the class correlated, the proposed algorithm strictly dominates AESSEA in all instances with $n = 100,\ldots, 500$. For instances with $n = 25$ and 50, Mem-MC-MST weakly dominates the other algorithm and for the instance with $n = 10$, the performance of the algorithms is incomparable. Runtimes of the proposed memetic algorithm are, in the majority, much smaller than the other algorithm. Moreover, the number of non-dominated solutions generated by the proposed algorithm is superior to the other. In average, the proposed algorithm finds six times more non-dominated solutions and runs in a computational time eleven times less than the other algorithm, regarding the class of correlated instances.

**Table 2.** Comparison of Mem-MC-MST and AESSEA on correlated instances

| Instance | $I_{\varepsilon+}(A,B)$ | $I_{\varepsilon+}(B,A)$ | Mem-MC-MST | | AESSEA | |
| | | | Time (s) | #solutions | Time (s) | #solutions |
| --- | --- | --- | --- | --- | --- | --- |
| 10 | 0.0000 | 0.0000 | 2.97 | 24 | 4.42 | 24 |
| 25 | 0.0000 | 0.0107 | 7.40 | 53 | 4.92 | 51 |
| 50 | 0.0000 | 0.1381 | 26.68 | 412 | 6.70 | 104 |
| 100_1 | -0.1439 | 0.6222 | 45.83 | 620 | 7.95 | 119 |
| 100_2 | -0.0609 | 0.3335 | 55.48 | 669 | 7.55 | 99 |
| 200_1 | -0.2362 | 1.0316 | 88.31 | 710 | 50.90 | 155 |
| 200_2 | -0.1485 | 0.6006 | 90.12 | 667 | 50.92 | 84 |
| 300_1 | -0.3516 | 1.7338 | 150.60 | 753 | 556.02 | 79 |
| 300_2 | -0.2566 | 1.3532 | 140.94 | 616 | 525.62 | 77 |
| 400_1 | -0.4813 | 2.2925 | 212.41 | 707 | 3185.00 | 84 |
| 400_2 | -0.2596 | 1.6500 | 212.62 | 595 | 3193.55 | 50 |
| 500_1 | -0.3576 | 2.3219 | 308.34 | 707 | 5666.54 | 159 |
| 500_2 | -0.3774 | 1.2539 | 309.11 | 608 | 5675.69 | 101 |

For the anti-correlated instances, the proposed algorithm strictly dominates AESSEA in seven instances, is weakly dominated by the other algorithm on the instance with $n = 10$, and for the remaining instances the performance of both algorithms is incomparable. In average, the proposed algorithm finds three times more non-dominated solutions than the other algorithm and spends 50% less time.

**Table 3.** Comparison of Mem-MC-MST and AESSEA on anti-correlated instances

| Instance | $I_{\varepsilon+}(A,B)$ | $I_{\varepsilon+}(B,A)$ | Mem-MC-MST | | AESSEA | |
|---|---|---|---|---|---|---|
| | | | Time (s) | #solutions | Time (s) | #solutions |
| 10 | 0.0275 | 0.0000 | 33.15 | 109 | 5.31 | 111 |
| 25 | 0.1499 | 0.1923 | 144.03 | 587 | 12.94 | 414 |
| 50 | 0.0419 | 0.6827 | 274.16 | 944 | 23.05 | 412 |
| 100_1 | -0.2058 | 1.9306 | 325.51 | 980 | 23.21 | 376 |
| 100_2 | -0.1052 | 1.9886 | 578.59 | 1062 | 28.26 | 393 |
| 200_1 | -0.1082 | 2.6343 | 726.97 | 1215 | 69.95 | 305 |
| 200_2 | 0.0168 | 3.2005 | 1136.79 | 1307 | 79.31 | 476 |
| 300_1 | -0.2191 | 4.3289 | 991.59 | 1243 | 542.96 | 366 |
| 300_2 | 0.0818 | 4.9293 | 1456.53 | 1327 | 604.51 | 357 |
| 400_1 | -0.7082 | 5.4709 | 1351.85 | 1442 | 2472.03 | 326 |
| 400_2 | 0.0219 | 6.6840 | 1946.63 | 1381 | 2488.57 | 431 |
| 500_1 | -0.6249 | 6.5998 | 1330.17 | 1440 | 6712.68 | 436 |
| 500_2 | -0.0962 | 6.6972 | 1999.35 | 1543 | 6456.46 | 461 |

Finally, for the concave instances, the proposed algorithm strictly dominates AESSEA in all instances with $n$ =200,...,500, weakly dominates the other algorithm on the instance 100_1, and for the remaining instances the performance of both algorithms is incomparable. In average, the proposed algorithm finds 65% more non-dominated solutions than the other algorithm and spends 3 times less computational time.

**Table 4.** Comparison of Mem-MC-MST and AESSEA on concave instances

| Instance | $I_{\varepsilon+}(A,B)$ | $I_{\varepsilon+}(B,A)$ | Mem-MC-MST | | AESSEA | |
|---|---|---|---|---|---|---|
| | | | Time (s) | #solutions | Time (s) | #solutions |
| 10 | 0.0000 | 0.0000 | 4.63 | 46 | 4.61 | 46 |
| 25 | 0.0078 | 0.0132 | 21.97 | 322 | 6.68 | 239 |
| 50 | 0.0063 | 0.0315 | 25.74 | 415 | 11.57 | 284 |
| 100_1 | 0.0000 | 0.0083 | 37.94 | 392 | 15.88 | 313 |
| 100_2 | 0.0023 | 0.0816 | 34.13 | 426 | 14.32 | 248 |
| 200_1 | -0.0329 | 0.2231 | 69.75 | 343 | 56.83 | 177 |
| 200_2 | -0.0005 | 0.0315 | 86.03 | 319 | 57.71 | 238 |
| 300_1 | -0.0354 | 0.2271 | 120.64 | 360 | 557.82 | 170 |
| 300_2 | -0.0278 | 0.1545 | 117.65 | 272 | 563.52 | 208 |
| 400_1 | -0.0462 | 0.2860 | 171.00 | 340 | 2155.66 | 125 |
| 400_2 | -0.0610 | 0.4353 | 174.06 | 352 | 2246.88 | 261 |
| 500_1 | -0.0519 | 0.2627 | 241.23 | 333 | 5990.03 | 172 |
| 500_2 | -0.0747 | 0.3523 | 247.26 | 355 | 6387.97 | 104 |

# 5  Conclusions

This paper presented a new memetic algorithm for the biobjective minimum spanning tree problem. A randomized greedy method is proposed to generate part of the initial population and tabu search procedure is introduced as an intensification tool for the evolutionary algorithm. A variant of the known binary tournament selection scheme is utilized in the algorithm.

The solution set generated by the proposed algorithm is compared with the solution set generated by a known algorithm, AESSEA, with basis on a general indicator, the binary additive ε-indictor.

The results of the computational experiments show that the proposed algorithm performs better than the other algorithm, regarding solution quality, mainly in large instances. In average, the algorithm also finds a higher number of solutions and runs in less time than the comparison algorithm.

As future works, the authors intend to investigate the use of other neighborhood structures for the tabu search methods, examines extensions of the selection scheme to problems with more objectives and extend the proposed approach to the degree constrained multiobjective minimum spanning tree problem.

# Acknowledgements

The authors want to thank Dr Joshua Knowles who kindly furnished the AESSEA code. This research was partially funded by CNPq.

# References

1. Aggarwal, V., Aneja, Y., Nair, K.: Minimal Spanning Tree Subject to a Side Constraint. Computers and Operations Research Vol. 9 (1982) 287-296
2. Bazlamaçci, C.F., Hindi, K.S.: Minimum-weight Spanning Tree Algorithms A Survey and Empirical Study. Computers and Operations Research Vol. 28 (2001) 767-785
3. Chartrand, G., Oellermann, O.R.: Applied and Algorithmic Graph Theory. McGraw-Hill (1993)
4. Coello, C. A.: A Comprehensive Survey of Evolutionary-based Multiobjective Optimization Techniques. Knowledge and Information Systems Vol.1 (1999) 269-308
5. Ehrgott, M.: Approximation Algorithms for Combinatorial Multicriteria Optimization Problems. International Transactions in Operational Research Vol. 7(2000) 5-31
6. Ehrgott, M., Gandibleux, X.: A Survey and Annotated Bibliography of Multiobjective Combinatorial Optimization. OR Spektrum Vol. 22 (2000) 425-460
7. Ehrgott, M., Gandibleux, X.: Approximative Solution Methods for Multiobjective Combinatorial Optimization. Top Vol. 12, N. 1 ( 2004) 1-89
8. Feo, T. A., Resende, M. G. C.: Greedy Randomized Adaptive Search Procedures. Journal of Global Optimization Vol. 6 (1995) 109-133
9. Garey, M. R., Johnson, D. S.: Computers and Intractability: A Guide to the Theory of NP-completeness. Freeman, New York (1979)

10. Goldberg, D. e Deb, K.: A Comparative Analysis of Selection Schemes Used in Genetic Algorithms, In: Rawlins, G.J.E (Ed.) Foundations of Genetic Algorithms. Morgan Kaufmann Publishers (1991) 69-93

11. Graham, R.L., Hell, P.: On the History of the Minimum Spanning Tree Problem. Ann. History of Comp. Vol. 7 (1985) 43-57

12. Gutin, G., Punnen, A. P.: Traveling Salesman Problem and Its Variations, In: Kluwer Academic Publishers (2002)

13. Hakami, S. L.: Steiner's Problem in Graphs and Its Implications. Networks Vol. 1 (1971) 113-133

14. Knowles, J.D.: Local-Search and Hybrid Evolutionary Algorithms for Pareto Optimization. PhD Thesis. Department of Computer Science, University of Reading, Reading, UK (2002)

15. Knowles, J. D., Corne, D. W.: Approximating the Nondominated Front Using the Pareto Archived Evolution Strategy. Evolutionary Computation Vol. 8, Issue 2 (2000) 149-172

16. Knowles, J. D., Corne, D. W.: A Comparison of Encodings and Algorithms for Multiobjective Spanning Tree Problems. Proceedings of the 2001 Congress on Evolutionary Computation (CEC01) (2001) 544-551

17. Knowles, J. D., Corne, D. W.: Enumeration of Pareto Optimal Multi-criteria Spanning Trees – A Proof of the Incorrectness of Zhou and Gen's Proposed Algorithm. European Journal of Operational Research Vol. 143 (2002) 543-547

18. Kruskal, J.B.: On the Shortest Spanning Subtree of a Graph and the Travelling Salesman Problem. Pric. AMS 7 (1956) 48-50

19. Merz, P., Freisleben, B.: Fitness Landscape Analysis and Memetic Algorithms for the Quadratic Assignment Problem. IEEE Transactions on Evolutionary Computation Vol. 4, N. 4 (2000) 337-352

20. Moscato, P.: On Evolution, Search, Optimization, Genetic Algorithms and Martial Arts: Towards Memetic Algorithms. In: Caltech Concurrent Computation Program, C3P Report 826 (1989)

21. Prim, R.C.: Shortest Connection Networks and Some Generalizations. Bell Systems Techn. J. 36 (1957) 1389-1401

22. Prüfer, H.: Neuer Beweis Eines Satzes Uber Permutationen. Arch. Math. Phys. Vol. 27 (1918) 742-744

23. Ramos, R. M., Alonso, S., Sicilia, J., Gonzáles, C.: The Problem of the Biobjective Spanning Tree. European Journal of Operational Research Vol. 111 (1998) 617-628

24. Raidl, G.R.: An Efficient Evolutionary Algorithm for the Degree-constrained Minimum Spanning Tree Problem. Proceedings of the 2000 Congress on Evolutionary Computation (CEC 2000), IEEE Press (2000) 104-111

25. Raidl, G., Julstrom, B.A.: Edge Sets: An Efficient Evolutionary Coding of Spanning Trees. IEEE Transactions on Evolutionary Computation Vol 7, N. 3 (2003) 225-239.

26. Steiner, S., Radzik, T.: Solving the Biobjective Minimum Spanning Tree Problem using a $k$-best Algorithm. Technical Report TR-03-06, Department of Computer Science, King's College London (2003)

27. Zhou G., Gen, M.: Genetic Algorithm Approach on Multi-Criteria Minimum Spanning Tree Problem. European Journal of Operational Research Vol. 114 (1999) 141-152

28. Zitzler E., Thiele, L., Laumanns, M., Fonseca, C. M., Fonseca, V.G.: Performance Assessment of Multiobjective Optimizers: An Analysis and Review. IEEE Transactions on Evolutionary Computation Vol. 7, N. 2 (2003) 117-132

# A Comparative Study of Ant Colony Optimization and Reactive Search for Graph Matching Problems

Olfa Sammoud[1,2], Sébastien Sorlin[2],
Christine Solnon[2], and Khaled Ghédira[1]

[1] SOIE, Institut Supérieur de Gestion de Tunis,
41 rue de la Liberté, Cité Bouchoucha, 2000 Le Bardo, Tunis
{olfa.sammoud, khaled.ghedira}@isg.rnu.tn
[2] LIRIS, CNRS UMR 5205, bât. Nautibus, University of Lyon I,
43 Bd du 11 novembre, 69622 Villeurbanne cedex, France
{sammoud.olfa, sebastien.sorlin, christine.solnon}@liris.cnrs.fr

**Abstract.** Many applications involve matching two graphs in order to identify their common features and compute their similarity. In this paper, we address the problem of computing a graph similarity measure based on a multivalent graph matching and which is generic in the sense that other well known graph similarity measures can be viewed as special cases of it. We propose and compare two different kinds of algorithms: an Ant Colony Optimization based algorithm and a Reactive Search. We compare the efficiency of these two algorithms on two different kinds of difficult graph matching problems and we show that they obtain complementary results.

## 1 Introduction

Graphs are often used to model structured objects: vertices represent object components while edges represent binary relations between components. For example, graphs are used to model images [3, 5], design objects [10], molecules or proteins [1], course timetables [9]. In this context, object recognition, classification and identification involve comparing graphs, *i.e.*, matching graphs to identify their common features [11]. This may be done by looking for an exact graph or subgraph isomorphism in order to show graph equivalence or inclusion. However, in many applications, one looks for similar objects and not identical ones and exact isomorphisms cannot be found. As a consequence, error-tolerant graph matchings such as the maximum common subgraph and the graph edit distance have been proposed [6, 7, 8, 11]. Such matchings drop the condition that the matching must preserve all vertices and edges: the goal is to find a "best" matching, *i.e.*, one which preserves a maximum number of vertices and edges.

Most recently, three different papers ([3, 5, 10]) proposed to go one step further by introducing multivalent matchings, where a vertex in one graph may be matched with a set of vertices of the other graph in order to associate one single

J. Gottlieb and G.R. Raidl (Eds.): EvoCOP 2006, LNCS 3906, pp. 234–246, 2006.

component of an object to a set of components of another object. This allows one to compare objects described at different levels of granularity such as under- or over- segmented images [3], or a model of an image having a schematic aspect with real and over-segmented images [5].

We more particulary focus on the multi-labeled graph similarity measure of [10] as it has been shown in [20] that it is more general than the two other ones proposed in [3, 5]. Indeed, it is parameterized by similarity functions that allow one to easily express problem-dependent knowledge and constraints.

In section 2, we briefly present the generic graph similarity measure of [10]. In section 3 and 4, we propose two algorithms based on two different approaches for computing this graph similarity measure. The first one is based on Ant Colony Optimization (ACO) and the second one is based on Reactive Search (RS). In section 5, we experimentally compare the two proposed algorithms on two kinds of graph matching problems. Finally, we conclude on the complementarity of these two algorithms and we discuss some further work.

## 2   A Generic Similarity Measure for Multi-labeled Graphs

**Definition of Multi-labeled Graphs.** A directed graph is defined by a couple $G = (V, E)$, where $V$ is a finite set of vertices and $E \subseteq V \times V$ is a set of directed edges. Vertices and edges may be associated with labels that describe their properties. Without loss of generality, we assume that each vertex and each edge has at least one label. Given a set $L_V$ of vertex labels and a set $L_E$ of edge labels, a multi-labeled graph is defined by a triple $G = \langle V, r_V, r_E \rangle$ such that:

- $V$ is a finite set of vertices,
- $r_V \subseteq V \times L_V$ is a relation associating labels to vertices, *i.e.*, $r_V$ is the set of couples $(v_i, l)$ such that vertex $v_i$ is labeled by $l$,
- $r_E \subseteq V \times V \times L_E$ is a relation associating labels to edges, *i.e.*, $r_E$ is the set of triples $(v_i, v_j, l)$ such that edge $(v_i, v_j)$ is labeled by $l$. Note that the set of edges of the graph can be defined by $E = \{(v_i, v_j) | \exists l, (v_i, v_j, l) \in r_E\}$.

**Similarity Measure.** We now briefly describe the graph similarity measure introduced in [10]. We refer the reader to the original paper for more details.

The similarity measure is computed with respect to a matching of the vertices of the two graphs. We consider here a multivalent matching, *i.e.*, each vertex of one graph is matched with a –possibly empty– set of vertices of the other graph. More formally, a multivalent matching of two multi-labeled graphs $G = \langle V, r_V, r_E \rangle$ and $G' = \langle V', r_{V'}, r_{E'} \rangle$ is a relation $m \subseteq V \times V'$ which contains every couple $(v, v') \in V \times V'$ such that vertex $v$ is matched with vertex $v'$.

Once a multivalent matching is defined, the next step is to identify the set of features that are common to the two graphs with respect to this matching. This set contains all the features from both $G$ and $G'$ whose vertices (resp. edges) are matched by $m$ to at least one vertex (resp. edge) that has the same label. More formally, the set of common features $G \sqcap_m G'$ of two graphs $G = \langle V, r_V, r_E \rangle$ and $G' = \langle V', r_{V'}, r_{E'} \rangle$, with respect to a matching $m \subseteq V \times V'$, is defined as follows:

$$G \sqcap_m G' \doteq \{(v,l) \in r_V \mid \exists (v,v') \in m, \, (v',l) \in r_{V'}\}$$
$$\cup \{(v',l) \in r_{V'} \mid \exists (v,v') \in m(v), \, (v,l) \in r_V\}$$
$$\cup \{(v_i,v_j,l) \in r_E \mid \exists (v_i,v_i') \in m, \exists (v_j,v_j') \in m \, (v_i',v_j',l) \in r_{E'}\}$$
$$\cup \{(v_i',v_j',l) \in r_{E'} \mid \exists (v_i,v_i') \in m, \exists (v_j,v_j') \in m \, (v_i,v_j,l) \in r_E\}$$

Given a multivalent matching $m$, we also have to identify the set of split vertices, *i.e.*, the set of vertices that are matched to more than one vertex, each split vertex $v$ being associated with the set $s_v$ of its matched vertices:

$$splits(m) = \{(v,s_v) \mid v \in V, \; s_v = \{v' \in V' \mid (v,v') \in m\}, |s_v| \geq 2\}$$
$$\cup \{(v',s_{v'}) \mid v' \in V', \, s_{v'} = \{v \in V \mid (v,v') \in m\}, \; |s_{v'}| \geq 2\}$$

The similarity of two graphs $G = \langle V, r_V, r_E \rangle$ and $G' = \langle V', r_{V'}, r_{E'} \rangle$ with respect to a matching $m$ is then defined by:

$$sim_m(G,G') = \frac{f(G \sqcap_m G') - g(splits(m))}{f(r_V \cup r_{V'} \cup r_E \cup r_{E'})} \tag{1}$$

where $f$ and $g$ are two functions that are defined to weight features and splits, depending on the considered application.

Finally, the similarity $sim(G,G')$ of two graphs $G = \langle V, r_V, r_E \rangle$ and $G' = \langle V', r_{V'}, r_{E'} \rangle$ is the greatest similarity with respect to all possible matchings, *i.e.*,

$$sim(G,G') = \max_{m \subseteq V \times V'} sim_m(G,G')$$

Note that the denominator of formula (1) does not depend on the matching $m$ —this denominator is introduced to normalize the similarity value between zero and one. Hence, it will be sufficient to find the matching $m$ that maximizes the *score* function below:

$$score(m) = f(G \sqcap_m G') - g(splits(m))$$

**Using this Graph Similarity Measure to Solve Different Graph Matching Problems.** Thanks to the functions $f$ and $g$ of formula (1), the graph similarity measure of [10] is generic. [20] shows how this graph similarity measure can be used to solve many different graph matching problems such as the (sub)graph isomorphism problem, the graph edit distance, the maximum common subgraph problem and the multivalent matching problems of [5] and [10].

The graph matching problem has been shown to be $NP$-hard in [20]. A complete algorithm has been proposed for computing the matching which maximizes formula (1) in [10]. This kind of algorithm, based on an exhaustive exploration of the search space [16] combined with pruning techniques, guarantees solution optimality. However, this algorithm is limited to very small graphs (having less than 10 vertices in the worst case). Therefore, incomplete algorithms, that do not guarantee optimality but have a polynomial time complexity, appear to be good alternatives.

# 3   ACO for the Graph Matching Problem

The ACO (Ant Colony Optimization) meta-heuristic is a bio-inspired approach [12, 13] that has been used to solve many hard combinatorial optimization problems. The main idea is to model the problem to solve as a search for an optimal path in a graph –called the construction graph– and to use artificial ants to search for 'good' paths. The behavior of artificial ants mimics the behavior of real ones: *(i)* ants lay pheromone trails on the components of the construction graph to keep track of the most promising components, *(ii)* ants construct solutions by moving through the construction graph and choose their path with respect to probabilities which depend on the pheromone trails previously laid, and *(iii)* pheromone trails decrease at each cycle simulating in this way the evaporation phenomena observed in the real world.

In order to solve graph matching problems, we have proposed in [18] a first ACO algorithm called ANT-GM (ANT-Graph Matching). However, if this algorithm appeared to be competitive with tabu search on sub-graph isomorphism problems, it was clearly outperformed on multivalent graph matching problems. The algorithm presented bellow called ANT-GM'06 improves ANT-GM with respect to the following points: *(i)* we consider a new heuristic function in the definition of the transition probability, *(ii)* we consider a new pheromonal strategy, and *(iii)* we introduce a local search procedure to improve solutions constructed by ants.

**Algorithmic Scheme.** At each cycle, each ant constructs a complete matching in a randomized greedy way. Once every ant has generated a matching, a local search procedure takes place to improve the quality of the best matching of the cycle. Pheromone trails are updated according to this improved matching. This process stops iterating either when an ant has found an optimal matching, or when a maximum number of cycles has been performed.

Contrary to the algorithm introduced in [18], ANT-GM'06 follows the *Max-Min Ant System*[21]: we explicit impose lower and upper bounds $\tau_{min}$ and $\tau_{max}$ on pheromone trails (with $0 < \tau_{min} < \tau_{max}$). The goal is to favor a larger exploration of the search space by preventing the relative differences between pheromone trails from becoming too extreme during processing. Also and in order to achieve a higher exploration of the search space at the first cycles, pheromone trails are set to $\tau_{max}$ at the beginning.

**Construction Graph.** The construction graph is the graph on which artificial ants lay pheromone trails. Vertices of this graph are solution components that are selected by ants to generate solutions. In our graph matching application, ants build matchings by iteratively selecting couples of vertices to be matched. Hence, given two attributed graphs $G = \langle V, r_V, r_E \rangle$ and $G' = \langle V', r_{V'}, r_{E'} \rangle$, the construction graph is the complete non-directed graph that associates a vertex to each couple $(u, u') \in V \times V'$.

**Pheromone Trails.** A key point when developing any ACO algorithm is to decide where pheromone trails should be laid and how they should be exploited and updated. In our case, we may consider two different possibilities:

- the first one turns into laying pheromone on the vertices of the construction graph. So, the amount of pheromone on a vertex $(u, u')$ represents the learnt desirability to match $u$ with $u'$ when constructing matchings.
- the second consists in laying pheromone on the edges of the construction graph. The amount of pheromone on an edge $< (u, u'), (v, v') >$ represents the learnt desirability to match together $u$ with $u'$ and $v$ with $v'$ when constructing matchings.

Experimental results presented in [18] have been obtained with the second strategy, that appears to be the best-performing one on the maximum clique problem [19] and multiple knapsack problem [2]. Since then, we have compared this second strategy with the first one, and experiments showed us that when we choose to lay pheromone on vertices (instead of edges), better results are obtained, also the algorithm is much less time consuming (pheromone laying and evaporation has a linear complexity with respect to the number of vertices of the construction graph and not a quadratic one).

So, in `ANT-GM'06`, pheromone is laid on vertices and not on edges like in the initial version of `ANT-GM`. The amount of pheromone on a vertex $(u, u')$ will be noted $\tau(u, u')$.

**Matching Construction by Ants.** At each cycle, each ant constructs a complete matching: starting from an empty matching $m = \emptyset$, by iteratively adding couples of vertices that are chosen within the set $cand = \{(u, u') \in V \times V' - m\}$. As usually in ACO algorithm, the choice of the next couple to be added to $m$ is done with respect to a probability that depends on pheromone and heuristic factors. More formally, given a matching $m$ and a set of candidates $cand$, the probability $p_m(u, u')$ of selecting $(u, u') \in cand$ is:

$$p_m(u, u') = \frac{[\tau(u, u')]^\alpha \cdot [\eta_m(u, u')]^\beta}{\displaystyle\sum_{(v,v') \in cand} [\tau(v, v')]^\alpha \cdot [\eta_m(v, v')]^\beta} \tag{2}$$

where:

- $\tau(u, u')$ is the pheromone factor (when choosing the first couple, $\tau_m(u, u') = 1$, so that the probability only depends on the heuristic factor), and
- $\eta_m(u, u')$ is a heuristic factor that aims at favoring couples that most increase the score function, *i.e.*, $\eta_m(u, u') = score(m \cup \{(u, u')\}) - score(m)$.
- $\alpha$ and $\beta$ are two parameters that determine the relative importance of the two factors.

Ants stop adding new couples to the matching when the addition of every candidate couple decreases the score function or when the score function has not been increased since the last three iterations.

**Local Search Procedure.** The best performing ACO algorithms for many combinatorial problems are hybrid algorithms that combine probabilistic solution construction by a colony of ants with local search. In a same perspective, we have tried to improve the performance of ANT-GM'06 by coupling it with a local search procedure. In our case, we have chosen a local search which achieves a 'good' compromise between quality and time consuming. Once every ant has constructed a matching, we try to improve the quality of the best matching constructed during the cycle as follows: the three worse couples of the matching are removed from it, and the resulting matching is completed in a greedy way, *i.e.*, by iteratively adding couples of vertices that most increase the *score* function. This local search process is iterated until no more improvement is obtained.

Note that the *'goodness'* of a couple is judged according to its contribution in our *score* function, and couples to be removed at each step of a local search improvement are couples which have not already been removed.

**Pheromone Updating Step.** Once each ant has constructed a matching, and the best of these matchings has been improved by local search, pheromone trails are updated according to the *Max-Min Ant System*. First, evaporation is simulated by multiplying every pheromone trail by a pheromone persistence rate $\rho$ such that $0 \leq \rho \leq 1$. Then, the best ant of the cycle deposits pheromone. More precisely, let $m_k$ be the best matching (with respect to the score function) built during the cycle and improved by local search, and $m_{best}$ be the best matching built since the beginning of the run (including the current cycle), the quantity of pheromone laid is inversely proportional to the gap of score between $m_k$ and $m_{best}$, *i.e.* it is equal to $1/(1 + score(m_{best}) - score(m_k))$.

As, we have chosen to put pheromone on the vertices of the construction graph, the quantity of pheromone to be added is deposited on each couple of vertices $(u, u')$ in $m_k$.

## 4   Reactive Search for the Graph Matching Problem

**Greedy Algorithm.** [10] proposed a greedy algorithm to solve the graph matching problem. We briefly describe it because it is used as a starting point of our Reactive Search algorithm. More information can be found in [10]. The algorithm starts from an empty matching $m = \emptyset$, and iteratively adds to $m$ couples of vertices chosen within the set of candidate couples $cand = V \times V' - m$. This greedy addition of couples to $m$ is iterated until $m$ is locally optimal, *i.e.*, until no more couple addition can increase the similarity. At each step, the couple to be added is randomly chosen within the set of couples that most increase the *score* function. This greedy algorithm has a polynomial time complexity of $\mathcal{O}((|V| \times |V'|)^2)$, provided that the computation of the $f$ and $g$ functions have linear time complexities with respect to the size of the matching. As a counterpart of this rather low complexity, this algorithm never backtracks and is not complete.

**Local Search.** The greedy algorithm of [10] returns a "locally optimal" matching in the sense that adding or removing one couple of vertices to this matching cannot improve it. However, it may be possible to improve it by adding and/or removing more than one couple to this matching. A local search [14, 15] tries to improve a solution by locally exploring its neighborhood: the neighbours of a matching $m$ are the matchings which can be obtained by adding or removing one couple of vertices to $m$:

$$\forall m \in \wp(V \times V'), neighbourhood(m) = \{m \cup \{(v, v')\} | (v, v') \in (V \times V') - m\}$$
$$\cup \{m - \{(v, v')\} | (v, v') \in m\}$$

From an initial matching, computed by the greedy algorithm, the search space is explored from neighbour to neighbour until the optimal solution is found (when the optimal value is known) or until a maximum number of moves have been performed. The next neighbour to move on at each step is selected according to the Tabu meta-heuristic.

**Tabu Meta-Heuristic.** *Tabu* search [14, 17] is one of the best known heuristic to choose the next neighbour to move on. At each step, one chooses the best neighbour with respect to the *score* function. To avoid staying around locally optimal matchings by always performing the same moves, a Tabu list is used. This list has a length $k$ and memorizes the last $k$ moves (*i.e.*, the last $k$ added/removed couples) in order to forbid backward moves (*i.e.*, to remove/add a couple recently added/removed). An exception named "aspiration" is added: if a forbidden move reaches a better matching than the best known matching, the move is nevertheless done.

**Reactive Search.** The length $k$ of the tabu list is a critical parameter that is hard to set: if the list is too long, search diversification is too strong so that the algorithm converges too slowly; if the list is too short, intensification is too strong so that the algorithm may be stuck around local maxima and fail in improving the current solution. To solve this parameter tuning problem, [4] introduced *Reactive Search* where the length of the Tabu list is dynamically adapted during the search. To make the Tabu algorithm reactive, one must evaluate the need for diversifying the search. When the same matching is explored twice, the search must be diversified. In order to detect such redundancies, a hashing key is memorized for each explored matching. When a collision occurs in the hash table, the list length is increased. On the contrary, when there is no collision during a fixed number of moves, thus indicating that search is diversified enough, one can reduce the list length. Hashing keys are incrementally computed so that this method has a negligible added cost.

**Iterated Reactive Search.** This reactive search process is iterated from different starting points: the total number of allowed moves $maxMoves$ is divided by $k$ and $k$ executions of reactive search having each one $maxMoves/k$ allowed moves are launched. Finally, we keep the best matching found during the $k$ executions.

# 5   Experimental Comparison Results of RS and ACO

**Problem Instances.** We compare our two algorithms on two different sets of multivalent matching problems: a randomly generated one and a set of seven instances introduced in [5].

*Test suite 1.* We have used a random graph generator to generate "similar" pairs of graphs: it randomly generates a first graph and applies some vertex splitting/merging and some edge and vertex insertion/suppression to build a second graph which is similar to the first one. When graph components have many different labels, the best matching is trivially found as nearly all vertices/edges have different labels. Therefore, to obtain harder instances, we have generated 100 graphs such that all vertices and edges have the same label. These graphs have between 80 and 100 vertices and between 200 and 360 edges. The second graph is obtained by doing 5 vertex merging/splitting and 10 edge or vertex insertion/suppression. We define function $f$ of formula (1) as the cardinality function and function $g$ as a weighted sum:

$$g(S) = w * \sum_{(v,s_v) \in S} (|s_v| - 1) \text{ where } w \text{ is the weight of a split.}$$

The chosen weight $w$ can drastically change the hardness of instances: with a null weight, problem is trivially solved (one can make as many splitted vertices as needed to recover labels), whereas, with a high weight, optimal solutions do not split vertices and problem turns into an univalent graph matching problem. With intermediate weights, the problem is harder: optimal solutions must do a balancing between the number of splitted vertices and the number of recovered labels. We display experimental results obtained for two different "intermediate" split weights in order to compare the capacity of our algorithms to deal with splitted vertices. We first consider instances where $w = 1$, so that optimal solutions may contain several splits. We also consider instances where $w = 3$, so that optimal solutions contain less splitted vertices. We keep only the 13 hardest instances (*i.e.*, the ones that cannot be solved by the iterated greedy algorithm of [10]).

*Test suite 2.* A non-bijective graph matching problem was introduced in [5] to find the best matching between models and over-segmented images of brains. Given a model graph $G=(V, E)$ and an image graph $G'=(V', E')$, a matching is defined as a function $\phi : V \to \wp(V')$ which associates to each vertex of the model graph $G$ a non empty set of vertices of $G'$, and such that *(i)* each vertex of the image graph $G'$ is associated to exactly one vertex of the model graph $G$, *(ii)* for some forbidden couples $(v, v') \in V \times V'$, $v'$ must not belong to $\phi(v)$, and *(iii)* the subgraph induced by every set $\phi(v)$ must be connected. A weight $s^v(v_i, v'_i)$ (resp. $s^e(e_i, e'_i)$) is associated with each couple of vertices $(v_i, v'_i) \in V \times V'$ (resp. of edges $(e_i, e'_i) \in E \times E'$). The goal is to find the matching which maximizes a function depending on these weights of matched vertices and edges.

One can define functions $f$ and $g$ so that the matching which maximizes formula (1) corresponds to the best matching as defined in [5]. We refer the reader to [20] for more details on the definition of these two functions. We have taken the 7 instances of the non-bijective graph matching problem of [5]. Scheme graphs have between 10 and 50 vertices while image graphs have between 30 and 250 vertices. For these instances, we compare our two algorithms with $LS+$, a randomized construction algorithm proposed by [5] that quickly computes a set of possible non-bijective graph matchings and improves the best of these matchings with a local search algorithm until a locally optimal point is reached. For more details on these instances and on the $LS+$ algorithm, please refer to [5].

**Experimental Settings for ACO.** For ANT-GM'06, we have set the pheromone factor weight $\alpha$ to 1 and 2 respectively on test suite 1 and test suite 2, the pheromone persistance rate $\rho$ to 0.98, the heuristic factor weight $\beta$ to 10, the maximum number of cycles $MaxCycle$ respectively to 1000 and 2000, the number of ants $nbAnts$ to 20, the pheromone lower and upper bound $\tau_{min}$ and $\tau_{max}$ to 0.01 and 6.

To evaluate the benefit of integrating local search within ANT-GM'06, we display results obtained without and with local search. We respectively call these algorithms ANT-GM'06 and ANT-GM'06+LS.

**Experimental Settings for RS.** Reactive Search needs 5 parameters: the minimum ($min$) and the maximum ($max$) length of the list, the length of extension (and shortening) $diff$ of the list when a reaction process occurs, the frequency $freq$ of reduction of the list and the maximum number of allowed moves $nbMoves$. The initial length of the list is always set to $min$. For the two test suites, $max$ is set to 50, $diff$ is set to 15 and the number of moves is set to 50000. In order to obtain better results, the two others parameters must be chosen depending on the considered problem. On instances of test suite 1, $min$ is set to 15 and $freq$ is set to 5000 whereas on instances of test suite 2, $min$ is set to 10 and $freq$ is set to 1000. As 50000 moves of RS are performed much quicker than 1000 cycles of ANT-GM'06, we iterated RS from different calls to the greedy algorithm. The number of iterations of RS is setted in such a way that both algorithms spend the same time. Note however that execution of our reactive local search is deterministic on instances of test suite 2: the weights used are real numbers and as a consequence couples of vertices are never chosen randomly. As a consequence, it is useless to use iterated version of RS and we perform only one execution of RS.

We made at least 20 executions of each algorithm on each instance.

**Results.** Table 1 displays results on the 13 instances of test suite 1 with splits weighted to 1. First, we can see that our 3 algorithms seems to be robust in the sense that their average results are close to their best results on 20 executions (for 9 of the 13 instances, the average result is equal to the best result).

**Table 1.** Results on multivalent graph matching problems with splits weighted to 1. For each instance, the table reports the number of vertices and edges of the two graphs, the CPU time limit $L$ for one run of each algorithm (on a Pentium IV 1.7 GHz) and, for each algorithm, the best *score* and the average *score* found over at least 20 runs and the average time needed (in seconds) to get the best *score*.

| Problem | | | | RS | | | ANT-GM'06 | | | ANT-GM'06+LS | | |
|---|---|---|---|---|---|---|---|---|---|---|---|---|
| Nbr ($\mid V_1 \mid, \mid E_1 \mid$) | ($\mid V_2 \mid, \mid E_2 \mid$) | L | Best | Avg | T. | Best | Avg | T. | Best | Avg | T. |
| 1 | (80, 200) | (74, 186) | 1512 | 511 | 511.00 | 57 | 511 | 511.00 | 131 | **512** | **511.10** | 140 |
| 2 | (80, 240) | (82, 261) | 1415 | 644 | 644.00 | 60 | 644 | 644.00 | 266 | 644 | 644.00 | 239 |
| 3 | (80, 320) | (83, 362) | 1445 | 821 | 820.97 | 279 | 821 | 820.50 | 498 | **822** | **821.20** | 660 |
| 4 | (80, 340) | (72, 302) | 1174 | 753 | 753.00 | 55 | 753 | 753.00 | 111 | 753 | 753.00 | 130 |
| 5 | (80, 360) | (77, 367) | 1139 | **856** | **855.97** | 187 | 855 | 855.00 | 321 | 855 | 855.00 | 249 |
| 6 | (80, 360) | (78, 367) | 1196 | 863 | 863.00 | 21 | 863 | 863.00 | 187 | **864** | **863.94** | 565 |
| 7 | (90, 300) | (91, 307) | 1670 | 762 | 762.00 | 98 | 762 | 762.00 | 326 | 762 | 762.00 | 213 |
| 8 | (90, 320) | (87, 310) | 1611 | 780 | 780.00 | 51 | 780 | 780.00 | 572 | 780 | 780.00 | 409 |
| 9 | (90, 320) | (90, 339) | 1716 | 816 | **816.00** | 69 | 816 | 815.45 | 546 | 816 | 815.45 | 602 |
| 10 | (100, 260) | (96, 263) | 2093 | 697 | 696.63 | 628 | 697 | 696.90 | 976 | 697 | **697.00** | 812 |
| 11 | (100, 300) | (100, 304) | 2078 | 780 | 780.00 | 148 | 780 | 780.00 | 278 | 780 | 780.00 | 279 |
| 12 | (100, 320) | (98, 331) | 2080 | 828 | 828.00 | 46 | 828 | 828.00 | 286 | 828 | 828.00 | 218 |
| 13 | (100, 360) | (99, 371) | 2455 | 915 | 915.00 | 90 | 915 | 915.00 | 267 | 915 | 915.00 | 152 |

**Table 2.** Results on multivalent graph matching problems with splits weighted to 3. For each instance, the table reports the number of vertices and edges of the two graphs, the CPU time limit $L$ for one run of each algorithm (on a Pentium IV 1.7 GHz) and, for each algorithm, the best *score* and the average *score* found over at least 20 runs and the average time needed (in seconds) to get the best *score*.

| Problem | | | | RS | | | ANT-GM'06 | | | ANT-GM'06+LS | | |
|---|---|---|---|---|---|---|---|---|---|---|---|---|
| Nbr ($\mid V_1 \mid, \mid E_1 \mid$) | ($\mid V_2 \mid, \mid E_2 \mid$) | L | Best | Avg | T. | Best | Avg | T. | Best | Avg | T. |
| 1 | (80, 200) | (74, 186) | 659 | 496 | 496.00 | 28 | 496 | 496.00 | 132 | 496 | 496.00 | 121 |
| 2 | (80, 240) | (82, 261) | 798 | 624 | 624.00 | 26 | 624 | 624.00 | 108 | 624 | 624.00 | 88 |
| 3 | (80, 320) | (83, 362) | 896 | 801 | 801.00 | 17 | 801 | 801.00 | 213 | 801 | 801.00 | 218 |
| 4 | (80, 340) | (72, 302) | 737 | 732 | 732.00 | 27 | 732 | 732.00 | 185 | 732 | 732.00 | 194 |
| 5 | (80, 360) | (77, 367) | 852 | 846 | 846.00 | 198 | 846 | 846.00 | 116 | 846 | 846.00 | 77 |
| 6 | (80, 360) | (78, 367) | 855 | 840 | 840.00 | 36 | 840 | 840.00 | 94 | 840 | 840.00 | 67 |
| 7 | (90, 300) | (91, 307) | 1140 | 748 | 748.00 | 82 | 748 | 748.00 | 186 | 748 | 748.00 | 150 |
| 8 | (90, 320) | (87, 310) | 1079 | 766 | 766.00 | 44 | 766 | 766.00 | 187 | 766 | 766.00 | 187 |
| 9 | (90, 320) | (90, 339) | 1127 | 802 | 802.00 | 70 | 802 | 802.00 | 167 | 802 | 802.00 | 163 |
| 10 | (100, 260) | (96, 263) | 1346 | 683 | 683.00 | 114 | 683 | 682.75 | 556 | 683 | 683.00 | 354 |
| 11 | (100, 300) | (100, 304) | 1466 | 769 | 769.00 | 358 | 769 | 769.00 | 274 | 769 | 769.00 | 285 |
| 12 | (100, 320) | (98, 331) | 1463 | 814 | 814.00 | 51 | 814 | 814.00 | 241 | 814 | 814.00 | 201 |
| 13 | (100, 360) | (99, 371) | 1528 | 900 | 900.00 | 54 | 900 | 900.00 | 245 | 900 | 900.00 | 243 |

Also, we can note that RS performs better than ANT-GM'06: it obtains better result on 1 instance and is always as fast as ANT-GM'06. Integrating local search within ANT-GM'06 actually improves the solution process so that ANT-GM'06

**Table 3.** Results on non-bijective graph matching of [5]. For each instance, the table reports its name, the number of vertices of the two graphs, the CPU time limit $L$ for one run of each algorithm (on a Pentium IV 1.7 GHz), the similarities of the best solutions obtained by LS+[5], RS, ANT-GM'06 without LS and with LS (the best solution, the average solution found over 20 runs and the average time in seconds).

| Problem | | | LS+ | RS | | ANT-GM'06 | | | ANT-GM'06+LS | | |
|---|---|---|---|---|---|---|---|---|---|---|---|
| GM-i ($\mid V_1 \mid$, $\mid V_2 \mid$) | | L | Sim | Sim | T. | Sim | Avg | T. | Sim | Avg | T. |
| 5 | $(10, 30)$ | 18 | .5474 | .5481 | 0.9 | .5601 | .5598 | 16 | **.5608** | .5604 | 15 |
| 5a | $(10, 30)$ | 19 | .5435 | .5529 | 4.6 | .5638 | .5638 | 10 | **.5645** | .5641 | 7 |
| 6 | $(12, 95)$ | 269 | .4248 | .4213 | 0.0 | **.4252** | .4251 | 211 | **.4252** | .4251 | 215 |
| 7 | $(14, 28)$ | 13 | .6319 | .6333 | 2.1 | .6369 | .6369 | 7 | **.6376** | .6369 | 5 |
| 8 | $(30, 100)$ | 595 | .5186 | .5210 | 1.3 | .5229 | .5226 | 462 | **.5232** | .5228 | 229 |
| 8a | $(30, 100)$ | 595 | .5222 | .5245 | 1.3 | .5263 | .5261 | 456 | **.5269** | .5264 | 241 |
| 9 | $(50, 250)$ | 6018 | .5187 | .5199 | 81.7 | .5201 | .5201 | 4133 | **.5203** | .5202 | 2034 |

obtains better (resp. worse) results than RS on 3 (resp. 1) instances, whereas they obtain same results on 9 instances. On these instances, RS and ANT-GM'06+LS obtain complementary results: ANT-GM'06+LS outperforms RS more frequently than RS outperforms ANT-GM'06+LS but RS is much quicker than ANT-GM'06+LS, even when the two algorithms obtain the same results. Finally, note that if one cycle of ANT-GM'06+LS is more time consuming than one cycle of ANT-GM'06, ANT-GM'06+LS does not generally need more CPU time than ANT-GM'06 to find a solution: the local search procedure speed up the convergence of ANT-GM'06 and less cycles are needed to find the best solution.

Table 1 does not show results for the first ANT-GM algorithm described in [18] but one should note that the new ACO algorithm ANT-GM'06 clearly outperforms this first one. Actually, on all the considered instances, ANT-GM'06 computes much better solutions in less CPU time than ANT-GM. For example, on instance 1, the best *score* found by ANT-GM is 505 and is found in 8648 seconds, whereas ANT-GM'06 finds a *score* of 511 in 131 seconds.

Table 2 displays results on the 13 instances of test suite 1 with splits weighted to 3. On each instance, our three algorithms always find the same best score and the same average score (except for ANT-GM'06 on one instance). However, RS finds the solution in shorter times than ANT-GM'06 and ANT-GM'06+LS except for only two instances. These results show that on these instances, one clearly has to use our RS algorithm.

On the 7 instances of test suite 2, our three algorithms obtain better results than $LS+$, the reference algorithm of [5] (6 instances on 7 are better solved by RS and 7 instances on 7 are better solved by ACO algorithms). Note that, because the considered weights are real numbers, an execution of RS is deterministic. As a consequence, RS is less randomized than for multivalent matching problems, it quickly converges to "good" matchings but can easily be trapped into local optimum. So, results show that, as for non-bijective graph matching problems, ACO gives better results than RS but needs much more time.

In conclusion, the local search procedure helps ANT-GM'06 to improve the quality of the results. As a consequence, ANT-GM'06+LS usually obtains better results but is slower than RS. The time limits allowed to our three algorithms have been set depending on the ACO algorithms which generaly need a long time to converge. So, these time limits penalize RS which, within a shorter time, can generaly find a better solution then ACO algorithms. ANT-GM'06+LS and RS are complementary: if we need to compute quickly a "good" solution of hard instances or if instances are easy, we can use RS but if we have more time to spend on computation or if we want to solve very hard instances, we can use ANT-GM'06+LS.

# 6   Conclusion and Further Work

In this paper, we address the problem of computing the generic graph similarity measure of [10]. We propose and compare two different kinds of algorithms: an Ant Colony Optimization (ACO) based algorithm boosted with local search and a Reactive Search based on a tabu local search heuristic. We compare the efficiency of these two algorithms on two different kinds of difficult graph matching problems. We show that ACO usually obtains better results but is slower than Reactive Search. These two algorithms are complementary: if we need to compute quickly a "good" solution of hard instances or if instances are easy, we can use RS but if we have more time to spend on computation or if we want to solve very hard instances, we can use ACO.

In further work, we would like to compare these algorithms on some other graph matching problems such as maximum common subgraph problems. For ACO, we would like to speed up the convergence of the algorithm. This could be done by using a better local search strategy to repair built matchings. For RS, we would like to diversify the search. This could be done by using an other strategy than the elitist greedy algorithm of [10] to choose the starting points and then, start shorter tabu searches from many starting points.

## References

1. Tatsuya Akutsu. Protein structure alignment using a graph matching technique. Technical report, Bioinformatics Center, Institute for Chemical Research, Kyoto University. Uji, Kyoto 611-0011, Japan, 1995.
2. I. Alaya, C. Solnon, and K. Ghédira. Ant algorithm for the multi-dimensional knapsack problem. *International Conference on Bioinspired Optimization Methods and their Applications (BIOMA 2004)*, pages 63–72, 2004.
3. R. Ambauen, S. Fischer, and H. Bunke. Graph edit distance with node splitting and merging and its application to diatom identification. *4th IAPR-TC15 Wk on Graph-based Representations in Pattern Recognition*, LNCS 2726-Springer:95–106, 2003.
4. R. Battiti and M. Protasi. Reactive local search for the maximum clique problem. In Springer-Verlag, editor, *Algorithmica*, volume 29, pages 610–637, 2001.

5. M. Boeres, C. Ribeiro, and I. Bloch. A randomized heuristic for scene recognition by graph matching. In *Wkshp on Experimental and Efficient Algorithms (WEA 2004)*, pages 100–113, 2004.

6. H. Bunke. On a relation between graph edit distance and maximum common subgraph. *PRL: Pattern Recognition Letters*, 18:689 694, 1997.

7. H. Bunke and X. Jiang. *Graph Matching and Similarity*, volume Teodorescu, H.-N., Mlynek, D., Kandel, A., Zimmermann, H.-J. (eds.): Intelligent Systems and Interfaces, chapter 1. Kluwer Academic Publishers, 2000.

8. H. Bunke and K. Shearer. A graph distance metric based on the maximal common subgraph. *Pattern recognition letters*, 19(3–4):255–259, 1998.

9. E.K Burke, B. MacCarthy, S. Petrovic, and R. Qu. Case based reasoning in course timetabling: An attribute graph approach. *Proc. 4th Int. Conf. on Case-Based Reasoning (ICCBR-2001)*, LNAI 2080-Springer:90–104, 2001.

10. P.-A. Champin and C. Solnon. Measuring the similarity of labeled graphs. In *5th Int. Conf. on Case-Based Reasoning (ICCBR 2003)*, volume LNAI 2689-Springer, pages 80–95, 2003.

11. D. Conte, P. Foggia, C. Sansone, and M. Vento. 30 years of graph matching in pattern recognition. *Int. Jour. of Pattern Recogn. and AI*, 18(3):265–298, 2004.

12. M. Dorigo and G. Di Caro. The ant colony optimization meta-heuristic. *In D. Corne, M. Dorigo, and F. Glover, editors, New Ideas in Optimization. McGraw Hill, London, UK*, pages 11–32, 1999.

13. M. Dorigo and T. Stützle. Ant colony optimization. *MIT Press*, 2004.

14. F. Glover. Tabu search - part I. *Journal on Computing*, pages 190–260, 1989.

15. S. Kirkpatrick, S. Gelatt, and M. Vecchi. Optimisation by simulated annealing. In *Science*, volume 220, pages 671–680, 1983.

16. B.T. Messmer and H.Bunke. Efficient subgraph isomorphism detection: a decomposition approach. *IEEE Trans. on Knowledge and Data Engineering*, 12-2:307–323, 2000.

17. Sanja Petrovic, Graham Kendall, and Yong Yang. A tabu search approach for graph-structured case retrieval. In IOS Press, editor, *Proc. of the STarting Artificial Intelligence Researchers Symposium (STAIRS 2002)*, pages 55–64, 2002.

18. O. Sammoud, C. Solnon, and K. Ghédira. An ant algorithm for the graph matching problem. *5th European Conference on Evolutionary Computation in Combinatorial Optimization (EvoCOP 2005)*, LNCS 3448 - Springer:213–223, 2005.

19. C. Solnon and S. Fenet. A study of aco capabilities for solving the maximum clique problem. *To appear in Journal of Heuristics- Springer*, 2006.

20. S. Sorlin and C. Solnon. Reactive tabu search for measuring graph similarity. *5th IAPR Workshop on Graph-based Representations in Pattern Recognition (GbR 2005)*, LNCS 3434 - Springer:172–182, 2005.

21. T. Stützle and H.H. Hoos. Max-min Ant System. *Journal of Future Generation Computer Systems*, 16:889–914, 2000.

# Divide-and-Evolve: A New Memetic Scheme for Domain-Independent Temporal Planning

Marc Schoenauer[1], Pierre Savéant[2], and Vincent Vidal[3]

[1] Projet TAO, INRIA Futurs, LRI, Bt. 490, Université Paris Sud, 91405 Orsay, France
[2] Thales Research & Technology France, RD 128, F-91767 Palaiseau, France
[3] CRIL & Université d'Artois, rue de l'université - SP16, 62307 Lens, France

**Abstract.** An original approach, termed *Divide-and-Evolve* is proposed to hybridize Evolutionary Algorithms (EAs) with Operational Research (OR) methods in the domain of Temporal Planning Problems (TPPs). Whereas standard Memetic Algorithms use local search methods to improve the evolutionary solutions, and thus fail when the local method stops working on the complete problem, the *Divide-and-Evolve* approach splits the problem at hand into several, hopefully easier, sub-problems, and can thus solve globally problems that are intractable when directly fed into deterministic OR algorithms. But the most prominent advantage of the *Divide-and-Evolve* approach is that it immediately opens up an avenue for multi-objective optimization, even though the OR method that is used is single-objective. Proof of concept approach on the standard (single-objective) Zeno transportation benchmark is given, and a small original multi-objective benchmark is proposed in the same Zeno framework to assess the multi-objective capabilities of the proposed methodology, a breakthrough in Temporal Planning.

## 1 Introduction

*Artificial Intelligence Planning* is a form of general problem solving task which focuses on problems that map into *state models* that can be defined by a state space $S$, an initial state $s_0 \subseteq S$, a set of goal states $S_G \subseteq S$, a set of actions $A(s)$ applicable in each state $S$, and a transition function $f(a, s) = s'$ with $a \in A(s)$, and $s, s' \in S$. A solution to this class of models is a sequence of applicable actions mapping the initial state $s_0$ to a goal state that belongs to $S_G$.

An important class of problems is covered by Temporal Planning which extends classical planning by adding a duration to actions and by allowing concurrent actions in time [8]. In addition, other metrics are usually needed for real-life problems to qualify a good plan, for instance a cost or a risk criterion. A usual approach is to aggregate the multiple criteria, but this relies on highly problem-dependent features and is not always meaningful. A better solution is to compute the set of optimal non-dominated solutions – the so-called Pareto front.

Because of the high combinatorial complexity and the multi-objective features of Temporal Planning Problems (TPPs), Evolutionary Algorithms are good general-purpose candidate methods.

J. Gottlieb and G.R. Raidl (Eds.): EvoCOP 2006, LNCS 3906, pp. 247–260, 2006.
© Springer-Verlag Berlin Heidelberg 2006

However, there has been very few attempts to apply Evolutionary Algorithms to planning problems and, as far as we know, not any to Temporal Planning. Some approaches use a specific representation (e.g. dedicated to the battlefield courses of action [15]). Most of the domain-independent approaches see a plan as a program and rely on Genetic Programming and on the traditional blocks-world domain for experimentation (starting with the Genetic Planner [17]). A more comprehensive state of the art on Genetic Planning can be found in [1] where the authors experimented a variable length chromosome representation. It is important to notice that all those works search the space of (partial) plans.

It is also now well-known that Evolutionary Algorithms (EAs) can rarely efficiently solve Combinatorial Optimization Problems on their own, i.e. without being hybridized, one way or another, with local search *ad hoc* techniques. The most successful of such hybridizations use Operational Research methods to locally improve any offspring that was born from EA variation operators (crossover and mutation): such algorithms have been termed "Memetic Algorithms" or "Genetic Local Search" [14]. Those methods are now the heart of a whole research field, as witnessed by the series of WOMA's (Workshops on Memetic Algorithms) organized every year now, Journal Special Issues and edited books [10].

However, most memetic approaches are based on finding local improvements of candidate solutions proposed by the evolutionary search mechanism using dedicated local search methods that have to tackle the complete problem. In some combinatorial domains such as Temporal Planning, this simply proves to be impossible when reaching some level of complexity.

This paper proposes *Divide-and-Evolve*, borrowing to the Divide-and-Conquer paradigm for such situations: the problem at hand is sliced into a sequence of problems that are hopefully easier to solve by OR or other local methods. The solution to the original problem is then obtained by a concatenation of the solutions to the different sub-problems.

Next section presents an abstract formulation of the *Divide-and-Evolve* scheme, and starting from its historical (and pedagogical) root, the TGV paradigm. Generic representation and variation operators are also introduced. Section 3 introduces an actual instantiation of the *Divide-and-Evolve* scheme to TPPs. The formal framework of TPPs is first introduced, then the TPP-specific issues for the *Divide-and-Evolve* implementation are presented and discussed. Section 4 is devoted to experiments on the transportation Zeno benchmark for both single and multi-objective cases. The last section opens a discussion highlighting the limitations of the present work and giving hints about on-going and future work.

## 2     The *Divide-and-Evolve* Paradigm

### 2.1     The TGV Metaphor

The *Divide-and-Evolve* strategy springs from a metaphor on the route planning problem for the French high-speed train (TGV). The original problem consists in computing the shortest route between two points of a geographical landscape with strong bounds on the curvature and slope of the trajectory. An evolutionary algorithm was designed [4] based on the fact that the only local search algorithm at hand was a greedy deterministic algorithm that could solve only very simple (i.e. short distance) problems. The evolutionary algorithm looks for a split of the global route into small consecutive segments such that a local search algorithm can easily find a route joining their extremities. Individuals represent sets of intermediate train stations between the station of departure and the terminus. The convergence toward a good solution was obtained with the definition of appropriate variation and selection operators [4]. Here, the state space is the surface on which the trajectory of the train is defined.

**Generalization.** Abstracted to Planning, the route is replaced by a sequence of actions and the "stations" become intermediate states of the system. The problem is thus divided into sub-problems and "to be close" becomes "to be easy to solve" by some local algorithm $\mathcal{L}$. The evolutionary algorithm plays the role of an oracle pointing at some imperative states worth to go trough.

### 2.2 Representation

The problem at hand is an abstract AI Planning problem as described in the introduction. The representation used by the evolutionary algorithm is a variable length list of states: an individual is thus defined as $(s_i)_{i \in [1,n]}$, where the length $n$ and all the states $s_i$ are unknown and subject to evolution. States $s_0$ and $s_{n+1} \equiv s_G$ will represent the initial state and the goal of the problem at hand, but will not be encoded in the genotypes. By reference to the original TGV paradigm, each of the states $s_i$ of an individual will be called a *station*.

**Requirements.** The original TGV problem is purely topological with no temporal dimension and reduces to a planning problem with a unique action: moving between two points. The generalization to a given planning domain requires to be able to:

1. define a distance between two different states of the system, so that $d(S, T)$ is somehow related to the difficulty for the local algorithm $\mathcal{L}$ to find a plan mapping the initial state $S$ to the final state $T$;
2. generate a chronological sequence of virtual "stations", i.e. intermediate states of the system, that are close to one another, $s_i$ being close to $s_{i+1}$;
3. solve the resulting "easy" problems using the local algorithm $\mathcal{L}$;
4. "glue" the sub-plans into an overall plan of the problem at hand.

### 2.3 Variation Operators

This section describes several variation operators that can be defined for the general *Divide-and-Evolve* approach, independently of the actual domain of application (e.g. TPPs, or the original TGV problem).

**Crossover.** Crossover operators amounts to exchanging stations between two individuals. Because of the sequential nature of the fitness, it seems a good idea to try to preserve sequences of stations, resulting in straightforward adaptations to variable-length representation of the classical 1- or 2-point crossover operators.

Suppose you are recombining two individuals $(s_i)_{1 \leq ?n}$ and $(T_i)_{1 \leq ?m}$. The 1-point crossover amounts to choosing one station in each individual, say $s_a$ and $T_b$, and exchanging the second part of the lists of stations, obtaining the two offspring $(s_1, \ldots, s_a, T_{m+1}, \ldots T_b)$ and $(T_1, \ldots, T_b, s_{n+1}, \ldots, s_n)$ (2-point crossover is easily implemented in a similar way). Note that in both cases, the length of each offspring is likely to differ from those of the parents.

The choice of the crossover points $s_a$ and $T_b$ can be either uniform (as done in all the work presented here), or distance-based, if some distance is available: pick the first station $s_a$ randomly, and choose $T_b$ by e.g. a tournament based on the distance with $s_a$ (this is on-going work).

**Mutation.** Several mutation operators can be defined. Suppose individual $(s_i)_{1 \leq ?n}$ is being mutated:

- **At the individual level**, the *Add* mutation simply inserts a new station $s_{new}$ after a given station $(s_a)$, resulting in an $n + 1$-long list, $(s_1, \ldots, s_a, s_{new}, s_{a+1}, \ldots, s_n)$. Its counterpart, the *Del* mutation, removes a station $s_a$ from the list.

  Several improvements on the pure uniform choice of $s_a$ can be added and are part of on-going work, too: in case the local algorithm fails to successfully join all pairs of successive stations, the last station that was successfully reached by the local algorithm can be preferred for station $s_a$ (in both the *Add* and *Del* mutations). If all partial problems are solved, the most difficult one (e.g. in terms of number of backtracks) can be chosen.

- **At the station level**, the definition of each station can be modified – but this is problem-dependent. However, assuming there exists a station-mutation operator $\mu_S$, it is easy to define the individual-mutation $M_{\mu_S}$ that will simply call $\mu_S$ on each station $s_i$ with a user-defined probability $p_{\mu_S}$. Examples of operators $\mu_S$ will be given in section 3, while simple Gaussian mutation of the $(x, y)$ coordinates of a station were used for the original TGV problem [4].

# 3   Application to Temporal Planning

## 3.1   Temporal Planning Problems

Domain-Independent planners rely on the Planning Domain Definition Language (PDDL) [13], inherited from the STRIPS model [5], to represent a planning problem. In particular, this language is used for a competition[1] which is held every two years since 1998. The language has been extended for representing Temporal Planning Problems in PDDL2.1 [7]. For the sake of simplicity, the temporal model is often simplified as explained below [18].

---

[1] http://ipc.icaps-conference.org/

A *Temporal PDDL Operator* is a tuple $o = \langle pre(o), add(o), del(o), dur(o) \rangle$ where $pre(o)$, $add(o)$ and $del(o)$ are sets of ground atoms that respectively denote the preconditions, add effects and del effects of $o$, and $dur(o)$ is a rational number that denotes the *duration* of $o$. The operators in a PDDL input can be described with variables, used in predicates such as (at ?plane ?city).

A *Temporal Planning Problem* is a tuple $P = \langle A, I, O, G \rangle$, where $A$ is a set of atoms representing all the possible facts in a world situation, $I$ and $G$ are two sets of atoms that respectively denote the initial state and the problem goals, and $O$ is a set of ground PDDL operators.

As is common in Partial Order Causal Link (POCL) Planning [19], two dummy actions are also considered, *Start* and *End* with zero durations, the first with an empty precondition and effect $I$; the latter with precondition $G$ and empty effects. Two actions $a$ and $a'$ interfere when one deletes a precondition or positive effect of the other. The simple model of time in [16] defines a valid plan as a plan where interfering actions do not overlap in time. In other words, it is assumed that the preconditions need to hold until the end of the action, and that the effects also hold at the end and cannot be deleted during the execution by a concurrent action.

A *schedule* $P$ is a finite set of actions occurrences $\langle a_i, t_i \rangle$, $i = 1, \ldots, n$, where $a_i$ is an action and $t_i$ is a non-negative integer indicating the starting time of $a_i$ (its ending time is $t_i + dur(a_i)$). $P$ must include the *Start* and *End* actions, the former with time tag 0. The same action (except for these two) can be executed more than once in $P$ if $a_i = a_j$ for $i \neq j$. Two action occurrences $a_i$ and $a_j$ *overlap* in $P$ if one starts before the other ends; namely if $[t_i, t_i + dur(a_i)] \cap [t_j, t_j + dur(a_j)]$ contains more than one time point.

A schedule $P$ is a *valid plan* iff interfering actions do not overlap in $P$ and for every action occurrence $\langle a_i, t_i \rangle$ in $P$ its preconditions $p \in pre(a)$ are true at time $t_i$. This condition is inductively defined as follows: $p$ is true at time $t = 0$ iff $p \in I$, and $p$ is true at time $t > 0$ if either $p$ is true at time $t - 1$ and no action $a$ in $P$ ending at $t$ deletes $p$, or some action $a'$ in $P$ ending at $t$ adds $p$. The *makespan* of a plan $P$ is the time tag of the *End* action.

### 3.2 CPT: An Optimal Temporal Planner

An optimal temporal planner computes valid plans with minimum makespan. Even though an optimal planner was not mandatory (as discussed in section 5), we have chosen *CPT* [18], a freely-available optimal temporal planner, for its temporal dimension and for its constraint-based approach which provide a very useful data structure when it comes to gluing the partial solutions (see section 2.2). Indeed, since in Temporal Planning actions can overlap in time, the simple concatenation of sub-plans, though providing a feasible solution, obviously might produce a plan that is not optimal with respect to the total makespan, even if the sequence of actions is the optimal sequence. However, thanks to the causal links and order constraints maintained by CPT, an improved global plan can be obtained by shifting sub-plans as early as possible in a final state of the algorithm.

## 3.3   Rationale for Using *Divide-and-Evolve* for Temporal Planning

The reasons for the failure of standard OR methods addressing TPPs come from the exponential complexity of the number of possible actions when the number of objects involved in the problem increases. It is known for a long time that taking into account the interactions between sub-goals can decrease the complexity of finding a plan, in particular when these sub-goals are independent [12]. Moreover, computing an ideal ordering on sub-goals is as difficult as finding a plan (PSPACE-hard), as demonstrated in [11]. The basic idea when using the *Divide-and-Evolve* approach is that each local sub-plan ("joining" stations $s_i$ and $s_{i+1}$) should be easier to find than the global plan (joining the station of departure $s_0$ and the terminus $s_{n+1}$). This will be now demonstrated on the Zeno transportation benchmark (see http://ipc.icaps-conference.org/).

Table 1 illustrates the decomposition of a relatively difficult problem in the Zeno domain (zeno14 from IPC-3 benchmarks), a transportation problem with 5 planes (plane1 to plane5) and 10 persons (person0 to person9) to travel among 10 cities (city0 to city9).

**Table 1.** State Decomposition of the Zeno14 Instance. (The new location of moved objects appears in bold.).

| Objects | Init (station 0) | Station 1 | Station 2 | Station 3 | Station 4 | Goal (station 5) |
|---------|---------|-----------|-----------|-----------|-----------|------|
| plane 1 | city 5 | **city 6** | city 6 | city 6 | city 6 | city 6 |
| plane 2 | city 2 | city 2 | **city 3** | city 3 | city 3 | city 3 |
| plane 3 | city 4 | city 4 | city 4 | **city 9** | city 9 | city 9 |
| plane 4 | city 8 | city 8 | city 8 | city 8 | **city 5** | city 5 |
| plane 5 | city 9 | city 9 | city 9 | city 9 | city 9 | **city 8** |
| person 1 | city 9 | city 9 | city 9 | city 9 | city 9 | city 9 |
| person 2 | city 1 | city 1 | city 1 | city 1 | city 1 | **city 8** |
| person 3 | city 0 | city 0 | **city 2** | city 2 | city 2 | city 2 |
| person 4 | city 9 | city 9 | city 9 | **city 7** | city 7 | city 7 |
| person 5 | city 6 | city 6 | city 6 | city 6 | city 6 | **city 1** |
| person 6 | city 0 | **city 6** | city 6 | city 6 | city 6 | city 6 |
| person 7 | city 7 | city 7 | city 7 | city 7 | **city 5** | city 5 |
| person 8 | city 6 | city 6 | city 6 | city 6 | city 6 | **city 1** |
| person 9 | city 4 | city 4 | city 4 | city 4 | **city 5** | city 5 |
| person 0 | city 7 | city 7 | city 7 | **city 9** | city 9 | city 9 |
| Makespan | | 350 | 350 | 280 | 549 | 522 |
| Backtracks | | 1 | 0 | 0 | 195 | 32 |
| Search time | | 0.89 | 0.13 | 0.52 | 4.34 | 1.64 |
| Total time | | 49.10 | 49.65 | 49.78 | 54.00 | 51.83 |
| | | **Compression** | | | **Global Search** | |
| Makespan | | 772 | | | 476 | |
| Backtracks | | 0 | | | 606,405 | |
| Search time | | 0.01 | | | 4,155.41 | |
| Total time | | 0.02 (total : 254.38) | | | 4,205.40 | |

Analyzing the optimal solution found by CPT-3 it was easy to manually divide the optimal "route" of this solution in the state space into four intermediate stations between the initial state and the goal. It can be seen that very few moves (plane or person) occur between two consecutive stations (the ones in bold in each column of Table 1). Each sub-plan is easily found by CPT, with a maximum of 195 backtracks and 4.34 seconds of search time. It should be noted that most of the time spent by CPT is for pre-processing: this operation is actually repeated each time CPT is called, but could be factorized at almost no cost.

Note that the final step of the process is the compression of the five sub-plans (see section 2.2): it is here performed in 0.02 seconds without any backtracking, and the overall makespan of the plan is 772, much less than the sum of the individual makespans of each sub-plan (2051).

To summarize, the recomposed plan, with a makespan of 772, required a total running time of 254.38 seconds (including only 7.5s of pure search) and 228 back-tracks altogether, whereas a plan with the optimal makespan of 476 is found by CPT in 4,205 seconds and 606,405 backtracks. Section 5 will discuss this issue.

## 3.4   Description of the State Space

**Non-temporal States.** A natural state space for TPPs, as described at the beginning of this section, would be the actual space of all possible time-stamped states of the system. Obviously, the size of such a space is far too big and we simplified it by restricting the stations to non-temporal states. However, even with this simplification, not all "non-temporal" states can be considered in the description of the "stations".

**Limiting the Possible States.** First, the space of all possible states grows exponentially with the size of the problem. Second, not all states are consistent w.r.t. the planning domain. For instance, an object cannot be located at two places at the same time in a transportation problem – and inferring such state invariants is feasible but not trivial [6]. Note also that determining plan existence from a propositional STRIPS description has been proved to be PSPACE-complete [2].

A possible way to overcome this difficulty would be to rely on the local algorithm to (rapidly) check the consistency of a given situation, and to penalize unreachable stations. However, this would clearly be a waste of computational resources.

On the other hand, introducing domain knowledge into EAs has been known for long as the royal road toward success in Evolutionary Computation [9]. Hence, it seems a more promising approach to add state invariants to the description of the state space in order to remove the inconsistent states as much as possible. The good thing is that it is not necessary to remove *all* inconsistent states since, in any case, the local algorithm is there to help the EA to spot them – inconsistent stations will be given poor fitness, and will not survive next selection steps. In particular, only state invariants involving a single predicate have been implemented in the present work.

## 3.5   Representation of Stations

It was hence decided to describe the stations using **only the predicates that are present in the goal** of the overall problem, and to maintain the state invariants based on the semantics of the problem.

A good example is given in Table 1: the goal of this benchmark instance is to move the persons and planes in cities listed in the last column. No other predicate than the corresponding (at objectN cityM) predicates is present in the goal. Through a user-supplied file, the algorithm is told that only the at predicates will be used to represent the stations, with the syntactic restrictions that within a given station, the first argument of an at predicate can appear only once (at is said to be *exclusive* with respect to its first argument). The state space that will be explored by the algorithm thus amounts to a vector of 15 fluents (instantiated predicates) denoting that an item is located in a city (a column of table 1). In addition, the actual implementation of a station includes the possibility to "remove" (in fact, comment out) a predicate of the list: the corresponding object will not move during this sub-plan.

**Distance.** The distance between two stations should reflect the difficulty for the local algorithm to find a plan joining them. At the moment, a purely syntactic domain-independent distance is used: the number of different predicates not yet reached. The difficulty can then be estimated by the number of backtracks needed by the local algorithm. It is reasonable to assume that indeed most local problems where only a few predicates need to be changed from the initial state to the goal will be easy for the local algorithm - though this is certainly not true in all cases.

## 3.6   Representation-Specific Operators

**Initialization.** First, the number of stations is chosen uniformly in a user-supplied interval. The user also enters a maximal distance $d_{max}$ between stations. A matrix is then built, similar to the top lines of table 1: each line corresponds to one of the goal predicates, each column is a station. Only the first and last columns (corresponding to initial state and goal) are filled with values. A number of "moves" is then randomly added in the matrix, at most $d_{max}$ per column, and at least one per line. Additional moves are then added according to another user-supplied parameter, and without exceeding the $d_{max}$ limit per column. The matrix is then filled with values, starting from both ends (init and goal), constrained column-wise by the state invariants. A final sweep on all predicates comments out some of the predicates with a given probability.

**Station Mutation.** Thanks to the simplified representation of the states (a vector of fluents with a set of state invariants), it is straightforward to modify one station randomly: with a given probability, a new value for the non-exclusive arguments is chosen among the possible values respecting all constraints (including the distance constraints with previous and next stations). In addition, each predicate might be commented out from the station with a given probability, like in the initialization phase.

# 4 First Experiments

## 4.1 Single Objective Optimization

Our main playground to validate the *Divide-and-Evolve* approach is that of transportation problems, and started with the `zeno` domain as described in section 3.3. As can be seen in table 1, the description of the stations in `zeno` domain involves a single predicate, `at`, with two arguments. It is *exclusive* w.r.t. its first argument. Three instances have been tried, called `zeno10`, `zeno12` and `zeno14`, from the simplest to the hardest.

**Algorithmic Settings.** The EA that was used for the first implementation of the *Divide-and-Evolve* paradigm use standard algorithmic settings at the population level: a $(10, 70) - ES$ evolution engine (10 parents give birth to 70 children, and the best 10 among the children become the next parents), the children are created using 25% 1-point crossover (see section 2.3) and 75% mutation (individual level), out of which 25% are the *Add* (resp. *Del*) generic mutations (section 2.3). The remaining 50% of the mutations call the problem-specific station mutation. Within a station mutation, a predicate is randomly changed in 75% of the cases and a predicate is removed (resp. restored) in each of the remaining 12.5% cases. (see section 3.6). Initialization is performed using initial size in $[2, 10]$, maximum distance of 3 and probability to comment out a predicate is set to 0.1. Note that no lengthy parameter tuning was performed for those proof-of-concept experiments, and the above values were decided based upon a very limited set of initial experiments.

**The Fitness.** The target objective is here the total makespan of a plan – assuming that a global plan can be found, i.e. that all problems $(s_i, s_{i+1})$ can be solved by the local algorithm. In case one of the local problems could not be solved, the individual is declared *infeasible* and is penalized in such a way that all unfeasible individuals were worse than any feasible one. Moreover, this penalty is proportional to the number of remaining stations after the failure, in order to provide a nice slope of the fitness landscape toward feasibility. For feasible individuals, an average of the total makespan and the sum of the makespans of all partial problems is used: when only the total makespan is used, some individuals start bloating, without much consequence on the total makespan thanks to the final compression that is performed by CPT, but nevertheless slowing down the whole run because of all the useless repeated calls to CPT.

**Preliminary Results.** The simple `zeno10` (resp. `zeno12`) instance can be solved very easily by CPT-2 alone, in less than 2s (resp. 125s), finding the optimal plans with makespan 453 (resp. 549) using 154 (resp. 27560) backtracks.

For `zeno10`, all runs found the optimal solution in the very first generations (i.e. the initialization procedure always produced a feasible individual that CPT could compress to the optimal makespan. For `zeno12`, all runs found a suboptimal solution with makespan between 789 and 1222. Note that this final solution was found after 3 to 5 generations, the algorithm being stuck to this solution thereafter. The CPU time needed for 10 generations is around 5 hours.

A more interesting case is that of `zeno14`. First of all, it is worth mentioning that the present *Divide-and-Evolve* EA uses as local algorithm CPT version 2, and this version of CPT was unable to find a solution to `zeno14`: the results given in table 1 have been obtained using the (yet experimental and not usable from within the EA) new version of CPT. But whereas it proved unable to solve the full problem, CPT-2 could nevertheless be used to solve the hopefully small instances of `zeno14` domain that were generated by the *Divide-and-Evolve* approach – though taking a huge amount of CPU time for that. Setting a limit on the number of backtracks allowed for CPT was also mandatory to force CPT not to explore the too complex cases that would have resulted in a never-returning call.

However, a feasible individual was found in each of the only 2 runs we could run – one generation (70 evaluations) taking more than 10 hours. In the first run, a feasible individual was found in the initial population, with makespan 1958, and the best solution had a makespan of 773. In the other run, the first feasible solution was found at generation 3 – but the algorithm never improved on that first feasible individual (makespan 1356).

Though disappointing with respect to the overall performances of the algorithm, those results nevertheless witness for the fact that the *Divide-and-Evolve* approach can indeed solve a problem that could not be solved by CPT alone (remember that the version of CPT that was used in all experiments is by far less efficient than the one used to solve `zeno14` in section 3.3, and was not able to solve `zeno14` at all.

## 4.2   A Multi-objective Problem

**Problem Description.** In order to test the feasibility of the multi-objective approach based on the *Divide-and-Evolve* paradigm, we extended the `zeno` benchmark with an additional criterion, that can be interpreted either as a cost, or as a risk: in the former case, this additional objective is an additive measure, whereas in the latter case (risk) the aggregation function is the `max` operator.

The problem instance is shown in Figure 1: the only available routes between cities are displayed as edges, only one transportation method is available (plane), and the duration of the transport is shown on the corresponding edge. Risks (or costs) are attached to the cities (i.e., concern any transportation that either lands or takes off from that city). In the initial state, the 3 persons and the 2 planes are in `City 0`, and the goal is to transport them into `City 4`.

As can be easily computed (though there is a little trick here), there are 3 remarkable Pareto-optimal solutions, corresponding to traversing only one of the

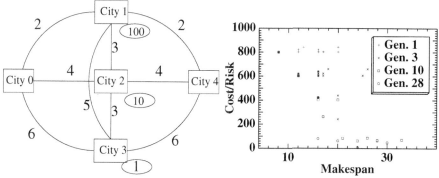

a) The instance: Durations are attached to edges, costs/risks are attached to cities (in gray circles)

b) The population at different generations for a successful run on the cost (additive) instance of the zeno miniproblem of Figure 1-a

**Fig. 1.** The multi-objective Zeno benchmark

3 middle cities. Going through City 1 is fast, but risky (costly), whereas going through City 3 is slow and safe and cheap.

When all persons go through respectively City 1, City 2 and City 3, the corresponding values of the makespans and costs in the additive case are (8, 800), (16, 80) and (24, 8), whereas they are, in the max case, (8, 100), (16, 10) and (24, 1).

**Problem Complexity** It is easy to compute the number of possible virtual stations: each one of the 3 persons can be in one of the 5 cities, or not mentioned (absent predicate). Hence there are $3^6 = 729$ possible combinations, and $729^n$ possible lists of length $n$. So even when $n$ is limited to 6, the size of the search space is approx. $10^{17}$ ...

**The Algorithm.** The EA is based on the standard NSGA-II multi-objective EA [3]: standard tournament selection of size 2 and deterministic replacement among parents + offspring, both based on the Pareto ranking and crowding distance selection; a population size of 100 evolves during 30 generations. All other parameters were those used for the single objective case.

**Fitnesses.** The problem has two objectives: one is the the total makespan (as in the single-objective case), the other is either the **risk** (aggregated using the **max** operator) or the **cost** (an **additive** objective). Because the global risk only takes 3 values, there is no way to have any useful gradient information when used as fitness in the max case. However, even in the additive case, the same arguments than for the makespan apply (section 4.1), and hence, in all cases, the second objective is the sum of the overall risk/cost and the average (not the sum) of the values for all partial problems – excluding from this average those partial problems that have a null makespan (when the goal is already included in the initial state).

**Results.** For the additive (**cost**) case, the most difficult Pareto optimum (going through city 3 only) was found 4 times out of 11 runs. However, the 2 other remarkable Pareto optima, as well as several other points in the Pareto front were also repeatedly found by all runs. Figure 1-b shows different snapshots of the population at different stages of the evolution for a typical successful run: at first ('+'), all individuals have a high cost (above 800); At generation 3 ('×'), there exist individuals in the population that have cost less than 600; At generation 10 (squares), many points have a cost less than 100. But the optimal (24,8) solution is only found at generation 28 (circles).

The problem in the **risk** context (the max case) proved to be, as expected, slightly more difficult. All three Pareto optima (there exist no other point of the true Pareto front in the max case) were found only in 2 runs out of 11. However, all runs found both the two other Pareto optima, as well as the slightly sub-optimal solutions that goes only through city 3 but did not find the little trick mentioned earlier, resulting in a (36,1) solution.

In both cases, those results clearly validate the *Divide-and-Evolve* approach for multi-objective TPPs – remember that CPT has no knowledge of the risk/cost in its optimization procedure - it only aggregates the values a posteriori, after having computed its optimal plan based on the makespan only – hence the difficulty to find the 3rd Pareto optimum going only through `city3`.

## 5    Discussion and Further Work

A primary concern is the existence of a decomposition for any plan with optimal makespan. Because of the restriction of the representation to the predicates that are in the goal, some states become impossible to describe. If one of these states is mandatory for all optimal plans, the evolutionary algorithm is unable to find the optimal solution. In the `zeno14` benchmark detailed in section 3.3, for instance, one can see from the optimal solution that the `in` predicate should be taken into account when splitting the optimal solution, in order to be able to link a specific person to a specific plane. The main difficulty, however, is to add the corresponding state invariant between `at` and `in` (a person is either `at` a location or `in` a plane). Future work will include state invariants involving pairs of predicates, to cope with such cases. Along the same line, we will investigate whether it might be possible to automatically infer some state invariants from the data structures maintained by CPT.

It is clear from the somehow disappointing results presented in section 4.1 that the search capabilities of the proposed algorithm should be improved. But there is a lot of space for improvements. First, and most immediate, the variation operators could use some domain knowledge, as proposed in section 2.3 – even if this departs from "pure" evolutionary blind search. Also, all parameters of the algorithm will be carefully fine-tuned.

Of course the *Divide-and-Evolve* scheme has to be experimented on more examples. The International Planning Competition provides many instances in several domains that are good candidates. Preliminary results on the `driver` problem showed very similar results that those reported here on the `zeno` domain. But other domains, such as the `depot` domain, or many real-world domains, involve (at least) 2 predicates in their goal descriptions (e.g., `in` and `on` for `depot`). It is hence necessary to increase the range of allowed expressions in the description of individuals.

Other improvements will result from the move to the new version of CPT, entirely rewritten in C. It will be possible to call CPT from within the EA, and hence to perform all grounding, pre-processing and CSP representation only once: at the moment, CPT is launched anew for each partial computation, and a quick look at table 1 shows that on `zeno14` problem, for instance, the run-time per individual will decrease from 250 to 55 seconds. Though this will not *per se* improve the quality of the results, it will allow us to tackle more complex problems than even `zeno14`. Along the same lines, other planners, in particular sub-optimal planners, will also be tried in lieu of CPT, as maybe the *Divide-and-Evolve* approach could find optimal results using sub-optimal planners (as done in some sense in the multi-objective case, see section 4.2).

A last but important remark about the results is that, at least in the single objective case, the best solution found by the algorithm was always found in the very early generations of the runs: it could be that the simple splits of the problem into smaller sub-problems that are done during the initialization are the main reasons for the results. Detailed investigations will show whether or not an Evolutionary Algorithm is useful in that context!

Nevertheless, we do believe that using Evolutionary Computation is mandatory in order to solve multi-objective optimization problems, as witnessed by the results of section 4.2, that are, to the best of our knowledge, the first ever results of Pareto optimization for TPPs.

# References

1. A. H. Brie and P. Morignot. Genetic Planning Using Variable Length Chromosomes. In *15th Intl Conf. on Automated Planning and Scheduling*, 2005.
2. T. Bylander. The Computational Complexity of Propositional STRIPS planning. *Artificial Intelligence*, 69(1-2):165–204, 1994.
3. K. Deb, S. Agrawal, A. Pratab, and T. Meyarivan. A Fast Elitist Non-Dominated Sorting Genetic Algorithm for Multi-Objective Optimization. In M. Schoenauer et al., editor, *PPSN'2000*, pages 849–858. Springer-Verlag, LNCS 1917, 2000.
4. C. Desquilbet. Détermination de trajets optimaux par algorithmes génétiques. Rapport de stage d'option B2 de l'Ecole Polytechnique. Palaiseau, France, Juin 1992. Advisor: Marc Schoenauer. In French.
5. R. Fikes and N. Nilsson. STRIPS: A New Approach to the Application of Theorem Proving to Problem Solving. *Artificial Intelligence*, 1:27–120, 1971.

6. M. Fox and D. Long. The Automatic Inference of State Invariants in TIM. *Journal of Artificial Intelligence Research*, 9:367–421, 1998.

7. M. Fox and D. Long. PDDL2.1: An Extension to PDDL for Expressing Temporal Planning Domains. *Journal of Artificial Intelligence Research*, 20:61–124, 2003.

8. H. Geffner. Perspectives on Artificial Intelligence Planning. In *Proc. AAAI-2002*, pages 1013–1023, 2002.

9. J. J. Grefenstette. Incorporating Problem Specific Knowledge in Genetic Algorithms. In Davis L., editor, *Genetic Algorithms and Simulated Annealing*, pages 42–60. Morgan Kaufmann, 1987.

10. W.E. Hart, N. Krasnogor, and J.E. Smith, editors. *Recent Advances in Memetic Algorithms*. Studies in Fuzziness and Soft Computing, Vol. 166. Springer Verlag, 2005.

11. J. Koehler and J. Hoffmann. On Reasonable and Forced Goal Orderings and their Use in an Agenda-Driven Planning Algorithm. *JAIR*, 12:338–386, 2000.

12. R. Korf. Planning as Search: A Quantitative Approach. *Artificial Intelligence*, 33:65–88, 1987.

13. D. McDermott. PDDL – The Planning Domain Definition language. At `http://ftp.cs.yale.edu/pub/mcdermott`, 1998.

14. P. Merz and B. Freisleben. Fitness Landscapes and Memetic Algorithm Design. In David Corne, Marco Dorigo, and Fred Glover, editors, *New Ideas in Optimization*, pages 245–260. McGraw-Hill, London, 1999.

15. J.L. Schlabach, C.C. Hayes, and D.E. Goldberg. FOX-GA: A Genetic Algorithm for Generating and Analyzing Battlefield Courses of Action. *Evolutionary Computation*, 7(1):45–68, 1999.

16. D. Smith and D. S. Weld. Temporal Planning with Mutual Exclusion Reasoning. In *Proceedings of IJCAI-99*, pages 326–337, 1999.

17. L. Spector. Genetic Programming and AI Planning Systems. In *Proc. AAAI 94*, pages 1329–1334. AAAI/MIT Press, 1994.

18. V. Vidal and H. Geffner. Branching and Pruning: An Optimal Temporal POCL Planner based on Constraint Programming. In *Proceedings of AAAI-2004*, pages 570–577, 2004.

19. D. S. Weld. An Introduction to Least Commitment Planning. *AI Magazine*, 15(4):27–61, 1994.

# A Variable Neighbourhood Search Algorithm for Job Shop Scheduling Problems

Mehmet Sevkli[1] and M. Emin Aydin[2,*]

[1] Fatih University, Dept. of Industrial Engineering, Buyukcekmece, Istanbul, Turkey
msevkli@fatih.edu.tr
[2] University of Luton, Department of Computing and Information Systems, Luton, UK
mehmetaydin@acm.org

**Abstract.** Variable Neighbourhood Search (VNS) is one of the most recent metaheuristics used for solving combinatorial optimization problems in which a systematic change of neighbourhood within a local search is carried out. In this paper, a variable neighbourhood search algorithm is proposed for Job Shop Scheduling (JSS) problem with makespan criterion. The results gained by VNS algorithm are presented and compared with the best known results in literature. It is concluded that the VNS implementation is better than many recently published works with respect to the quality of the solution.

## 1 Introduction

Metaheuristics are general strategies for designing heuristic procedures to solve an optimization problem by a search process on the solution space. The heuristic search procedures are generally based on transformations of the alternatives that determine a neighbourhood structure on the solution space. Variable neighbourhood search (VNS) is a recent metaheuristic for solving combinatorial and global optimization problems whose basic idea is systematic change of neighbourhood within a local search. It is based upon a simple principle: change the neighbourhood structure when the search is trapped on a local minimum, which is very likely in most of combinatorial and/or multi-model numerical optimisation problems. Especially, as search space grows fast with growing problem sizes, the likelihood of being trapped in local minima becomes inevitable. The main concern of research in this field is to recover trapped search or to put effort for preventing on-line. VNS offers a multiple neighbourhood structure with which one recovers the solutions trapped via the others. The main idea here is to choose heuristics (neighbourhood structures) complementary to each other.

Job Shop Scheduling (JSS) problems with the objective function of minimizing makespan ($C_{max}$) is one of the best known and strongly NP-hard [16] combinatorial optimization problems. Among the benchmarks within the literature, small size instances of the JSS problems can be solved in reasonable computational time by exact algorithms such as branch-and-bound approach [3, 10]. However, when the problem size increases, the computational time of the exact methods grows exponentially.

---

[*] Corresponding author.

J. Gottlieb and G.R. Raidl (Eds.): EvoCOP 2006, LNCS 3906, pp. 261–271, 2006.
© Springer-Verlag Berlin Heidelberg 2006

Therefore, the recent research on JSS problems is focused on heuristic algorithms such as Simulated Annealing (SA) [4, 21, 27], Genetic Algorithm (GA) [7, 14, 17, 18], Taboo Search (TS) [13, 23, 24], Ant Colony Optimization (ACO) [8, 12], Neural Network (NN) [26], Shifting Bottleneck Procedure [1, 19], Guided Local Search [5], Parallel Greedy Randomized Adaptive Search Procedure (GRASP) [2] and Constraint Propagation [14]. A comprehensive survey of the JSS problem can be found in [20].

In this paper, we propose a new implementation of VNS algorithm for JSS problems. The main idea of new implementation is to look for the best time and condition to switch to the other neighbourhood structure from one, whereas heuristics are utilised on a periodical bases in traditional VNS. That is to diversify the solution when it needs. The algorithms have been tested with several hard benchmark instances of the JSS problems. The novelty of this study comes from the success of modified VNS algorithm as it is the first implementation of VNS for JSS problems to our knowledge.

The organization of this paper is as follows. Section 2 introduces the foundations of VNS algorithm, job shop scheduling problems and the representation method exploited in this study. A novel implementation of VNS, namely Modified VNS, is elaborated in Section 3 while experimental results are presented and discussed in Section 4. Finally, Section 5 presents the concluding remarks.

## 2  Background

In the following sub-sections, the foundations of VNS, JSS and the way in which JSS problems are represented have been elaborated.

### 2.1  Variable Neighbourhood Search (VNS) Algorithms

VNS algorithm, one of very well-known local search methods [22], gets more attention day-by-day, because of its ease of use and accomplishments in solving combinatorial optimisation problems [25, 28]. Basically, a local search algorithm carries out exploration within a limited region of the whole search space. That facilitates a provision of finding better solutions without going further investigation. The VNS is a simple and effective search procedure that proceeds to a systematic change of neighbourhood. An ordinary VNS algorithm (Fig. 1) gets an initial solution, $s \in S$, where $S$ is the whole set of search space, than manipulates it through a two nested loop in which the core one alters and explores via two main functions so called *shake* and *localSearch*. The outer loop works as a refresher reiterating the inner loop, while the inner loop carries the major search. *localSearch* explores an improved solution within the local neighbourhood, whilst *shake* diversifies the solution by switching to another local neighbourhood. The inner loop iterates as long as it keeps improving the solutions, where an integer, $k$, controls the length of the loop. Once an inner loop is completed, the outer loop re-iterates unless the termination condition is not met. Since the complementariness of neighbourhood functions is the key idea behind VNS, the neighbourhood structure / heuristic functions should be chosen very rigorously so as to achieve an efficient VNS.

```
procedure VNS
   Generate initial solution, s∈S
   while termination condition not met do
      k ←1
      while k≤k_max do
         s'∈S ← shake( s )
         s"∈S ← localSearch(s')
         if f(s")< f(s) then
            s ← s"
            k ←1
         else k ←k+1
         end-if
      end-while
   end-while
end-procedure
```

Fig. 1. A pseudo code for VNS algorithm

## 2.2 Job Shop Scheduling Problems

Job Shop Scheduling (JSS) problems have been studied for a long time. Since it is not that easy to reach the optimal solutions within a short time, because of the NP-Hard nature, and there is no guarantee to switch to a better state from a feasible state, this problem type has been a very strong testbed for metaheuristics. Furthermore, they have never been dropped from scientific research.

The problem is comprised of a set of jobs ($J$) to be processed on a set of machines ($M$) subject to a number of technological constraints. Each job consists of $m$ operations, $O_j=\{o_{1j},...,o_{mj}\}$, each operation must be processed on a particular machine, and there is only one operation of each job to be processed on each machine. There is a predefined order of the operations of each particular job in which each operation has to be processed after its predecessor ($PJ_j$) and before its successor ($SJ_j$). In the end of the whole schedule, each machine completes processing $n$ operations in an order that is determined during the scheduling time, although there is no such order initially. Therefore, each operation processed on the $M_i$ has a predecessor ($PM_i$) and a successor ($SM_i$). A machine can process only one operation at a time. There are no set-up times, no release dates and no due dates.

Each operation has a processing time ($p_{ij}$) on related machine starting at the time of $r_{ij}$. The completion time of $o_{ij}$ is therefore: $c_{ij}=r_{ij}+p_{ij}$, where $i = (1,...,m)$, $j = (1,..., n)$ and $r_{ij} = \max(c_{iPJ_j}, c_{PM_{ij}})$. Machines and jobs have particular completion times, which are denoted and identified as: $C_{M_i} = c_{in}$ and $C_{J_j} = c_{in}$ where $c_{in}$ and $c_{jm}$ are the completion time of the last ($n^{th}$) operation on $i^{th}$ machine and the completion time of the last ($m^{th}$) operation of $j^{th}$ job, respectively. The overall objective is to minimise the completion time of the whole schedule (makespan), which is the maximum of machines' completion times, $C_{max} = max (C_{M1},...,C_{Mm})$. The representation is done via a disjunctive graph, as it is widely used.

## 2.3  Problem Representation

Schedules are represented in a set of integers, where each stands for an operation. It is also called chromosome of $n \times m$ gene represents a problem of $n$ jobs, $m$ machines. Since each integer does not represent a certain operation, but the last completed operation of corresponding job, each job is represented $m$ times within the chromosome. This way of representation prevents infeasibility, and always provide with a feasible active schedule.  For instance, we are given a chromosome of [2 1 2 2 1 3 1 3 3], where $\{1, 2, 3\}$ represents $\{ j_1, j_2, j_3 \}$ respectively. Obviously, there are totally 9 operations, but, 3 different integers, each is repeated 3 times. The integer on the first gene, 2, represents the first operation of the second job to be processed first on corresponding machine. Likewise, the integer on the second gene, 1, represents the first operation of the first job on corresponding machine.  Thus, the chromosome of [2 1 2 2 1 3 1 3 3] is understood as $[o_{21}, o_{11}, o_{22}, o_{23}, o_{12}, o_{31}, o_{13}, o_{32}, o_{33}]$ where $o_{ij}$ stands for the $i^{th}$ operation of $j^{th}$ job. More details can be found in [11].

## 2.4  Neighbourhood Structure

The neighbourhood structure with which the neighbouring solutions are determined to move to is one of the key elements of metaheuristics, as the performance of the meta-heuristic algorithm significantly depends on the efficiency of the neighbourhood structure. The following two neighbourhood structures are employed in this study:

- *Exchange* is a function used to move around in which any two randomly selected operations are simply swapped. For instance, suppose that we are given a state of [2 1 2 2 1 3 1 3 3] and the two random numbers derived are 2 and 8. After applying *Exchange*, the new state will be [2 1 3 2 1 3 1 3 2]. Obviously, the $2^{nd}$ and $8^{th}$ genes of the chromosome were 2 and 3, respectively. Applying *Exchange* function, the new $2^{nd}$ and $8^{th}$ genes were swapped and turned to 3 and 2, respectively.
- *Insert* is another fine-tuning function that inserts a randomly chosen gene in front or back of another randomly chosen gene. For instance, we are given the same state as before. In order to apply *Insert*, we also need to derive two random numbers; one is for determining the gene to be inserted and the other is for the gene that insertion to be done in front/back of it. Let us say those number are 3 and 5, where $3^{rd}$ gene is 2 and the $5^{th}$ one is 3. Consecutively, the new state will be [2 1 2 1 2 3 1 3 3].

Although there are other possible functions to be applied, we preferred the aforementioned functions due to their simplicity and ease of use alongside of a reasonable efficiency.

# 3  The Modified VNS Implementation for JSS

The VNS implementation for JSS problems has been carried out based on the aforementioned information provided. *Exchange* and *Insert* functions are adopted as local search and shake respectively, meaning that the solution undertaken is shaked by

*Insert* function and then submitted it to the local search that manipulates the solution by *Exchange*. Although *Insert* and *Exchange* are found very complementary to each other, the VNS developed ordinarily has not provided impressive results. For that purpose, we kept investigating for a better equipped VNS algorithm in which the problems can be solved straightforwardly.

Modified VNS algorithm is the newly developed VNS implementation based on complementary heuristics for better quality of solution within a shorter time. We realised that the main bottleneck of VNS lies in the pairing of the heuristics to shape up, *shake* and *localSearch* functions. In fact, they have to complete each other so as to develop an efficient algorithm. For this purpose, the core of VNS algorithm, which functions in the middle, has been shaped up with two complementary heuristics, exchange and insert, in a way that the heuristics runs as the shake function is invoked whenever it is needed subject to the conditions. Otherwise, the one functions as *localSearch* keeps running. The other significant novelty is the insert of a perturbation function into the algorithm. This is because of the need of periodical diversification in the search. In many cases, local optima blocks the way of search, especially in the algorithms as such.

```
procedure modified VNS
    Generate initial solution, s∈S
    while termination condition not met do
        k←1
        s' ∈S ← perturb ( s )
        while k≤k_max do
            if (k=1) then s" ∈S ←Exchange( s' )
            if (k=2) then s" ∈S ← Insert( s' )
            if f(s") < f(s') then
                s' ← s"
            else k←k+1
            end-if
        end-while
        if f(s") < f(s) then s← s"
    end-while
end-procedure
```

Fig. 2. Pseudo code for Modified VNS algorithm

Modified VNS algorithm is sketched in a pseudo code given by Fig. 2. As explained in the previous paragraph, it has been developed wrapping shake and local search functions into a combined function, which allows each working subject to certain conditions. In addition, it has been equipped with a perturbation function outside the inner loop. This algorithm has been tested as reported in the following section. The perturbation procedure consists of a combination of *Exchange* and *Insert* functions, where *Exchange* operates first, then *Insert* runs with its outcome and finally *Exchange* re-operates on the outcome of *Insert*. This is shown by Fig. 3.

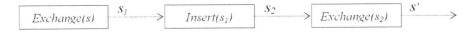

**Fig. 3.** The way perturbation function is shaped up

## 4  Experimental Results

In this paper, we propose a VNS implementation for job shop scheduling problems with a wide range of experimentation. In the following section, we provide with experimental results and relevant discussions in order to make clear the fundamentals of the implementation. The measures considered in this study are mainly about quality of solution and/or computational time. The success of the algorithm regarding the quality of solution has mainly been accounted with respect to the relative percentage of error (RPE) index, which is calculated as follows:

$$RPE = \frac{(bf - opt)}{opt} \times 100 \qquad (1)$$

where *bf* is the best makespan found and *opt* is either the optimum or the lowest boundary known for unknown optimum values. Obviously, RPE is calculated based on the mean, and also can be measured benchmark-by-benchmark. In order to review the results in a broader point of view, we developed a second index based on the latter RPE calculation averaged over the 30 repetitions. That is called ARPE standing for averaged relative percentage of error. The third index used is the hitting-ratio (HR) being calculated as the number of optimum found through the whole repetitions. This is needed as other indexes may not build a sufficient level of confidence with the results. The experimentation was carried out on a PC equipped with Intel Pentium IV 2.6 GHz processor and 256MB memory. The software coded in C programming language. The JSS benchmark problems, which are very well known within the field, were picked up from OR-Library [6].

In Table 1, the experimental results gained by VNS and Modified VNS algorithms with respect to the quality of solutions measured in RPE, ARPE and HT indexes, where first two are minimized and the last one (HR) is maximized. The experiments have been conducted over 31 benchmarks tackled; some are known moderately hard but some are very hard. The accomplishments of both algorithms are clearly reflected. Last three columns of Table 1 are assigned for the results gained by the ordinary VNS, while the middle three present the results by Modified VNS, which are significantly better and evidently improved. HR index by normal VNS remains 0 % for almost all of the benchmarks, which means that almost none of the trail of experiments provided with an optimum makespan for none of the benchmarks. On the other hand, HR by Modified VNS remains 0 % for only 7 of 31 benchmarks, which are known very difficult, where the 11 of 31 are 50 % or more and the rest provided with less than 50 % provision of optimum. The other two indexes by normal VNS are very solidly and significantly higher than the ones by Modified VNS.

**Table 1.** Results obtained from two VNS algorithms with respect to 3 indexes for quality of solutions

| Benchmarks | | Modified VNS | | | Ordinary VNS | | |
|---|---|---|---|---|---|---|---|
| Name | Opt. | RPE | ARPE | HR | RPE | ARPE | HR |
| ft10 | 930 | 0.00 | 0.55 | 0.60 | 3.02 | 7.47 | 0.00 |
| ft20 | 1165 | 0.00 | 0.54 | 0.50 | 1.27 | 2.70 | 0.00 |
| la16 | 945 | 0.00 | 0.30 | 0.50 | 0.11 | 2.26 | 0.00 |
| la19 | 842 | 0.00 | 0.09 | 0.93 | 2.77 | 3.57 | 0.00 |
| abz05 | 1234 | 0.00 | 0.10 | 0.60 | 0.00 | 0.86 | 0.03 |
| abz06 | 943 | 0.00 | 0.00 | 1.00 | 2.38 | 2.58 | 0.00 |
| orb01 | 1059 | 0.00 | 1.47 | 0.13 | 2.22 | 4.79 | 0.00 |
| orb02 | 888 | 0.11 | 0.33 | 0.00 | 0.89 | 2.92 | 0.00 |
| orb03 | 1005 | 0.00 | 2.60 | 0.10 | 6.25 | 7.85 | 0.00 |
| orb04 | 1005 | 0.00 | 0.62 | 0.40 | 1.66 | 3.48 | 0.00 |
| orb05 | 887 | 0.00 | 0.31 | 0.17 | 2.31 | 4.92 | 0.00 |
| orb06 | 1010 | 0.00 | 0.89 | 0.10 | 2.04 | 7.20 | 0.00 |
| orb07 | 397 | 0.00 | 0.27 | 0.80 | 2.46 | 3.31 | 0.00 |
| orb08 | 899 | 0.00 | 1.63 | 0.23 | 1.86 | 5.45 | 0.00 |
| orb09 | 934 | 0.00 | 0.78 | 0.13 | 0.53 | 2.94 | 0.00 |
| orb10 | 944 | 0.00 | 0.00 | 1.00 | 0.74 | 4.19 | 0.00 |
| abz07 | 656 | 0.46 | 2.01 | 0.00 | 4.93 | 6.45 | 0.00 |
| abz08 | 665 | 0.60 | 1.99 | 0.00 | 6.99 | 7.28 | 0.00 |
| abz09 | 679 | 0.15 | 2.17 | 0.00 | 6.34 | 7.16 | 0.00 |
| la21 | 1046 | 0.00 | 0.62 | 0.03 | 1.88 | 4.81 | 0.00 |
| la22 | 927 | 0.00 | 0.24 | 0.57 | 1.70 | 2.90 | 0.00 |
| la24 | 935 | 0.00 | 0.60 | 0.03 | 1.58 | 4.12 | 0.00 |
| la25 | 977 | 0.00 | 0.64 | 0.07 | 0.91 | 3.72 | 0.00 |
| la27 | 1235 | 0.08 | 0.91 | 0.00 | 2.76 | 4.08 | 0.00 |
| la28 | 1216 | 0.00 | 0.03 | 0.87 | 1.38 | 2.33 | 0.00 |
| la29 | 1152 | 0.95 | 1.80 | 0.00 | 5.88 | 7.39 | 0.00 |
| la36 | 1268 | 0.00 | 0.49 | 0.27 | 1.78 | 3.60 | 0.00 |
| la37 | 1397 | 0.00 | 0.73 | 0.37 | 3.72 | 5.62 | 0.00 |
| la38 | 1196 | 0.00 | 0.97 | 0.07 | 2.13 | 5.18 | 0.00 |
| la39 | 1233 | 0.00 | 0.43 | 0.30 | 1.83 | 3.49 | 0.00 |
| la40 | 1222 | 0.16 | 0.42 | 0.00 | 2.16 | 3.59 | 0.00 |

The investigation on fine-tuning the local search algorithms has been carried out over various combinations of aforementioned local search heuristics, namely *Exchange* and *Insert*. Table 2 reflects the performances of each case considered with respect to the three performance indexes. The cases considered are *Exchange* only, *Insert* only, *Exchange* with *Insert*, *Insert* with *Exchange* and *Exchange* with *Insert* without perturbation. Obviously, the worst case indicated in the lowermost row of Table 2 presents the performance of *Exchange* with *Insert* without Perturbation, which reflects the significance of perturbation in this algorithm. The other cases are not that significantly worse, where the performance of *Exchange* only and *Insert* only are remarkably distinguishable from other two. The rest two cases are not significantly different from each other, though *Exchange* with *Insert* seems the best in numbers as PRE and HR indexes are better, but APRE is worse than *Insert* with *Exchange*. All these experiments are carried out over 30 trails each.

**Table 2.** The performance of various heuristics with modified VNS algorithms with respect to all three indexes

| Heuristics | PRE | ARPE | HR |
|---|---|---|---|
| *Exchange* | 0.30 | 1.22 | 0.52 |
| *Insert* | 0.13 | 1.00 | 0.74 |
| *Exchange+Insert* | 0.08 | 0.79 | 0.77 |
| *Insert+Exchange* | 0.13 | 0.76 | 0.71 |
| Without *Perturbation* | 0.44 | 3.10 | 0.45 |

The impact of perturbation is obvious from Table 2 as the worst case indicated is the case with which perturbation operation has not been considered. As mentioned before, perturbation operation is a combined operation of *Exchange* and *Insert* functions, where *Exchange* operates once before and once after *Insert*. The idea behind this operation is to release the possible restrictions of some improved results via the heuristics; each functions as an independent hill climber. *Perturbation* refreshes the solution and provides some diversity for preventing further traps.

Table 3 presents results provided with various meta-heuristics recently published and Modified VNS algorithm with respect to the quality of the solutions in PRE index, which is the only measure provided in common. The benchmarks chosen are those which considered very hard among the list of 31 in Table 1. We are not able to provide a comparison for the whole list of benchmarks provided in Table 1 as the comparing works have mostly considered those listed in Table 3, but not the whole list of Table 1. These algorithms taken into account are listed as follows:

- Distributed evolutionary simulated annealing algorithm (dESA) Aydin and Fogarty [4].
- Ant colony optimization algorithm (ACO GSS) by Blum and Sampels [8].
- Parallel GRASP with path-relinking (GRASP) by Aiex et al.[2]
- A Hybrid Genetic Algorithm (HGA) by Goncalves et al.[18].
- A Tabu Search Method (TSSB) by Pezzella and Merelli [24].

Obviously, the column providing best results is the last one that presents the results with Modified VNS, where the others are competitive among one another. The last row presents the average PRE values of each column, where the minimum is also in the column of VNS. All these results prove the strength of this VNS implementation (Modified VNS) for JSS problems.

**Table 3.** A comparison among the meta-heuristics recently published with respect to the quality of the solution in RPE index

| Benchmarks | | PRE performance | | | | | |
|---|---|---|---|---|---|---|---|
| Name | Opt. | dESA [4] | ACO GSS [8] | GRASP [2] | HGA [18] | TSSB [24] | VNS |
| abz07 | 656 | 2.44 | 2.74 | 5.49 | NA | 1.52 | 0.46 |
| abz08 | 665 | 2.41 | 3.61 | 6.02 | NA | 1.95 | 0.60 |
| abz09 | 679 | 2.95 | 3.39 | 8.98 | NA | 2.06 | 0.15 |
| la21 | 1046 | 0.00 | 0.10 | 1.05 | 0.00 | 0.00 | 0.00 |
| la24 | 935 | 0.32 | 0.96 | 2.03 | 1.93 | 0.32 | 0.00 |
| la25 | 977 | 0.00 | 0.00 | 0.72 | 0.92 | 0.20 | 0.00 |
| la27 | 1235 | 0.40 | 0.65 | 2.75 | 1.70 | 0.00 | 0.08 |
| la29 | 1152 | 2.08 | 1.39 | 4.43 | 3.82 | 1.39 | 0.95 |
| la38 | 1196 | 0.42 | 2.59 | 1.84 | 1.92 | 0.42 | 0.00 |
| la40 | 1222 | 0.49 | 0.49 | 1.80 | 1.55 | 0.90 | 0.16 |
| Average | | 1.15 | 1.59 | 3.51 | 1.69 | 0.88 | 0.24 |

# 5   Conclusion

Job shop scheduling problem has been studied for far a long time. Because of its hardness and being representative for planning problems, many methods have been tested with this family of problems. In this paper, a novel VNS implementation has been introduced for classical job shop problems. The novelty of this implementation is provided with a combined local search algorithm and a perturbation function which frequently refreshes the solution undertaken. It has been shown that this method has done well and outperformed a number of recently published meta-heuristics. The research is ongoing on the course of parallelization of VNS algorithm to get better quality of solution within far shorter time.

# References

1. Adams, J., Balas, E., and Zawack, D. :The Shifting Bottleneck Procedure for Job Shop Scheduling. Management Science 34 (1988) 391-401.
2. Aiex, R. M., Binato, S., and Resende, M. G. C. :Parallel GRASP with Path-Relinking for Job Shop Scheduling. Parallel Computing 29 (2003) 393-430.
3. Applegate, D., and Cook, W. :A Computational Study of Job-Shop Scheduling. ORSA Journal on Computing 3(2) (1991) 149-156.
4. Aydin, M. E., and Fogarty, T. C.: A Distributed Evolutionary Simulated Annealing Algorithm for Combinatorial Optimisation Problems. Journal of Heuristics 10 (2004) 269-292.
5. Balas, E., and Vazacopoulos, A. :Guided Local Search with Shifting Bottleneck for Job Shop Scheduling. Management Science 44 (1998) 262-275.

6. Beasley, J.E. "Obtaining Test Problems via Internet." Journal of Global Optimisation 8, 429-433, http://people.brunel.ac.uk/~mastjjb/jeb/info.html.

7. Bierwith, C. :A Generalized Permutation Approach to Job Shop Scheduling with Genetic Algorithms. OR Spektrum 17 (1995) 87-92.

8. Blum, C., and Sampels, M. :An Ant Colony Optimization Algorithm for Shop Scheduling Problems. Journal of Mathematical Modelling and Algorithms 3 (2004) 285-308.

9. Bruce, K.B., Cardelli, L., Pierce, B.C.: Comparing Object Encodings. In: Abadi, M., Ito, T. (eds.): Theoretical Aspects of Computer Software. Lecture Notes in Computer Science, Vol. 1281. Springer-Verlag, Berlin Heidelberg New York (1997) 415–438

10. Carlier, J., and Pison, E. :An Algorithm for Solving the Job-Shop Problem, Management Science 35 (1989) 164-176.

11. Cheng, R. , Gen, M., and Tsujimura, Y. :A Tutorial Survey of Job Shop Scheduling Problems Using genetic Algorithms-I. Representation. Journal of Computers and Industrial Engineering 30(4) (1996) 983-997.

12. Colorni, A., Dorigo, M., Maniezzo, V., and Trubian, M. :Ant System for Job-Shop Scheduling. Belgian Journal of Operations Research, Statistics and Computer Science (JORBEL) 34(1) (1994) 39-53.

13. Dell'Amico, M., and Trubian, M. :Applying Tabu-Search to the Job-Shop Scheduling Problem. Annals of Operations Research 4 (1993) 231-252.

14. Dorndorf, U., and Pesch, E.: Evolution Based Learning in a Job Shop Scheduling Environment, Computers & Operations Research 22 (1995) June 26, 2005 0:30 International Journal of Production Research "A hybrid PSO for the JSSP"

15. Dorndorf, U., Pesch, E., and Phan-Huy, T.: Constraint Propagation and Problem Decomposition: A Preprocessing Procedure for the Job Shop Problem, Annals of Operations Research 115 (2002) 125-145.

16. Garey, M. Johnson, D., and Sethy, R. :The Complexity of Flow Shop and Job Shop Scheduling. Mathematics of Operations Research 1 (1976) 117-129.

17. Groce, F. D. Tadei, R., and Volta, G. :A Genetic Algorithm for the Job Shop Problem. Computers & Operations Research 22 (1995) 15-24.

18. Goncalves, J. F., Mendes, J. M., and Resende, M. :A hybrid genetic algorithm for the job shop scheduling problem, European Journal of Operations Research 167(1) (2004) 77-95.

19. Huang, W., and Yin, A. : An Improved Shifting Bottleneck Procedure for the Job Shop Scheduling Problem. Computers & Operations Research 31 (2004) 2093-2110.

20. Jain, A., and Meeran, S. : Deterministic Job-Shop Scheduling: Past, Present and Future. European Journal of Operational Research. 113: (1999) 390-434.

21. Kolonko, M. :Some New Results on Simulated Annealing Applied to the Job Shop Scheduling Problem. European Journal of Operational Research 113, (1999) 123-136.

22. Mladenovic, N., and Hansen, P. :Variable Neighborhood Search. Computers and Operations Research 24 (1997) 1097-1100.

23. Nowicki, E., and Smutnicki, C.: A Fast Taboo Search Algorithm for the Job Shop Problem. Management Science 42 (1996) 797-813.

24. Pezzella, F. and Merelli, E. :A Tabu Search Method Guided by Shifting Bottleneck for the Job Shop Scheduling Problem. European Journal of Operational Research 120:297-310, 2000.

25. Ribeiroa, C. C. and Souza, M., C.: Variable neighborhood search for the degree-constrained minimum spanning tree problem, Discrete Applied Mathematics 118 (2002) 43–54

26. Satake, T., Morikawa, K., Takahashi, K., and Nakamura, N. :Neural Network Approach for Minimizing the Makespan of the General Job- Shop. International Journal of Production Economics 33 (1994) 67-74.

27. Satake, T., Morikawa, K., Takahashi, K., and Nakamura, N. :Simulated Annealing Approach for Minimizing the Makespan of the General Job- Shop. International Journal of Production Economics 60 (1999) 515-522.

28. Urosevic,D., Brimberg,J. and Mladenovic, N., Variable neighborhood decomposition search for the edge weighted k-cardinality tree problem, Computers & Operations Research 31 (2004) 1205-1213

# An Efficient Hybrid Search Algorithm for Various Optimization Problems

Mario Vanhoucke[1,2]

[1] Ghent University, Hoveniersberg 24, 9000 Ghent, Belgium
[2] Vlerick Leuven Gent Management School, Reep 1, 9000 Ghent, Belgium
Mario.vanhoucke@ugent.be
Mario.vanhoucke@vlerick.be

**Abstract.** This paper describes a detailed study of a recursive search algorithm for different optimization problems. Although the algorithm has been originally developed for a project scheduling problem with financial objectives, we show that it can be extended to many other application areas and therefore, can serve as a sub-procedure for various optimization problems. The contribution of the paper is threefold. First, we present a hybrid recursive search procedure for the project scheduling problem with net present value maximization and compare it with state-of-the-art procedures by means of computational tests. Second, we show how the procedure can be adapted to two other application areas: project scheduling with work continuity minimization and the open pit mining problem. Last, we highlight some future research areas where this hybrid procedure might bring a promising contribution.

## 1 Introduction

In 1964, [1] introduced the idea of maximizing the net present value ($npv$) of the cash flows of a project as a financially highly relevant criterion. Since then, a large amount of algorithms have been presented in the literature under different assumptions with respect to network representation (activity-on-the-node versus activity-on-the-arc) and cash flows patterns (positive and/or negative, event-oriented or activity-based and time-dependent vs. time-independent). In their paper [17] examined the rationale behind this idea and gave an overview of the existing algorithms.

The basic problem under study involves the scheduling of project activities in order to maximize the net present value ($npv$) of the project in the absence of resource constraints. Assume that the project is represented by an activity-on-the-node (AoN) network $G = (N, A)$ where the set of nodes, $N$, represents activities and the set of arcs, $A$, represents the precedence constraints. The activities are numbered from the dummy start activity 1 to the dummy end activity $n$. The duration of an activity is denoted by $d_i$ ($1 \leq i \leq n$) and the performance of each activity involves a series of cash flow payments and receipts throughout the activity duration. Assume that $cf_{it}$ ($1 < i < n$) denotes the known deterministic cash flow of activity $i$ in period $t$ of its execution. A terminal value of each activity upon completion can be calculated by compounding the associated cash flow to the end of the activity as follows:

J. Gottlieb and G.R. Raidl (Eds.): EvoCOP 2006, LNCS 3906, pp. 272–283, 2006.
© Springer-Verlag Berlin Heidelberg 2006

$$c_i = \sum_{t=1}^{d_i} cf_{it} e^{\alpha(d_i - t)} \tag{1}$$

where $\alpha$ represents the discount rate and $c_i$ the terminal value of cash flows of activity $i$ at its completion. The nonnegative integer variables $s_i$ and $f_i$ ($1 \leq i \leq n$) denote the starting time and completion time, respectively, of activity $i$. The discounted value of activity $i$ at the beginning of the project is

$$c_i e^{-\alpha(s_i + d_i)} = c_i e^{-\alpha f_i} \tag{2}$$

A formulation of the problem can be given as follows:

$$\text{Maximize} \sum_{i=1}^{n} c_i e^{-\alpha(s_i + d_i)} \tag{3}$$

Subject to

$$s_i + l_{ij} \leq s_j, \forall (i, j) \in A \tag{4}$$

$$s_n \leq \delta_n \tag{5}$$

$$s_1 = 0 \tag{6}$$

The objective in Eq. (3) maximizes the net present value of the project. The constraint set given in Eq. (4) maintains the precedence relations with time-lag $l_{ij}$ among the activities. Eq. (5) limits the project duration to a negotiated project deadline $\delta_n$ and Eq. (6) forces the dummy start activity to start at time zero. The time-lag $l_{ij}$ and the different types of generalized precedence relations can be represented in a standardized form by reducing them to minimal start-start precedence relations as shown by [2].

The early research has focused on the case where activity cash flows are independent on the completion times of the corresponding activities. Depending on the type of precedence relations, we distinguish between problems where only minimal time-lags between the activities are considered and problems with generalized precedence constraints, i.e. where both minimal and maximal time-lags are taken into account. Following the classification scheme of [18], the first type of project scheduling problem can be categorized as $min, \delta_n, c_j | npv$ and is further denoted as the max-$npv$ problem. Exact algorithms for the max-$npv$ problem have been presented by [6], [10], [15], [16], [25] and [32]. The introduction of generalized precedence relations (*gpr*) transforms the problem into the max-$npv$-$gpr$ problem (problem $gpr, \delta_n, c_j | npv$) and is the topic of research done by [8], [20], [23], [26] and [27].

Recent research has focused on the case where activity cash flows are dependent on the completion times of the corresponding activities (e.g. the papers by [3], [4], [5], [12], [13], [21], [28], [29], [32], [33] and [34]). In this paper, however, we restrict our attention to the maximization of the net present value of an unconstrained project in which we assume that the cash flows are time-independent. We show that an efficient hybrid solution procedure for the cash-flow problem can be extended to many other applications areas, and therefore, might be a promising algorithm for future research purposes.

The organization of the paper is as follows. In section 2 we briefly review the literature for the max-*npv-gpr* problem. Furthermore, we present a hybrid search algorithm for the max-*npv-gpr* problem based on ideas from different sources in literature. In section 3 we report the results of a computational experiment set up to validate the hybrid procedure described in this paper. In section 4 we describe two applications where the efficient hybrid algorithm can be used as a sub-problem. We conclude in section 5 with some overall conclusions.

# 2  The Max-npv-gpr Procedures

In section 2.1 we give an overview of the procedures in literature for the max-*npv-gpr* problem. In section 2.2 we discuss a hybrid approach in order to improve efficiency. Section 2.3 presents a project example.

## 2.1  The Max-*npv-gpr* Procedures in Literature

[20] has presented an exact solution procedure for the max-*npv-gpr* problem based on the approach by [15]. The introduction of a new pivot rule extends this last procedure to generalized precedence relations. [6] propose an activity-oriented recursive search algorithm for the max-*npv* problem as described in Eqs. (3)-(6) with $l_{ij} = d_i$ for all $(i, j)$ ∈ A. [32] have updated this recursive search algorithm and incorporated it in a branch-and-bound algorithm for the resource-constrained max-*npv* problem. [8] have extended this procedure using the so-called distance matrix $D$ in order to cope with generalized precedence relations. [9] have embedded the procedure of [8] into a branch-and-bound algorithm for the resource-constrained project scheduling problem with discounted cash flows and generalized precedence constraints. [23] have adapted the procedure by [15] and have investigated different pivot rules. Finally, [26] and [27] have presented a steepest ascent algorithm and compared different solution procedures on two randomly generated test sets. Note that [22] have shown that the project scheduling problem with general cost functions can be transformed into a minimum-cut problem, and hence, can be solved by any minimum-cut solution procedure. In the next sub-sections, we present a hybridized method based on the recursive search principles taken from [6], [8], [26], [27] and [32].

## 2.2  Hybridization of Max-*npv-gpr* Procedures

In this section, we extend the recursive search procedure of [32] to cope with generalized precedence relations. Furthermore, we hybridize this adapted recursive search method with ideas from the steepest ascent approach of [26] and [27] and extend it with a forward/backward calculation principle.

### 2.2.1 Adapting the Recursive Search Procedure for the Max-*npv-gpr* Problem

Although the original recursive search procedure of [32] has been developed for minimal precedence relations with a time-lag of zero, it can be easily modified in order to cope with generalized precedence relations. The procedure builds an initial tree for the project, and then iteratively searches the tree for sets of activities to shift. These activities are then shifted forwards in time within a calculated displacement interval, aiming at an improvement of the total net present value. In the adapted procedure, the allowable displacement interval $v_{kl}$, which simply calculates the minimal distance over which an activity set $SA$ can be shifted forwards in time, need to take the generalized precedence relations into account, as follows:

$$v_{k*l*} = \min_{\substack{(k,l) \in A \\ k \in SA \\ l \notin SA}} \{s_l - s_k - l_{kl}\} \tag{7}$$

This interval simply calculates all time windows between all activities $k \in SA$ and $l \notin SA$, taking into account the time-lags $l_{kl}$, and selects the shortest time window for arc $(k*, l*)$. Note that [8] also has transformed the procedure of [32] to cope with generalized precedence relations, but uses the distance matrix $D$ as a basis for the tree calculation. Our adapted approach uses the original precedence relations as a basis for the tree calculations, and hence, avoids the time-consuming computation of this distance matrix.

In section 3, we refer to the modified procedure as "*RS-gpr*" and use the abbreviations "*DM-gpr*" to refer to [8].

### 2.2.2 A Hybrid Search Procedure

The steepest ascent approach [27] follows a similar approach than the recursive search method [32] but there are a number of differences. In the recursive search method, each time a set of activities $SA$ has been found which can be shifted forwards, a number of steps are successively performed. First, the allowable displacement interval $v_{kl}$ is calculated and the activities are shifted. Second, the search tree $ST$ is updated and finally, the recursion step is repeated to search for another set of activities to shift. The steepest ascent approach calculates all sets of activities in one step and shifts them one by one in a second step.

The hybrid approach which is the subject of this section divides the logic of the recursive search method into two parts and borrows ideas from both the recursive search algorithm of [32] and the steepest ascent procedure of [27]. The first part (denoted as the "recursion" sub-procedure) only searches for sets of activities $SA$ with a negative *npv* and which are, consequently, candidates to be shifted. The second step (referred to as the "shift_activities" sub-procedure) calculates an allowable displacement interval for each set $SA \in SS$ similar to Eq. (7).

The pseudocode of a hybrid approach is described hereafter. We use $SA$ to denote the set of activities to be shifted, $SS$ to denote the various sets $SA$ found in the recursion step, $Z$ to denote all the activities in $SS$ that are candidates to be shifted, $CA$ to denote the set of already considered activities during the search process and $DC$ to denote the discounted cash flows. The symbol "$\rightarrow$" refers to the output of the recursion sub-procedure.

**Procedure** Hybrid_recursive_search
  $CA$ = $SS$ = $\emptyset$
  **Do** Recursion(1)
  **If** $SS \neq \emptyset$ **then**
    Shift_activities($SS$) and **repeat** Recursion(1)
  **Else**
    Report the optimal solution ($DC'$)
**Return**

The pseudo-code of the two sub-procedures can be described as follows:

**Sub-procedure** Recursion($newnode$)
  $SA$ = {$newnode$}, $DC$ = $DC_{newnode}$ and $CA$ = $CA \cup$ {$newnode$}
  **Do** $\forall i \mid i \notin CA$ and $i$ succeeds newnode in the tree $ST$:
    Recursion($i$) $\rightarrow$ $SA'$, $DC'$
    **If** $DC' \geq 0$ **then** set $SA$ = $SA \cup SA'$ and $DC$ = $DC + DC'$
    **Else** $ST$ = $ST\backslash(newnode,i)$ and $SS$ = $SS \cup SA'$
  **Do** $\forall i \mid i \notin CA$ and $i$ precedes newnode in the tree $ST$:
    Recursion($i$) $\rightarrow$ $SA'$, $DC'$
    Set $SA$ = $SA \cup SA'$ and $DC$ = $DC + DC'$
**Return**

**Sub-procedure** Shift_activities($SS$)
Let $Z$ = {$i \in SA \mid SA \in SS$} be the set of activities which can possibly be delayed;
**While** $Z \neq \emptyset$ **do**
  Compute $v_{k*l*} = \min\limits_{\substack{(k,l) \in A \\ k \in Z \\ l \notin Z}} \{s_l - s_k - l_{kl}\}$
  **Do** $\forall i \in SA \mid k* \in SA$ : Set $s_i$ = $s_i + v_{k*l*}$ and $Z$ = $Z\backslash\{i\}$;
  Set $ST$ = $ST \cup (k*,l*)$;
**Return**

In section 3, we refer to this procedure as "*RS/SA-gpr*". In section 2.3, we illustrate the pseudocode on a project network.

### 2.2.3 Forward and Backward

In their paper, [32] have tested the impact of the percentage of negative cash flows in the project on the required CPU-time and have revealed that a higher percentage (except for 100% negative cash flows) of negative cash flows results in a more difficult problem. This is quite logic since a higher number of negative cash flows implies a larger number of shifts and consequently a more extensive search in the search tree. Inspired by these results, we further modified the hybrid search procedure of previous section for problem instances with more than 50% of the activities with a negative cash flow. In this case, we schedule the activities as late as possible (within the project deadline) and try to find sets of activities to shift backwards (toward time zero) in order to increase the net present value. We simply use the dummy end node as a basis of our recursive search instead of the dummy start node. Therefore, the net present value calculations of Eq. (2) need to be modified, in order to calculate the discounted value of activity $i$ at the end of the project, as follows:

$$c_i e^{\alpha(\delta_n - (s_i + d_i))} = c_i e^{\alpha(\delta_n - f_i)} \tag{8}$$

This will lead to a considerable reduction of the number of shifts in our search, as revealed in the computational results section. In section 3, we refer to this procedure as "*RS/SA(FB)-gpr*"

## 2.3 An Illustrative Example

The example network of Fig. 1 (a) contains 4 non-dummy activities and minimal and maximal time-lags between the activities. The activity duration is displayed above the node while the cash flow is given below the node. The discount rate is 1.6% and the project deadline $\delta_n$ is 10. The numbers associated with the arcs denote the generalized precedence relations. A minimal time-lag is denoted by an arc $(i, j)$ with a positive number while a maximal time-lag is represented by an arc $(j, i)$ with a negative number (e.g. arc (4,3)). For the sake of simplicity, all arcs are of the start-start type.

The initial search tree, which spans all activities at its earliest start schedule $s_1 = 0$, $s_2 = 1$, $s_3 = 0$, $s_4 = 1$, $s_5 = 6$ and $s_6 = 7$, is given in Fig. 1 (b).

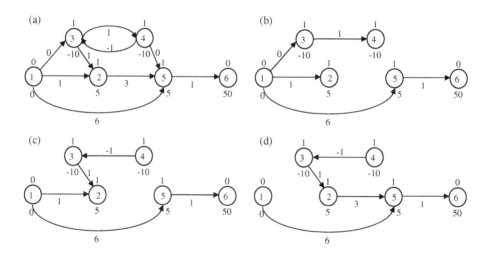

**Fig. 1.** An example project with generalized precedence relations and its corresponding trees

After the first run of the recursion sub-procedure, the sets $SS = \{(3), (4)\}$ and $Z = \{3, 4\}$ have been found. During this recursion search process, the arcs (1, 3) and (3, 4) have been removed from the tree. The activity sets $\{3\}$ and $\{4\}$ will be shifted in the shift_activities sub-procedure as follows: The sub-procedure calculates the minimal allowable displacement interval. The allowable displacement interval $v_{kl} = \min\{s_5 - s_4 - l_{45}; s_2 - s_3 - l_{32}\} = \{6 - 1 - 0; 1 - 0 - 1\}$ is minimal for $k^* = 3$ and $l^* = 2$ and hence, activity 3 is delayed in the search tree by adding arc (3, 2) and is removed from $Z$. Since $Z = \{4\}$ is not empty, a new allowable displacement interval $v_{kl} = $

$\min\{s_5 - s_4 - l_{45}; s_3 - s_4 - l_{43}\} = \{6 - 1 - 0; 0 - 1 + 1\}$ is minimal for $k^* = 4$ and $l^* = 3$, and hence, the sub-procedure adds arc (4, 3) in the search tree. Since both displacement intervals equal 0, the starting times remain unchanged. The search tree is displayed in Fig. 1 (c).

The second run of the recursion sub-procedure reports one set $SA = \{2, 3, 4\}$ and hence, $Z = SS = \{2, 3, 4\}$. During this recursion search process, the arc (1, 2) has been removed from the tree. The allowable displacement interval $v_{kl} = \min\{s_5 - s_4 - l_{45}; s_5 - s_2 - l_{25}\} = \{6 - 1 - 0; 6 - 1 - 3\}$ is minimal for $k^* = 2$ and $l^* = 5$, and hence, the sub-procedure adds arc (2, 5) in the search tree. Since $v_{kl} > 0$, the starting times of each activity of $SA$ is increased by 3 units to $s_2 = 3$, $s_3 = 2$ and $s_4 = 3$. The search tree is displayed in Fig. 1 (d).

The last run of the recursion sub-procedure does not report a non-empty set, and therefore, the optimal solution can be derived from the tree as displayed in Fig. 1 (d), i.e. $s_1 = 0$, $s_2 = 3$, $s_3 = 2$, $s_4 = 3$, $s_5 = 6$ and $s_6 = 7$. Note that the lump sum payment of 50 at the end of the project forces the algorithm to minimize the total project duration, and hence, the deadline of 10 is irrelevant.

## 3   Computational Experiences

In order to test the efficiency of all procedures, we have coded them in Visual C++ version 6.0 under Windows XP on a Toshiba personal computer, with a Pentium IV 2 GHz processor. We have used the 1,440 instances from [7] to validate the procedures for the max-*npv-gpr* problem. We have extended both datasets with cash flows generated from the interval [-500; 500], a discount rate of 1.6% and a project deadline which equals the critical path length increased by 100.

In the sequel of this section we use "% neg" to denote the percentage of negative cash flows in the problem instances and "*RS-gpr*" to refer to the extended recursive search method of [32] to cope with generalized precedence relations (see section 2.2.1). "*RS/SA-gpr*" and "*RS/SA(FB)-gpr*" refer to the adaptations as proposed in section 2.2.2 and section 2.2.3 respectively. "*SA-gpr*" refers to the steepest ascent procedure of [27]. The abbreviation "*DM-gpr*" is used to refer to the procedure of [8] that relies on the distance matrix.

Table 1 reports the results for the 15,840 (1,440 instances and 11 settings for the percentage of negative cash flows) instances for the max-*npv-gpr* problem. The columns labeled "CPU" display the average CPU-time in milliseconds needed to solve the problem instances. The columns labeled "*iter*" displays the average number of iterations for each procedure. For all recursive procedures, the number of iterations is defined as the number of times the recursion is called. For the steepest ascent approach, the number of iterations equals the number of times a steepest ascent direction has to be looked for. The results can be summarized as follows.

*RS-gpr* and *DM-gpr* are both adapted versions of the recursive search procedure of [32], but the former outperforms the latter. This is due to the fact that the latter needs the very time-consuming calculation of the distance matrix $D$ in order to solve the problem (see section 2.2.1).

The hybridization of the recursive search procedure and the steepest ascent approach, as discussed in section 2.2.2, is beneficial. The number of iterations for the *RS/SA-gpr, RS/SA(FB)-gpr* and *SA-gpr* procedures is significantly lower than the *RS-gpr* procedure, since these algorithms delay the activities of several disconnected subtrees at the same time.

The best performing procedure is the *RS/SA(FB)-gpr* procedure of section 2.2.3. For all other procedures, the CPU-times are positively correlated with the percentage of negative cash flows in the problem instances. In the forward/backward approach, problem instances can be resolved by using a backward recursive search procedure whenever the percentage of negative cash flows exceeds 50%.

**Table 1.** Computational performance for the different search procedures

| % neg | RS-gpr CPU | RS-gpr iter | RS/SA-gpr CPU | RS/SA-gpr iter | RS/SA(FB)-gpr CPU | RS/SA(FB)-gpr iter | SA-gpr CPU | SA-gpr iter | DM-gpr CPU | DM-gpr iter |
|---|---|---|---|---|---|---|---|---|---|---|
| 0% | 0.157 | 1.00 | 0.156 | 1.00 | 0.156 | 1.00 | 0.297 | 1.00 | 19.525 | 1.00 |
| 10% | 0.197 | 3.14 | 0.194 | 2.22 | 0.194 | 2.22 | 0.476 | 2.22 | 21.570 | 29.96 |
| 20% | 0.228 | 4.93 | 0.209 | 2.67 | 0.209 | 2.67 | 0.543 | 2.67 | 22.816 | 46.23 |
| 30% | 0.305 | 9.62 | 0.229 | 3.14 | 0.229 | 3.14 | 0.590 | 3.14 | 25.506 | 80.40 |
| 40% | 0.439 | 19.07 | 0.248 | 3.45 | 0.248 | 3.45 | 0.593 | 3.45 | 27.930 | 116.92 |
| 50% | 0.612 | 36.98 | 0.268 | 3.54 | 0.268 | 3.54 | 0.603 | 3.54 | 29.277 | 152.90 |
| 60% | 0.762 | 55.48 | 0.285 | 3.46 | 0.263 | 3.19 | 0.606 | 3.46 | 31.871 | 169.95 |
| 70% | 0.920 | 79.44 | 0.300 | 3.29 | 0.266 | 2.91 | 0.620 | 3.29 | 31.685 | 173.46 |
| 80% | 1.066 | 102.49 | 0.319 | 3.05 | 0.290 | 2.78 | 0.616 | 3.05 | 32.172 | 150.67 |
| 90% | 1.170 | 121.55 | 0.335 | 2.68 | 0.344 | 2.75 | 0.586 | 2.68 | 28.498 | 118.15 |
| 100% | 1.283 | 141.48 | 0.337 | 2.00 | 0.169 | 1.00 | 0.519 | 2.00 | 23.905 | 69.10 |
| Avg | 0.649 | 52.29 | 0.262 | 2.77 | 0.240 | 2.61 | 0.550 | 2.77 | 26.796 | 100.79 |
| Max | | 362 | | 18 | | 18 | | 18 | | 754 |

# 4   Other Application Areas

The excellent performance (the CPU-times are only fractions of milliseconds) of the hybrid recursive search procedure is extremely important, since the procedure will not be restricted to solve problem instances as proposed in Eqs (3)-(6). Instead, the procedure will often be embedded in larger, more complex, algorithms where the hybrid recursive search procedure acts as a sub-problem solver. A straightforward example is the incorporation of the original recursive search procedure for the max-*npv* problem into a branch-and-bound procedure to solve the resource-constrained project scheduling problem with discounted cash flows [32]. In this problem, the required computational effort by the recursive search procedure has to be extremely small to allow the efficient computation of upper bounds for the nodes in the branch-and-bound tree which may run in the thousands (even millions). In this section, we describe two application areas without cash-flow optimization, where the efficient use of the hybrid recursive search procedure is extremely important and acts as a sub-problem solver.

## 4.1   Work Continuity Constraints

Construction projects are often characterized by repeating activities that have to be performed from unit to unit. Highway projects, pipeline constructions and high-rise buildings, for example, commonly require resources to perform the work on similar activities that shift in stages. Indeed, construction crews perform the work in a sequence and move from one unit of the project to the next. This is mainly the result of the subdivision of a general activity (e.g. carpentry) into specific activities associated with particular units (e.g. carpentry at each floor of a high-rise building). Since the resources of these repeating activities have to be performed from unit to unit, it is crucial in scheduling these project types to ensure the uninterrupted usage of resources of similar activities between different units, as to enable timely movement of resources (crews) from one unit to the other, avoiding idle time. This feature is known as *work continuity constraints* [11].

The hybrid recursive search procedure is able to solve the project scheduling problem with work continuity constraints, to assure that idle time of resources is minimized. The basic philosophy of the net present value is to schedule activities with a negative cash flow as late as possible and activities with a positive cash flow as soon as possible. A similar philosophy can be applied to the work continuity constraint problem, since the idle time of resources can be minimized by minimizing the time window between the first and the last activity of the activity set that relies on the resource. Therefore, by assigning a fictive negative cash flow to the start activity of the activity set and a fictive positive cash flow to the last activity of the activity set, we simulate attraction between these two activities, and hence, the total time window (and consequently, the resource idle time) will be minimized.

[30] have illustrated the importance of work continuity constraints on the huge and complex "Westerschelde Tunnel" project. The incorporation of work continuity constraints reduces the total resource idle time in the schedule, which leads to a cost saving of more than 1 million euro. A detailed description of this project can be found in [30] or on http://www.westerscheldetunnel.nl.

## 4.2  Open Pit Mining

The open-pit mining problem aims at the determination of the contours of a mine, based on economic data and engineering feasibility requirement in order to yield maximum possible net income [19]. More precisely, blocks of earth are removed from the surface to retrieve the ore contained in them. Therefore, the entire volume is divided into blocks with a weight representing its estimated value. This value represents the value of the ore of the block, minus all the costs of excavating the block. The objective is to determine the set of blocks to remove, in order to maximize net benefits, and subject to several constraints. These constraints represent the precedence relations among blocks, preventing to remove blocks for which other blocks on top of them are not yet removed, and technological slope requirements constraints.

The open-pit mining problem can be represented by a graph $G = (N, A)$, where each node in $N$ represents a block with a weight equal to its net value. The arc set $A$ between the nodes represents the precedence relations between the blocks. The decision which blocks to extract is equivalent to the search of a maximum weight set of nodes such that all nodes in the set have their successor nodes in the set. This problem is known in the graph theory literature as the *maximum closure problem*, and such a set is called a maximum closure of graph $G$. [24] has shown that the maximal closure problem can be converted to a maximal flow problem. Consequently, any maximal flow algorithm can solve the problem with, as a by-product, a minimal cut that serves as the maximal closure.

The maximum closure problem can be solved by the hybrid recursive search procedure. Indeed, each node of the graph has a weight that can be either positive or negative. A closure of $G$ is a subset of nodes such that every successor of each node of the subset also belongs to that subset. Hence, a maximal closure of $G$ is a subset of nodes such that the sum of the node weights is at least as large as any other closure of $G$. The node weights can serve as cash flows for the hybrid search procedure, with a fictive project deadline equal to infinity. In doing so, the procedure splits the project activities into two parts: activities near the beginning of the projects and activities close to the deadline. This split corresponds to the closure of the project network and hence, determines to optimal blocks to extract for the open-pit mining problem.

## 5  Conclusions

In this paper we have compared five solution procedures for the unconstrained project scheduling problems with discounted cash flows with generalized precedence constraints (max-*npv-gpr* problem). We show that the hybrid adaptation of different procedure lead to the most promising results.

More precisely, we have compared three hybrid adaptations of a recursive search procedure of [32] with a steepest ascent procedure of [27] and a search procedure of [8]. Computational experience has revealed that the hybrid recursive search method outperforms the steepest ascent procedure. Additional tests have shown that both the hybrid recursive search procedure and the steepest ascent procedure outperform the procedure by [8]. The main reason is that the distance matrix is a very time-consuming activity in the last mentioned procedure.

Our future research intensions are two-fold. First, we want to improve the efficiency of the hybrid recursive search procedure to be able to cope with very large problem instances. Second, we want to use the procedure as a sub-problem solver in combinatorial optimization algorithms (such as branch-and-bound) to solve large, real-life problem instances. The application areas of the hybrid recursive search procedure are numerous. [14] mentioned possible applications for the transitive closure problem, such as portfolio selection, task scheduling in job shop environments, selection of freight handling terminals and many more. [31] have discussed a modified version for the resource-constrained project scheduling problem with JIT characteristics, but they do not incorporate generalized precedence relations. In the near future, we will embed the hybrid recursive search procedure in a branch-and-bound procedure to schedule R&D projects where quality restrictions are taken into account. The different procedures are available upon request, such that researchers will be motivated to use this efficient code as a part of their algorithms for complex, large-sized and realistic problems.

# References

[1] Battersby, A.: Network analysis for planning and scheduling. MacMillan (1964)

[2] Bartusch, M, Möhring, R.H., Radermacher, F.J.: Scheduling project networks with recource constraints and time windows. Annals of Operations Research. 16, 201-240 (1988)

[3] Dayanand, N., Padman, R..: The payment scheduling problem in project networks. Working Paper 9331, The Heinz School, CMU, Pittsburgh, PA 15213, U.S.A (1993)

[4] Dayanand, N., Padman, R.: Payments in projects: a constructor's model. Working Paper 9371, The Heinz School, CMU, Pittsburgh, PA 15213, U.S.A (1993)

[5] Dayanand, N., Padman, R.: On modeling payments in project networks. Journal of the Operational Research Society. 48, 906-918 (1997)

[6] Demeulemeester, E., Herroelen, W., Van Dommelen, P.: An optimal recursive search procedure for the deterministic unconstrained max-npv project scheduling problem. Research Report 9603, Department of Applied Economics, Katholieke Universiteit Leuven, Belgium (1996)

[7] De Reyck, B.: Scheduling Projects with Generalized Precedence Relations - Exact and Heuristic Procedures. Ph.D. Dissertation, Department of Applied Economics, Katholieke Universiteit Leuven, Belgium (1998)

[8] De Reyck, B., Herroelen, W.: An optimal procedure for the unconstrained max-npv project scheduling problem with generalized precedence relations. Research Report 9642, Department of Applied Economics, Katholieke Universiteit Leuven, Belgium (1996)

[9] De Reyck, B., Herroelen, W.: An optimal procedure for the resource-constrained project scheduling problem with discounted cash flows and generalized precedence relations. Computers and Operations Research. 25, 1-17 (1998)

[10] Elmaghraby, S.E., Herroelen, W.: The scheduling of activities to maximize the net present value of projects. European Journal of Operational Research. 49, 35-49 (1990)

[11] El-Rayes, K., Moselhi, O.: Resource-driven scheduling of repetitive activities. Construction Management and Economics. 16, 433-446 (1998)

[12] Etgar, R., Shtub, A., LeBlanc, L.J.: Scheduling projects to maximize net present value - The case of time-dependent, contingent cash flows. European Journal of Operational Research. 96, 90-96 (1996)

[13] Etgar, R., Shtub, A.: Scheduling project activities to maximize the net present value - The case of linear time dependent, contingent cash flows. International Journal of Production Research. 37, 329-339 (1999)

[14] Faaland, B., Kim, K., Schmitt, T.: A new algorithm for computing the maximal closure of a graph. Management Science. 36, 315-331 (1990)

[15] Grinold, R.C.: The payment scheduling problem. Naval Research Logistics Quarterly. 19, 123-136 (1972)

[16] Herroelen, W., Gallens, E.: Computational experience with an optimal procedure for the scheduling of activities to maximize the net present value of projects. European Journal of Operational Research. 65, 274-277 (1993)

[17] Herroelen, W., Demeulemeester, E., Van Dommelen, P.: Project network models with discounted cash flows: A guided tour through recent developments. European Journal of Operational Research. 100, 97-121 (1997)

[18] Herroelen, W., Demeulemeester, E., De Reyck, B.: A classification scheme for project scheduling problems. Weglarz J. (Ed.), Handbook on Recent Advances in Project Scheduling, Kluwer Academic Publishers. Chapter 1, 1-26 (1999)

[19] Hochbaum, D.S., Chen, A.: Performance analysis and best implementations of old and new algorithms for the open-pit mining problem. Operations Research. 48, 894-914 (2000)

[20] Kamburowski, J.: Maximizing the project net present value in activity networks under generalized precedence relations. Proceeding of 21st DSI Annual meeting, San Diego, 748-750 (1990)

[21] Kazaz, B., Sepil, C..: Project scheduling with discounted cash flows and progress payments. Journal of the Operational Research Society. 47, 1262-1272 (1996)

[22] Möhring, R.H., Schulz, A.S., Stork, F., Uetz, M.: On project scheduling with irregular starting time costs. Operations Research Letters. 28, 149-154 (2001)

[23] Neumann, K., Zimmermann, J.: Exact and heuristic procedures for net present value and resource levelling problems in project scheduling. European Journal of Operational Research. 127, 425-443 (2000)

[24] Picard, J.C.: Maximal closure of a graph and applications to combinatorial problems. Management Science. 22, 1268-1272 (1976)

[25] Russell, A.H.: Cash flows in networks. Management Science. 16, 357-373 (1970)

[26] Schwindt, C., Zimmermann, J.: Maximizing the net present value of projects subject to temporal constraints. WIOR-Report-536, Institut für Wirtschaftstheorie und Operations Research, University of Karlsruhe, Germany (1998)

[27] Schwindt, C., Zimmermann, J.: A steepest ascent approach to maximizing the net present value of projects. Mathematical Methods of Operations Research. 53, 435-450 (2001)

[28] Sepil, C., Ortaç, N.: Performance of the heuristic procedures for constrained projects with progress payments. Journal of the Operational Research Society. 48, 1123-1130 (1997)

[29] Shtub, A., Etgar, R..: A branch-and-bound algorithm for scheduling projects to maximize net present value: the case of time dependent, contingent cash flows. International Journal of Production Research. 35, 3367-3378 (1997)

[30] Vanhoucke, M.: Work continuity constraints in project scheduling. Journal of Construction Engineering and Management. 132, 1-12 (2006).

[31] Vanhoucke, M., Demeulemeester, E., Herroelen, W.: An exact procedure for the resource-constrained weighted earliness-tardiness project scheduling problem. Annals of Operations Research. 102, 179-196 (2000)

[32] Vanhoucke, M., Demeulemeester, E., Herroelen, W.: On maximizing the net present value of a project under renewable resource constraints. Management Science. 47, 1113-1121 (2001)

[33] Vanhoucke, M., Demeulemeester, E., Herroelen, W.: Scheduling projects with linearly time-dependent cash flows to maximize the net present value. International Journal of Production Research. 39, 3159-3181 (2001)

[34] Vanhoucke, M., Demeulemeester, E., Herroelen, W.: Progress payments in project scheduling problems. European Journal of Operational Research. 148, 604-620 (2003)

# A Hybrid VNS/Tabu Search Algorithm for Apportioning the European Parliament

Gabriel Villa[1], Sebastián Lozano[2], Jesús Racero[1], and David Canca[1]

[1] University of Seville, Camino de los Descubrimientos,
s/n Isla de la Cartuja,
E-41092 Sevilla, Spain
{gvilla, jrm, dco}@esi.us.es
[2] University of Seville, E.T.S. Ingeniería Informática,
Avda. Reina Mercedes, s/n
E-41012 Sevilla, Spain
slozano@us.es

**Abstract.** In a Proportional Representation (PR) electoral system it is assumed that seats are apportioned to the different electoral districts/states according to the corresponding voters' distribution. In a previous paper we proposed a MILP (Mixed Integer Linear Programming) model to apportion the seats in the European Parliament (EP). Since the exact solution to the problem is not computationally efficient, we have designed a hybrid metaheuristic algorithm based on Variable Neighborhood Search (VNS) and Tabu Search (TS). The proposed approach takes into account the existing situation, guaranteeing a minimum number of seats, independently of the population size of each member. The model is illustrated with actual data and its results are compared with the present apportionment. The results show that the proposed approach can significantly improve the proportionality of the present apportionment.

## 1 Introduction

PR is inspired in the principle of "one person, one vote". The application of this principle to the apportionment problem implies that the number of seats allocated to each electoral district should be as close as possible to what would correspond to its share of the electoral population. The idea is simple but its implementation is not trivial since seats are indivisible and hence while the distribution of the population is fractional the distribution of seats is discrete. This means that a perfectly proportional apportionment is usually not feasible. Since there are different ways to measure malapportionment, i.e. the deviation from perfect proportionality, ([1]) finding a "good" apportionment is a complex problem.

There are two classes of apportionment methods: divisor methods and quota methods. Among the divisor methods we find the Greatest Divisors or Jefferson method, the Smallest Divisors or Adams method, the Major Fractions or Webster method, the Harmonic Means or Dean method and the Equal Proportions, Hill or Huntington

J. Gottlieb and G.R. Raidl (Eds.): EvoCOP 2006, LNCS 3906, pp. 284–292, 2006.
© Springer-Verlag Berlin Heidelberg 2006

method [2]. Quota methods (a.k.a. Greatest Remainders methods) include the Hare method, the Droop method and the Imperiali method [3]. There are many real examples of significant malapportionment [4] and although there is an interesting debate about the comparative advantages of each of these methods (e.g. [2], [3], [5], [6], [7]), it cannot be concluded that any of them is better than the rest.

In the real world, apart from the fact that some malapportionment is unavoidable, there are also practical and political considerations in addition to pure proportionality that may have to be taken into account, e.g. establishing a lower limit of representation so that all electoral districts have a guaranteed minimum number of seats independently of its size.

In the case of the EP, the apportionment is the result of a complex, political, multilateral negotiation process. Table 1 shows the population and present number of seats for each EP member country. It can be seen in any of the figures in the Appendix that the present EP apportionment is not very proportional to population with larger countries getting a smaller share than smaller ones.

**Table 1.** Population distribution and present seats apportionment

| Country | Abrev | Population (millions) | Seats |
|---------|-------|----------------------|-------|
| Austria | at | 8,059 | 18 |
| Belgium | be | 10,348 | 24 |
| Cyprus | cy | 0,797 | 6 |
| Czech Rep | cz | 10,202 | 24 |
| Denmark | dk | 5,387 | 14 |
| Estonia | ee | 1,35 | 6 |
| Finland | fi | 5,21 | 14 |
| France | fr | 59,725 | 78 |
| Germany | de | 82,551 | 99 |
| Greece | el | 10,68 | 24 |
| Hungary | hu | 10,12 | 24 |
| Ireland | ie | 3,947 | 13 |
| Italy | it | 57,646 | 78 |
| Latvia | lv | 2,321 | 9 |
| Lithuania | lt | 3,454 | 13 |
| Luxembourg | lu | 0,448 | 6 |
| Malta | mt | 0,393 | 5 |
| Netherlands | nl | 16,215 | 27 |
| Poland | pl | 38,195 | 54 |
| Portugal | pt | 10,191 | 24 |
| Slovakia | sk | 5,381 | 14 |
| Slovenia | sl | 1,964 | 7 |
| Spain | es | 41,101 | 54 |
| Sweden | se | 8,956 | 19 |
| UK | uk | 59,28 | 78 |
| TOTAL | --- | 453,921 | 732 |

In the next section, we formulate a Mixed-Integer Linear Programming model for the apportionment problem. In section 3 we present the hybrid metaheuristic algorithm that we propose as efficient solution method. Computer results are reported in Section 4. Finally, in Section 5 some conclusions are drawn.

## 2   Quasi-Pure Proportional Apportionment Model

In ([8]) the authors have proposed a Quasi-Pure Proportional Apportionment (QPPA) model. Let:

| | |
|---|---|
| N | number of EP member countries |
| $p_r$ | population of EP member country r |
| $P=\Sigma_r p_r$ | sum of EP member countries population |
| $s_r$ | number of seats presently assigned to EP member country r |
| $S=\Sigma_r s_r$ | total number of seats in EP |
| smin | guaranteed minimum number of seats per EP member country |
| $\alpha^+_r$ | positive deviation (w.r.t. proportional ratio) of ratio of seats to population for country r |
| $\alpha^-_r$ | negative deviation (w.r.t. proportional ratio) of ratio of seats to population for country r |

The QPPA model is:

$$\text{Min} \quad \sum_{r=1}^{N}(\alpha_r^+ + \alpha_r^-)$$

s.t

$$s_r \geq s\min \qquad \forall r$$

$$\sum_{r=1}^{N} s_r = S \qquad\qquad\qquad (1)$$

$$\alpha_r^+ - \alpha_r^- = \frac{\dfrac{s_r}{S}}{\dfrac{p_r}{P}} - 1 \qquad \forall r$$

$$\alpha_r^+, \alpha_r^- \geq 0 \quad s_r \text{ integer}$$

This is a Mixed-Integer Linear Programming (MILP) with $2 \cdot N$ continuous variables and N integer variables. The model finds a feasible (i.e. taking into account the guaranteed minimum number of seats) allocation of the total number of seats minimizing the sum of deviations (either from above or from below) from proportional representation. For an ideal PR apportionment (generally not feasible) these deviations would be zero. Note that the third set of constraints are just the definitions of the variables that measure the deviation (positive or negative) of the ratio of seats to population for each country; these deviation variables are used in the objective function.

In ([8]) this combinatorial optimization problem was solved six times with smin varying from 0 to 5. Version 7.0 of the state-of-the-art CPLEX optimization software was used (http://www.cplex.com), running on an Intel Pentium IV 2.8 GHz microprocessor. As it often happens with MILP problems, finding the exact optimal solution can take an unpredictable (potentially large) amount of computing time. An upper limit of 5 hours of computing time was imposed so that CPLEX stopped after that time recording the best solution found. That is what happened in all six cases. Table 2 shows the results obtained as well as, again, the present apportionment.

**Table 2.** Solution provided by CPLEX and present seats apportionment

| Country | Present Appor. | smin | | | | | |
|---|---|---|---|---|---|---|---|
| | | 0 | 1 | 2 | 3 | 4 | 5 |
| Austria | 18 | 13 | 13 | 13 | 13 | 13 | 13 |
| Belgium | 24 | 17 | 17 | 17 | 17 | 17 | 17 |
| Cyprus | 6 | 1 | 1 | 2 | 3 | 4 | 5 |
| Czech Rep | 24 | 17 | 17 | 17 | 17 | 17 | 17 |
| Denmark | 14 | 9 | 9 | 9 | 9 | 9 | 9 |
| Estonia | 6 | 2 | 2 | 2 | 3 | 4 | 5 |
| Finland | 14 | 8 | 8 | 8 | 8 | 8 | 8 |
| France | 78 | 96 | 96 | 96 | 96 | 96 | 96 |
| Germany | 99 | 133 | 133 | 130 | 126 | 121 | 115 |
| Greece | 24 | 17 | 17 | 17 | 17 | 17 | 17 |
| Hungary | 24 | 16 | 16 | 16 | 16 | 16 | 16 |
| Ireland | 13 | 6 | 6 | 6 | 6 | 6 | 6 |
| Italy | 78 | 93 | 93 | 93 | 93 | 93 | 93 |
| Latvia | 9 | 4 | 4 | 4 | 4 | 4 | 5 |
| Lithuania | 13 | 6 | 6 | 6 | 6 | 6 | 6 |
| Luxembourg | 6 | 1 | 1 | 2 | 3 | 4 | 5 |
| Malta | 5 | 1 | 1 | 2 | 3 | 4 | 5 |
| Netherlands | 27 | 26 | 26 | 26 | 26 | 26 | 26 |
| Poland | 54 | 62 | 62 | 62 | 62 | 62 | 62 |
| Portugal | 24 | 16 | 16 | 16 | 16 | 16 | 16 |
| Slovakia | 14 | 9 | 9 | 9 | 9 | 9 | 9 |
| Slovenia | 7 | 3 | 3 | 3 | 3 | 4 | 5 |
| Spain | 54 | 66 | 66 | 66 | 66 | 66 | 66 |
| Sweden | 19 | 14 | 14 | 14 | 14 | 14 | 14 |
| UK | 78 | 96 | 96 | 96 | 96 | 96 | 96 |
| Objective Function Value | 30.5559 | 1.8096 | 1.8096 | 5.1283 | 9.1952 | 13.6424 | 18.4699 |

It can be seen that the present apportionment has a large value of the objective function (not unexpected since it is far from proportionality) and that the objective function decreases sharply (meaning a significant proportionality improvement) as the guaranteed minimum number of seats smin decreases. Those countries that gain more from the apportionment changes seem to be those whose population is more under-represented (e.g. Germany) while the ones that may lose seats would be some smaller, over-represented countries (e.g. Luxemburg).

## 3  The VNTS Proposed

Since optimally solving the above problem does not seem efficient, a search strategy must be designed to obtain better results. We have opted for a metaheuristic algorithm based on a hybrid of VNS ([9] and [10]) and TS ([11]). The initial solution is the present apportionment.

$$[s_1^o, s_2^o, ..., s_r^o, ..., s_N^o] \tag{2}$$

We have defined a set of pre-selected neighbourhood structures $N_k(x)$ that consist in adding and subtracting a total of k seats to some countries. The amount of seats added and subtracted has to be equal since the total number of seats (S) must be kept constant. For example, for neighborhood $N_2(x)$ a neighbor solution would be obtained taking one seat from each of two countries (which may coincide) and giving them to another two countries (which also may coincide)

$$[s_1 + 0, ..., s_i - 2, ..., s_j + 1, ..., s_k + 1, ..., s_N + 0] \tag{3}$$

Obviously, the proposed approach avoids unfeasible solutions such as those that would assign a negative number of seats or a number of seats less than smin to any country. The pseudocode of the proposed VNTS algorithm is:

Pseudo-code of proposed VNTS algorithm

```
Initialization. Select the set of neighborhood struc-
tures Nk, k=1,…, kmax, that will be used in the search;
find an initial solution x; choose a stopping condi-
tion;

Repeat the following sequence for a fixed number of it-
erations (ITER):
   1) Set k←1;
   2) Until k=kmax repeat the following steps:
        a) Exploration of neighborhood. Find the best
        neighbor  x'  of  x  (x'∈Nk(x))  using  Tabu
        Search;
```

> b) Move or not. If the solution thus obtained
> x' is better than x, set x ←x', and k←1;
> otherwise, set k←k+1;

end.

As for the tabu search mentioned in step 2a), it explores the neighborhood $N_k(x)$. The value of each movement, i.e. the variation of the objective function can be efficiently computed based on the change in the representation ratio of only the countries affected by the move. As adaptive memory mechanism, a tabu list has been proposed, where the previous value of the number of seats is recorded for the affected countries, so that for a number of T iterations (i.e. a tabu tenure) changing the number of seats of any country to its recorded value is prohibited. The tabu search algorithm method ends after a certain number of iterations (parameter ITERTB) without improving the best solution found.

# 4  Computational Experiences

In table 3 we show the solutions provided by the hybrid for smin = 0, 1, 2, 3, 4 and 5. We have used the following values for the different parameters: kmax=2; ITERTB=50; T=5; and ITER=5000. We obtained the similar solutions using different values which may show the robustness of the proposed approach. In particular, for kmax=3 the time taken by the metaheuristic was higher than for kmax=2, providing the same results. We tried to solve the model for kmax=4 as well, however, we had to interrupt the resolution after a couple of hours, obtaining again the same results as in the case of kmax=2.

The three last rows contain respectively the value of the objective function, the percentage error with respect to the solution obtained by CPLEX, measured as:

$$100*\left( \frac{Objective\ Function\ Value|_{VNTS} - Objective\ Function\ Value_{CPLEX}}{Objective\ Function\ Value|_{CPLEX}} \right) \tag{4}$$

and the CPU time used by the proposed hybrid approach.

We can observe from table 3 that the metaheuristic in all cases obtains better results than CPLEX (see the minus sign of the relative errors). Also, it is remarkable that, compared to CPLEX, the time taken to solve every case is insignificant.

Figure 1 (in the Appendix) graphically shows the solutions found by the proposed approach for smin varying from 0 through 5. Note that except for three member countries (namely Malta, Luxemburg and Cyprus) the seat apportionment of the rest of countries is very close to its proportional quota (the dashed line segment in each graphic) and for these three countries their deviation from their quotas is significantly smaller than at present. The range of the ratio %seats/%population would be reduced from the present (0.74, 8.31) to (0.78, 1.58) for smin=1 for example.

Finally, we can observe from table 4 how the value of the deviation ranges of the ratio %seats/%population (i.e. the difference between the maximum and minimum values) increases with smin.

**Table 3.** Solution provided by VNTS approach

| Country | Present Appor. | smin | | | | | |
|---|---|---|---|---|---|---|---|
| | | 0 | 1 | 2 | 3 | 4 | 5 |
| Austria | 18 | 13 | 13 | 13 | 13 | 13 | 13 |
| Belgium | 24 | 17 | 17 | 17 | 17 | 17 | 17 |
| Cyprus | 6 | 1 | 1 | 2 | 3 | 4 | 5 |
| Czech Rep | 24 | 16 | 16 | 16 | 16 | 16 | 16 |
| Denmark | 14 | 9 | 9 | 9 | 9 | 9 | 9 |
| Estonia | 6 | 2 | 2 | 2 | 3 | 4 | 5 |
| Finland | 14 | 8 | 8 | 8 | 8 | 8 | 8 |
| France | 78 | 97 | 97 | 96 | 96 | 96 | 96 |
| Germany | 99 | 133 | 133 | 133 | 129 | 124 | 118 |
| Greece | 24 | 17 | 17 | 17 | 17 | 17 | 17 |
| Hungary | 24 | 16 | 16 | 16 | 16 | 16 | 16 |
| Ireland | 13 | 6 | 6 | 6 | 6 | 6 | 6 |
| Italy | 78 | 93 | 93 | 93 | 93 | 93 | 93 |
| Latvia | 9 | 4 | 4 | 4 | 4 | 4 | 5 |
| Lithuania | 13 | 6 | 6 | 6 | 6 | 6 | 6 |
| Luxembourg | 6 | 1 | 1 | 2 | 3 | 4 | 5 |
| Malta | 5 | 1 | 1 | 2 | 3 | 4 | 5 |
| Netherlands | 27 | 26 | 26 | 26 | 26 | 26 | 26 |
| Poland | 54 | 62 | 62 | 61 | 61 | 61 | 61 |
| Portugal | 24 | 16 | 16 | 16 | 16 | 16 | 16 |
| Slovakia | 14 | 9 | 9 | 9 | 9 | 9 | 9 |
| Slovenia | 7 | 3 | 3 | 3 | 3 | 4 | 5 |
| Spain | 54 | 66 | 66 | 66 | 66 | 66 | 66 |
| Sweden | 19 | 14 | 14 | 14 | 14 | 14 | 14 |
| UK | 78 | 96 | 96 | 95 | 95 | 95 | 95 |
| Objective Function Value | 30.5559 | 1.8076 | 1.8076 | 5.1051 | 9.1719 | 13.6191 | 18.4466 |
| Relative Error (%) | -- | -0.1083 | -0.1083 | -0.4533 | -0.2531 | -0.1706 | -0.1263 |
| CPU time (s) | -- | 28.67 | 22.58 | 19.94 | 18.12 | 16.42 | 14.83 |

**Table 4.** Minimum, maximum and deviation range of the ratio %seats/%population

| smin | min | max | max-min |
|---|---|---|---|
| 0 | 0.78 | 1.58 | 0.80 |
| 1 | 0.78 | 1.58 | 0.80 |
| 2 | 0.92 | 3.16 | 2.24 |
| 3 | 0.94 | 4.73 | 3.79 |
| 4 | 0.93 | 6.31 | 5.38 |
| 5 | 0.89 | 7.89 | 7.00 |

# 5  Conclusions

In this article we have presented an efficient solution approach for the European Parliament apportionment problem that is able to allocate quasi-proportionally to their population the total number (732) of seats. Since the problem has a combinatorial structure exact methods are ineffective which makes necessary the use of metaheuristic approaches such as VNS or tabu search. We have introduced a new hybrid of these two metaheuristics that solves the problem in a matter of seconds. Experiments with various values of smin have been carried out giving in all cases a more proportional apportionment than the present one.

# References

1. Taagepera, R., Grofman, B.: Mapping the indices of seats-votes disproportionality and inter-election volatility. Party Politics, 9, 6 (2003) 659-677
2. Ernst, L.: Apportionment Methods for the House of Representatives and the Court Challenges. Management Sciences, 40, 10 (1994) 1207-1227
3. Riedwyl, H., Steiner, J.: What is Proportionality Anyway?. Comparative Politics, 27, 3 (1995) 357-369
4. Samuels, D., Snyder, R.: The Value of a Vote: Malapportionment in Comparative Perspective. British Journal of Political Science, 31 (2001) 651-671
5. Loosemore, J., Hanby, V.J.: The Theoretical Limits of Maximum Distortion: Some Analytical Expressions for Electoral Systems. British Journal of Political Science, 1, 4 (1971) 467-477
6. Benoit, K.: Which Electoral Formula Is the Most Proportional? A New Look with New Evidence. Political Analysis, 8, 4 (2000) 381-388
7. Schuster, K., Pukelsheim, F., Drton, M., Draper, N.R.: Seat biases of apportionment methods for proportional representation. Electoral Studies, 22 (2003) 651-676
8. Lozano, S., Villa, G.: Apportioning the European Parliament using Data Envelopment Analysis. Paper submitted to European Journal of Political Research
9. Hansen, P., Mladenovic, N.: Variable Neighborhood search: Principles and applications, European Journal of Operational Research, 130 (2001) 449-467
10. Hansen, P., Mladenovic, N., Pérez-Brito, D.: Variable Neighborhood Decomposition search. Journal of Heuristics, 7 (2001) 335-350
11. Glover, F., Laguna M.: Tabu Search. Kluwer Academic Publishers, Boston (1997)

# Appendix

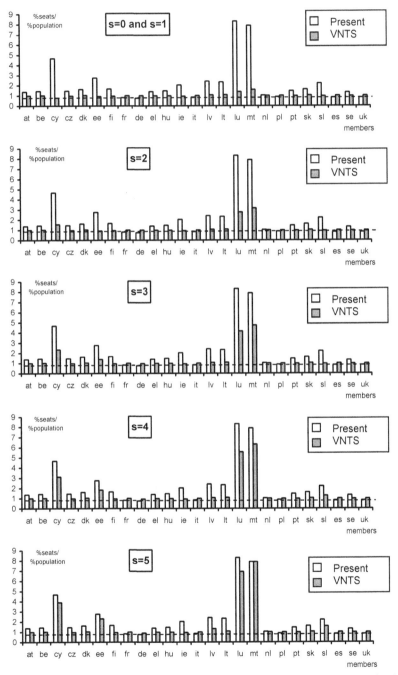

**Fig. 1.** Proportionality profile of present vs solution provided by VNTS approach (smin=0, 1, 2, 3, 4 and 5)

# Author Index

# Lecture Notes in Computer Science

For information about Vols. 1–3818

please contact your bookseller or Springer